Bioenergy Research

Bioenergy Research

Evaluating Strategies for Commercialization and Sustainability

Edited by

Neha Srivastava
IIT (BHU)
Chemical Engineering & Technology
221005 Varanasi
India

Manish Srivastava
IIT (BHU)
Chemical Engineering & Technology
221005 Varanasi
India

WILEY

This edition first published 2021

Registered Offices
John Wiley & Sons, Inc., 111 River Street, Hoboken, NJ 07030, USA
John Wiley & Sons Ltd, The Atrium, Southern Gate, Chichester, West Sussex, PO19 8SQ, UK

Editorial Office
The Atrium, Southern Gate, Chichester, West Sussex, PO19 8SQ, UK

For details of our global editorial offices, customer services, and more information about Wiley products visit us at www.wiley.com.
Wiley also publishes its books in a variety of electronic formats and by print-on-demand. Some content that appears in standard print versions of this book may not be available in other formats.

For general information on our other products and services or for technical support, please contact our Customer Care Department within the United States at (800) 762-2974, outside the United States at (317) 572-3993 or fax (317) 572-4002.
Wiley also publishes its books in a variety of electronic formats. Some content that appears in print may not be available in electronic formats. For more information about Wiley products, visit our web site at ww.wiley.com.

Library of Congress Cataloging-in-Publication Data

Name: Srivastava, Neha, 1981- editor. | Srivastava, Manish, editor.
Title: Bioenergy research : evaluating strategies for commercialization and sustainability / edited by Neha Srivastava, IIT (BHU) Chemical Engineering & Technology, Manish Srivastava, IIT (BHU) Chemical Engineering & Technology.
Description: First edition. | Hoboken, NJ : Wiley, 2021. | Includes bibliographical references and index.
Identifiers: LCCN 2020051125 (print) | LCCN 2020051126 (ebook) | ISBN 9781119772095 (cloth) | ISBN 9781119772101 (adobe pdf) | ISBN 9781119772118 (epub)
Subjects: LCSH: Biomass energy. | Renewable energy sources.
Classification: LCC TP339 .B533 2021 (print) | LCC TP339 (ebook) | DDC 662/.88–dc23
LC record available at https://lccn.loc.gov/2020051125
LC ebook record available at https://lccn.loc.gov/2020051126

Cover Design: Wiley
Cover Image: © esemelwe/iStock/Getty Images

Set in 9.5/12.5pt STIXTwoText by SPi Global, Chennai, India
Printed and bound by CPI Group (UK) Ltd, Croydon, CR0 4YY

C9781119772095_050421

Contents

List of Contributors

Komal Agrawal
Bioprocess and Bioenergy Laboratory
Department of Microbiology
Central University of Rajasthan
Ajmer
Rajasthan
India

Mayara C.S. Barcelos
Laboratory of Food Biotechnology
Institute of Science and Technology
UFVJM Diamantina
MG
Brazil

Nisha Bhardwaj
Bioprocess and Bioenergy Laboratory
Department of Microbiology
Central University of Rajasthan
Ajmer Rajasthan
India

S.M. Bhatt
Agriculture Department
ACET Amritsar
Amritsar Group of Colleges
Punjab
India

Sashita Bindu Ekka
Department of Environmental Science
Indira Gandhi National Tribal University
Amarkantak
MP
India

N.B. Chandrasekhar
Research and Development Center
Department of Biotechnology
Shridevi Institute of Engineering and Technology
Tumakuru
India

S. Dheiver
Institute of Computing
Federal University of Alagoas (UFAL)
Maceio
Brazil

Stephen Edward Chattree
Galgotias University
Greater Noida
UP
India

S K. Godlaveeti
Centre for Nanoscience and Nanotechnology
Sathayabama Institute of Science and Technology
Chennai
India

Rahul Kumar Goswami
Bioprocess and Bioenergy Laboratory
Department of Microbiology
Central University of Rajasthan
Ajmer
Rajasthan
India

Quyet Van Le
Faculty of Natural Sciences
Duy Tan University
Danang
Vietnam

P K. Mishra
Department of Chemical Engineering and Technology
Indian Institute of Technology (BHU)
Varanasi
India

Gustavo Molina
Laboratory of Food Biotechnology
Institute of Science and Technology
UFVJM Diamantina
MG
Brazil

Thi Hong Chuong Nguyen
Institute of Research and Development
Duy Tan University
Da Nang
Vietnam

and

Faculty of Natural Sciences
Duy Tan University
Danang
Vietnam

Lílian A. Pantoja
Institute of Science and Technology
Federal University of Jequitinhonha and Mucuri Valleys UFVJM Diamantina
MG
Brazil

Veena Paul
Department of Dairy Science and Food Technology
Institute of Agricultural Sciences
Banaras Hindu University
Varanasi
UP
India

Enosh Phillips
Department of Biotechnology
St. Aloysius College (Autonomous)
Jabalpur
Madhya Pradesh
India

Ram Prasad
Department of Botany
Mahatma Gandhi Central University
Motihari
Bihar
India

Spriha Raven
Diatom Research Laboratory
Amity Institute of Biotechnology
Amity University
Noida
UP
India

Lipi Rina
Department of Agronomy
SHUATS
Prayagraj
UP
India

Shivani Smita Sadanand
Department of Agriculture
SHUATS
Prayagraj
UP
India

Arla Sai Kumar
Department of Materials Science and Nanotechnology
Yogi Vemana University
Kadapa
Andhra Pradesh
India

Sana Siva Sankar
School of Chemical Engineering and Technology
North University of China
Taiyuan
China

Alexandre S. Santos
Department of Basic Science
Federal University of Jequitinhonha and Mucuri Valleys UFVJM Diamantina
MG
Brazil

Alexandre Soares dos Santos
Department of Basic Science
Federal University of Jequitinhonha and Mucuri Valleys - UFVJM – Diamantina
MG
Brazil

Sreedevi Sarsan
Department of Microbiology
St. Pious X Degree and PG College
Hyderabad
India

Abhishek Saxena
Diatom Research Laboratory
Amity Institute of Biotechnology
Amity University
Noida
Uttar Pradesh
India

Akshay Shrivastav
Department of Chemical Engineering
Madan Mohan Malaviya University of Technology
Gorakhpur
India

Rajeev Singh
Department of Chemical Engineering and Technology
Indian Institute of Technology (BHU)
Varanasi
India

Neha Srivastava
Department of Chemical Engineering and Technology
Indian Institute of Technology (BHU)
Varanasi
India

K.R. Srivastava
Department of Chemical Engineering and Technology
Indian Institute of Technology (BHU)
Varanasi
India

Manish Srivastava
Department of Chemical Engineering and Technology
Indian Institute of Technology (BHU)
Varanasi
India

Archana Tiwari
Diatom Research Laboratory
Amity Institute of Biotechnology
Amity University
Noida
UP
India

Abhishek Dutt Tripathi
Department of Dairy Science and Food Technology
Institute of Agricultural Sciences
Banaras Hindu University
Varanasi
UP
India

K. Vindhya Vasini Roy
Department of Microbiology
St. Pious X Degree and PG College
Hyderabad
India

Pradeep Verma
Bioprocess and Bioenergy Laboratory
Department of Microbiology
Central University of Rajasthan
Ajmer
Rajasthan
India

Kele A. C. Vespermann
Laboratory of Food Biotechnology
Institute of Science and Technology
UFVJM Diamantina
MG
Brazil

Ashok Kumar Yadav
Food Processing and Management
DDU Kaushal Kendra
Rajiv Gandhi South Campus
Banaras Hindu University
Barkachha
Mirzapur
UP
India

Foreword

To discontinue the chain of harmful impact of environmental pollution, fossil fuel issues must be resolved in a sustainable manner. Production of renewable energy from renewable resources is one of the most effective alternatives toward the replacement of fossil fuels. For decades, research on renewable energy production has been studied for its commercialization purpose in an environment-friendly manner; the various downsides keep it stuck, which hampers its commercial implementation effectively and uniformly. To date, many renewable energy options have been explored and analyzed for their commercial sustainability. Bioenergy production is the most sustainable alternative to replace fossil fuels, and rigorous research has been done on bioenergy topics as the most sustainable solution. Although the number of bioenergy options are available and detailed research has been performed, the sustainable bioenergy options are still far away from commercialization and practical viability. There is an urgent need to summarize and critically evaluate the available bioenergy options for the practical implementation for the replacement of fossil fuels.

Publication of *Bioenergy Research: Evaluating Strategies for Commercialization and Sustainability* is a useful approach in this area. I am writing this message with joy as a working researcher in this area. This book holds 13 chapters, presented as an evaluation of different bioenergy options with complete details and up-to-date research as well as development. The book expands on different bioenergy options, such as biohydrogen, bioethanol, and biomass to fuel production and algal biofuels, and so on. Additionally, the book expands upon the feasibility of these biofuels on the basis of their properties, advantages, drawbacks, research updates, and existing hurdles still in the way of their hopeful implementation. In my introduction, I talk about how this book will serve as a preliminary introduction to anyone working in the relevant areas, both academic and research institutions as well as industrial.

I appreciate the efforts of Dr. Neha Srivastava and Dr. Manish Srivastava for *Bioenergy Research: Evaluating Strategies for Commercialization and Sustainability*. The efforts taken to build this book will surely cover the gap between the academic research benches to a practical industrial-scale study. I congratulate the editors for their hard work in bringing a final shape to this book.

Dr Vijai Kumar Gupta
Center for Safe and Improved Food &
Biorefining and Advanced Biomaterials Research Center
Scotland's Rural College (SRUC), UK

Acknowledgments

The editors are thankful to all the academicians and scientists whose contributions have enriched this volume. We also express our deep sense of gratitude to our parents whose blessings have always prompted us to pursue academic activities deeply. It is quite possible that in a work of this nature, some mistakes might have crept in text inadvertently and for these we owe undiluted responsibility. We are grateful to all authors for their contribution to present book. We are also thankful to Wiley for giving this opportunity to editors and Department of Chemical Engineering & Technology, IIT (BHU) Varanasi, U.P., India for all technical support. We thank them from the core of our heart.

Biofuels Production Technologies: Recent Advancement

Neha Srivastava and Manish Srivastava

Department of Chemical Engineering and Technology, Indian Institute of Technology (BHU), Varanasi, 221005, India

1 Introduction

Fossil fuels are the hydrocarbon, coal, or natural gas that is produced from the dead remains of living organisms. The combustion of these fuels by humans causes the emission of carbon dioxide, sulfur dioxide, nitrous oxide, methane, etc., which causes harmful effects and contributes to global warming and pollutes the environment (Basu et al. 2020). The excessive utilization of fossil fuels has also led to its shortage. Fossil fuels are essential to fulfill global energy needs, accounting for more than 80% of the world's primary energy consumption (Karnauskas et al. 2020).

In order to reduce fossil fuel usage, other renewable and sustainable sources of energy, like solar power, wind energy, wave energy, and thermal energy from the earth's crust, as well as biofuels may be taken into consideration (Jia et al. 2018). Biofuels are one of the best green alternatives, which have the potential to replace fossil fuels because they are ecofriendly, easily available, completely combustible, and nontoxic. A biofuel is defined as a source of energy or a fuel, which is obtained using a biological carbon fixation process. It is the most productive form of renewable energy source, as it is generated from renewable sources such as biomass (Milano et al. 2016). It is widely acknowledged that biofuels combustion does not contribute to greenhouse effect. Biofuel not only reduces the dependency of the world toward oil but also decreases the emission of harmful gases; moreover, its production can also provide new employment and income options in rural and urban areas. Low emission, nontoxic, environmentally friendly, and safer are some of the green and unique characteristics of biofuels, which make them potential candidates for their effective utilization (Kumarappan and Joshi 2011). Further, the biofuels are categorized as first-generation, second-generation, third-generation, and fourth-generation biofuel based on feedstocks from which they are produced (Bhatia et al. 2017).

Biofuel production from the first generation requires food crops that are commonly grown on arable lands as a substrate. This process utilizes the complete food crops only for producing fuel and not anything else. The process used in this generation was yeast fermentation or transesterification for production of ethanol or biodiesel (Rodionova et al. 2017). The second generation of biofuel production requires lignocellulosic biomass, human waste, woody chips, and agricultural by-products as a substrate. The benefit of this process is that it utilizes any kind of biomass for the production of biofuel, but unlike the other processes, there are some difficulties in extraction of fuels. This process requires some extra steps, such as physical, chemical pretreatment just to obtain proper yield. The third generation of biofuel production utilizes oil-rich algae as a substrate for production of alcohol. The types of biofuel production in the third generation varies because of the two different characteristics of microbes. First, the oil produced by algae can be refined to biodiesel or some content of gasoline, whereas in the second, genetic manipulation leads to production of ethanol, butanol, gasoline, and even biodiesel (Leong et al. 2018). In the fourth generation of biofuel, the destruction of biomass is not required and biofuels are produced from nonarable land. This category includes biofuels, such as photobiological solar fuels and electrofuels (Abdullah et al. 2019). There are different types of biofuels, such as bioethanol, methanol, biodiesel, biogas, bio-butanol, and bio-hydrogen, which exist in nature. Different types of biofuels can be produced by varying fermentation techniques and microorganisms. This variation and comparisons between the different types of biofuel are described in Table 1. There are currently two main types of biofuels used: ethanol and biodiesel.

Table 1 Comparisons of different types of biofuel.

	Bioethanol	**Biodiesel**	**Biohydrogen**	**Biogas**
Substrate	Sugarcane, corn, and sugar beat	Vegetable oils	Food waste	Municipal solid waste
Process involved	Pretreatment Hydrolysis Fermentation Product upgradation	Pretreatment Hydrolysis Fermentation Lipid extraction Transesterification	Pretreatment Hydrolysis Fermentation	Pretreatment Hydrolysis Acidogenesis Acetogenesis Methanogenesis
Pre-treatment preferred	Steam explosion	Biological pre-treatment	Heating of inocula	Ultrasonification
Fermentation preferred	Solid state fermentation	Solid state fermentation	Dark fermentation	Anaerobic, fermentation
Microbes preferred	Bacteria and yeast *(Saccharomyces cerevisiae* and *Zymomonas mobilis)*	Fungi (white rot fungi and brown rot fungi)	Bacteria (Lactobacillus, *Thermoanaerobaterium*, Clostridium)	Bacteria (Methanogens and acidogens)
Current research	Downstream process and separation of biomass after pre-treatment	Cost-effective microalgae and biophotoreactor	Fermentation condition and catalyst	Bioreactors, process of anaerobic digestion and application of high pressure

2 Bioethanol

The process of ethanol production is classified into four different categories, (i) pretreatment, (ii) hydrolysis, (iii) fermentation, and (iv) product upgradation. Pretreatments can be done by physical means (microwave radiation, pyrolysis, sonication, and spry drying), chemical means (hydrolysis, oxidation, ozonation, and alkaline pretreatment), and biological means (bacterial and enzymatic treatment). In the process of ethanol production, steam explosion is mostly used a pretreatment process at the commercial level (Trinh et al. 2019). Pretreatment of lignocellulosic biomass helps the exposure of cellulose and hemicelluloses and removes lignin as it ruptures the complete structure (Figure 1) (Karuppiah and Azariah 2019). After the pretreatment process, enzymatic or chemical hydrolysis takes place, which converts cellulose and hemicelluloses hexose, pentose, and glucose as a simple fermentable carbohydrate. These simple carbohydrates get fermented by the help of yeast, fungi, or bacteria using two different mechanisms of fermentation, (i) solid state fermentation and (ii) submerged fermentation (Brexó and Sant'Ana 2017). These C-5 and C-6 carbohydrates can be fermented to ethanol by the help of yeast and bacteria (*Saccharomyces cerevisiae* and *Zymomonas mobilis* are the most common species used at industrial level currently) (Gallagher et al. 2018; Nakamura and Shima 2018; Karuppiah and Azariah 2019) (Figure 2). At the end, the produced bioethanol has to go through a dehydration and distillation process for producing ethanol with 99.9% purity. Earlier hydrolysis and fermentation are carried out in two different chambers; nevertheless, the process can be carried out in the same chamber. Using these techniques, fermentation and saccharification, direct conversion, prehydrolysis, and simultaneous fermentation and saccharification, as well as simultaneous saccharification and cofermentation can be occurred simultaneously (Zabed et al. 2017; Chen and Xiaoguo 2016). In the process of ethanol production, the solid state fermentation technique is mostly used and this technique performs two different steps in the same chamber.

The application of lignocellulosic biomass, such as sugarcane, corn, and kitchen waste as a substrate for bioethanol production in currently trending across the world. Microbes, such as *Codium tomentosum, Myceliophthora thermophila, S. cerevisiae,* and *moringaolefira* are currently used at different levels to produce and test bioethanol (Konur 2020). Countries like the UK and the USA are working on the production of bioethanol using marine yeast and sea water as a media. In 2019, the United States produced the maximum of ethanol, which is around 15.8 billion gallons, while Brazil was in second position with 8.6 billion gallons. On this list, India is at the fifth position with 530 million gallons in 2019 (Aparicio et al. 2020). The most common substrate for ethanol production is starch released from sugar cane, corn, and sugar beets. The United States is the largest producer of corn in the world, which produced about 345.89 million metric tons in 2019. Therefore, the United States utilizes corn as a feedstock for ethanol production because of its abundance and low cost, whereas Brazil is the largest producer of sugar cane in the world and utilizes it as a substrate because of its easily availability and low cost (Mayumi 2020). Bioethanol productions are still not being used at commercial level in many countries and the problem may be due to their feed and food purposes. One of the major disadvantages with the production of bioethanol in developing countries is a lack of availability of feed stocks for production. As the production of bioethanol continues to grow, it is the compulsion of producing

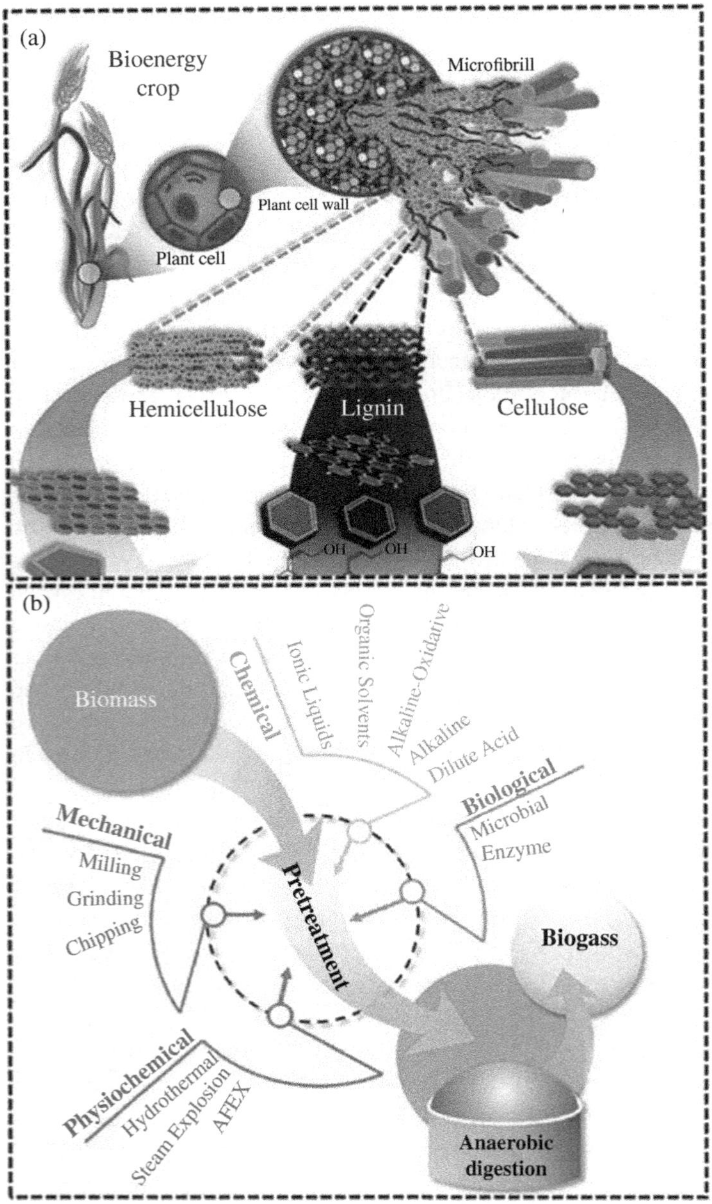

Figure 1 (a) Structure of lignocellulosic biomass and its biopolymers; cellulose, hemicellulose, and lignin [Hernández-Beltrán et al. (2019), CCBY 4.0]. (b) Pretreatment methods to increase the bioavailability of lignocellulosic biomass. [Hernández-Beltrán et al. (2019), CCBY 4.0].

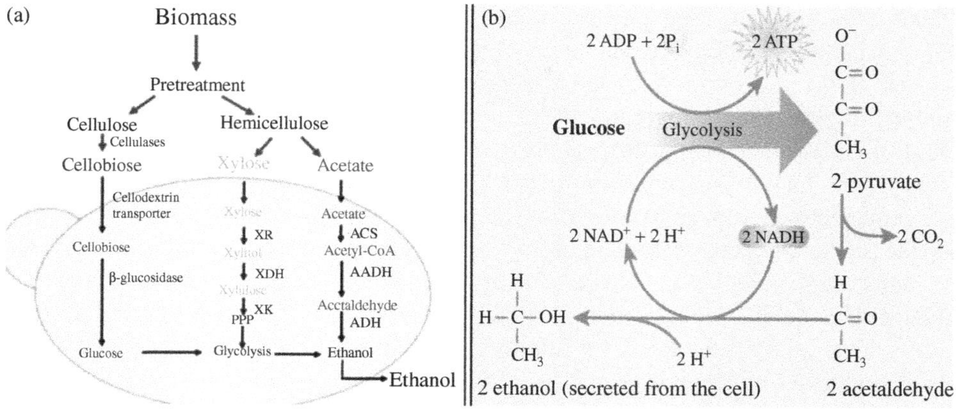

Figure 2 Schematic representation [adopted with permission from Wei et al. (2015)]. (a) and metabolic pathway of fermentation (b) for producing ethanol. Source: El-Dalatony et al. (2017). Licensed under CC BY 4.0.

bioethanol using renewable substrates which are still not found. Currently, the downstream process in ethanol production and separation of biomass constituents after pretreatment are focused, and several points of research are under process for developing an economical procedure (Mohanty and Swain 2019).

3 Biodiesel

Biodiesel can be produced using substrate like oils from different seeds (rapeseeds, soyabeans, palm), animal waste (animal fats, fish oil, poultry oils), as well as other sources, such as almond, barley, and lignocellulosic biomasses (Chen et al. 2018). The process of biodiesel production starts with pretreatment process in which most favorable is biological techniques using different microorganisms (white rot fungi such as *Irpexlacteus, P. floridensis, Ceriporiopsis subvermispora, Punctularia* spp., *Phlebia brevispora*, brown rot fungi such as *Meruliporia incrassate, Serpula lacrymans, Laetoporus sulphurous, Coniophora puteana*) (Rudakiya and Gupte 2017; Hegnar et al. 2019; Wang et al. 2017). After the process of pretreatment, th e process of enzymatic hydrolysis starts at a mild climatic condition (pH 4.5–5.0 and temperature 40–50 °C). Fermentation techniques depend upon the state of substrate, whether it is liquid (submerged fermentation) or solid (solid state fermentation), whereas solid state technique is mostly preferred in biodiesel production. Biodiesel can be produced using microorganisms which have the ability to accumulate concentration of lipids of more than 20% (Sohedein et al. 2020). These microbes can be bacteria (Arthrobacter sp., *Rhodococcus opacus, Acinetobacter calcoaceticus*), yeasts (*Cryptococcus albidus, Candida curvata, Rhodotorula glutinis*), microalgae (*Cylindrotheca* sp, *Schizochytrium* sp, *Botryococcus braunii*), and fungi (*Mortierella isabellina, Aspergillus oryzae, Mortierella vinacea*) (Bardhan et al. 2020). The produced lipids can be extracted using process like organic solvent, microwave, bead beatings, etc., and the extracted lipids then have go through a direct transesterification process for the production of biodiesel.

Biodiesel is generally used in normal diesel engines alone or as a blend with petrodiesel. Different countries utilize different feedstock as per their availability; for example, different parts of Europe utilize sunflower and rapeseeds as a substrate, while soybean is commonly used in the United States, and canola oil is used in Canada, and in tropical countries, palm oils are used (Yesilyurt et al. 2020). Application of these food oils have raised an issue related to food versus fuel. To overcome this problem, production of biodiesels for different second- (pongamia, jatropha, karanja) and third- (microalgae and other microorganisms) generation feedstocks are currently trending with lots of research, both at the lab scale as well as at the industrial scale (Eiben et al. 2020). In 2019, the United States produced the maximum amount of biodiesel, which is about 6.5 billion liters, and Brazil is a little behind the United States with annual production of 5.9 billion liter of biodiesel in 2019. India produced about 2.5 billion liters of biodiesel in year 2018–2019. United States is the largest producer of soybeans in the world and produced about 120.52 million metric tons in 2018–2019 (Bognar et al. 2020). Therefore, the United States utilizes soybeans as a feedstock for biodiesel production because of its abundance and low cost. Brazil overtook the United States in soybean production in 2019–2020 with 124 million metric tons (Vunnava and Singh 2020). Currently, different research is going on for the preparation of cost-effective microalgae and developing photobioreactors, which will increase the production of biodiesel (Atabani et al. 2017). Some important factors of using microalgae are high production, low cost, as well as easy availability as described in Figure 3, whereas some of the drawbacks in the process are low oil content, low concentration of biomass in the culture, and the small size of microalgae which increases the cost of harvest (Duong et al. 2012).

4 Biohydrogen

There are numerous types of pretreatments methods, such as microwave, heating, etc. However, the most important pretreatment method which is widely adopted by researchers is the heating of inocula for the production of biohydrogen (Prabakar et al. 2018). After the pretreatment process of biomass, enzymes can penetrate easily to cell walls and convert cellulose and hemicellulose to sugars that can be fermented easily via enzymatic hydrolysis (Algapani et al. 2016). Further, biohydrogen can be produced by different methods such as photo-fermentation, indirect biophotolysis, dark fermentation, and direct biophotolysis (Figure 4) (Chandrasekhar et al. 2015). Among these, dark fermentative biohydrogen production has been reported to be a potential economic and effective method due to the high yield and efficiency using a variety of substrate (cellulose, sugary waste water, sugarcane juice, corn pulp, food waste). Microbes (especially bacteria), such as *Lactobacillus*, *Thermoanaerobaterium*, *Clostridium*, *Rhodopseudomonas*, *Citrobacter*, and *Enterobacter* are mainly used as fermentative microorganisms in the biohydrogen production process (Kumar et al. 2019). There are three different pathways of conversion of organic waste in biohydrogen using dark fermentation, favored by the thermodynamics process. These pathways are as follows: (i) acetic acid from hexose, (ii) butyric acid from hexose, and (iii) ethanol from acetate, and among these pathways, the acetate ethanol pathway has been found to be more stable as compared to the others. The microbes grow on biodegradable wastes, having a high quantity of sugar in order to produce pyruvate through glycolysis

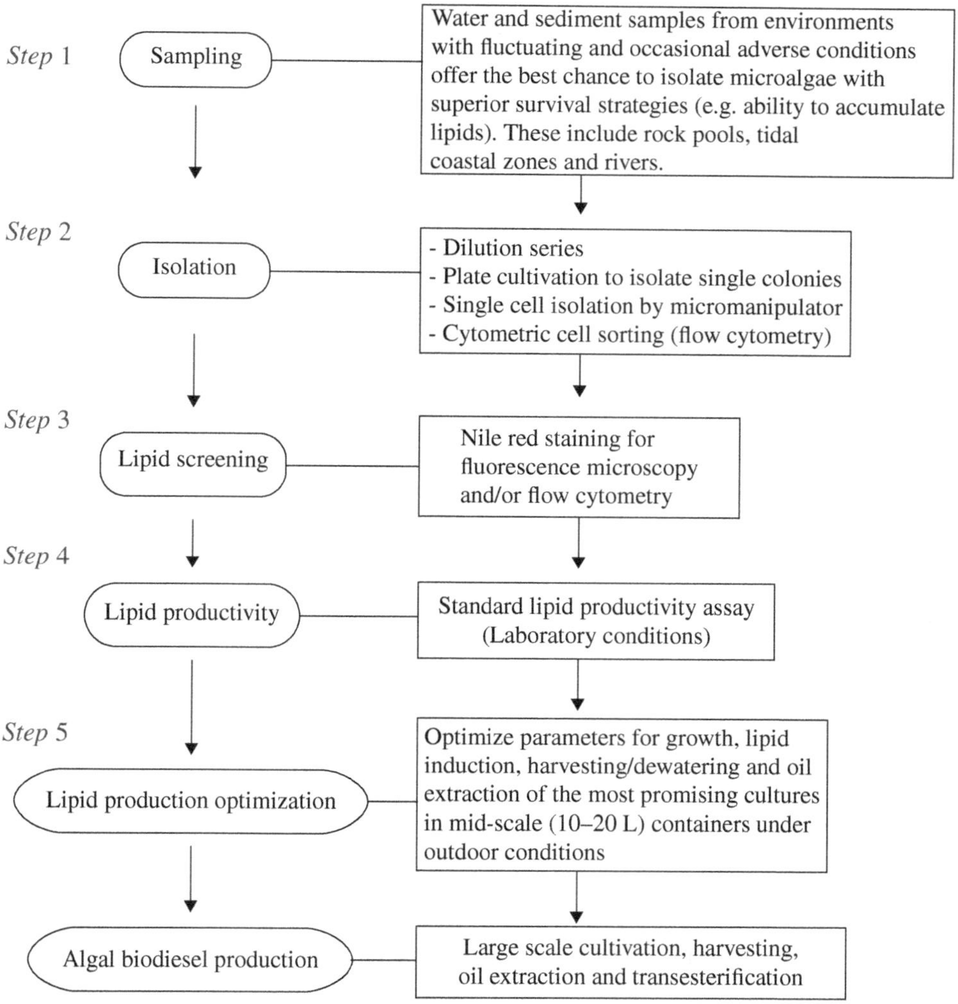

Figure 3 Microalgae selection and biodiesel production. Source: Duong et al. (2012). Licensed under CC BY 3.0.

techniques, which get further oxidized into acetyl CoA through the reduction of ferredoxin while producing adenosine di-phosphate (ATP) and acetate. Finally, ferredoxin is oxidized into hydrogen with the help of hydrogenase enzymes. This enzyme is responsible for the production of biohydrogen using dark fermentation (Liu et al. 2017).

In recent years, the production of biohydrogen using organic waste as a substrate is mostly found in developed countries like the USA, the UK, and Germany. In these countries, the application of two-stage bioreactors for producing both hydrogen as well as methane are focused on. Research is ongoing on the utilization of food waste as a substrate for producing hydrogen (Hassan et al. 2019). As per the Food and Agriculture Organization (FAO), about one third of the world foods are wasted, which is approximately1.3 billion tons per

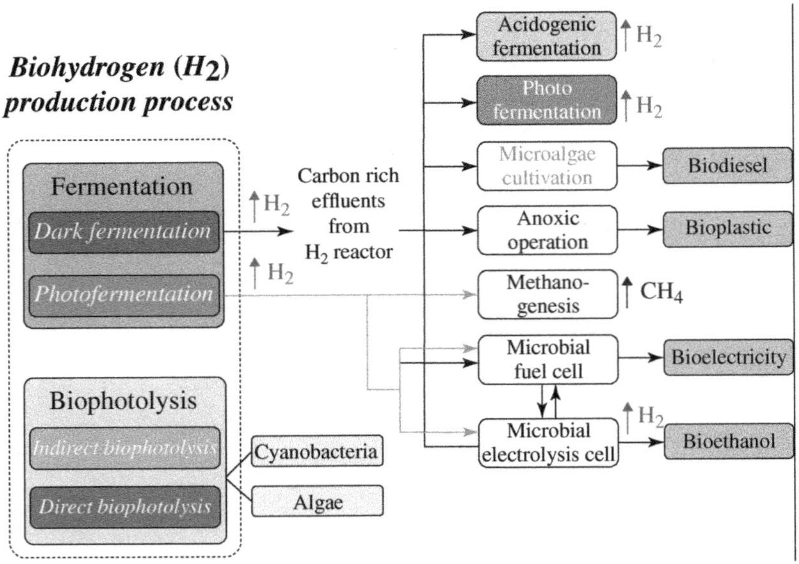

Figure 4 Different process and route for biohydrogen production. Source: Chandrasekhar et al. (2015). Licensed under CC BY 4.0.

year. According to the study performed by the Economist Intelligence Unit, food sustainability index (FSI), and as per the FSI 2017 report, Australia produced the highest quantity of food waste per capita (361 kg), whereas other countries, such as Sweden (200 kg), the USA (287 kg), Russia (56 kg), and China (44 kg) also produce a high amount of waste. Moreover, India also produces about 51 kg of kitchen waste per capita (Boliko 2019). Improvements as well as advancements are made, such as variation in different fermentation conditions like temperature and pH and concentration of substrate (carbon to nitrogen ratio), plays a significant role to increase the final yield. Apart from all these advancements and improvements, there are two major problems that act as a barrier in the path of biohydrogen production for being used at commercial scale. These problems are low yield and have a high operation cost, which acts as a serious issue in different countries (Ishangulyyev et al. 2019). Some latest advancements, such as application of nanoparticles, metal oxides and ions, and immobilization of microbes, are appealing in their prospects to achieve high yields as well as a profitable efficiency of the process (Yun et al. 2018).

5 Biogas

Biogas is a type of biofuel produced by anaerobic digestion of organic waste, such as biomass, cow dung, agricultural residue, green waste, sugar cane, and cassava, etc. The production of biogas is classified into different steps which occurs in an anaerobic reactor (Plugge 2017). These steps are as follows: pretreatment, hydrolysis, acidogenesis, acetogenesis, and methanogenesis (Campanari et al. 2017). There are lots of pretreatment processes which can be used in the process of biogas production, such as milling, preheating, steam explosion, and liquid hot water, etc. However, the most important pretreatment method which was widely adopted at the industrial level is the preheating by ultrasonification of substrate (Deepanraj et al. 2017). Hydrolysis is the primary step in the process of biogas production. This step is achieved by breaking down and solubilizing organic complex compounds to simple soluble compounds. Enzymatic hydrolysis is one of the most preferred processes which accelerates the process through oxidation of organic matter (Tongco et al. 2020). After the hydrolysis process, the substrate is available for transportation of cells and bacteria which will ferment the substrate during acidogenesis. In this process, the produced organic acids get transformed into hydrogen, carbon dioxide, acetic acid, and acid derivatives. In this process, the main role is being performed by microorganisms which are classified into two different groups based on the product they produce (Fernandez 2018). The two main groups of these microbes are the methane-producing bacteria (methanogens –*Methanoculleus, Methanosarcinales, Methanobacteriales*) and acid-producing bacterial (acidogens – *Moorella thermoacetica, Clostridium formiaceticum, Acetobacter woodii, Clostridium termo autotrophicum*). Methanogensis is one of the critical steps in the process of the acidogenesis because about 70% of methane used in anaerobic digestion is produced in this step only, whereas during acetogenesis ethanol, volatile fatty acids (VFAs) with more than two carbons get converted by acetate-forming bacteria into carbon dioxide and hydrogen (main product) and acetate (Mulat et al. 2016). In this step, only methanogens convert hydrogen (oxidizing) and carbon dioxide (reducing) to methane, as well as acetolactic methanogens which convert acetate to methane. Therefore, the produced biogas contains 1–5% other gases, including hydrogen, carbon dioxide 35–40%, and methane 55–60%.

Recently, lots of work continues in the development of bioreactors, the separation of anaerobic digestion into two different steps, and the high-pressure digestion which can increase the production of biogas. Developing countries have higher amounts of municipal solid waste, and about 74% of municipal solid waste generated in Indonesia, 75% generated in Bangladesh, 76.4% generated in Sri Lanka, 42% generated in India, and 52.6% generated in China is organic. China is the world's largest producer of municipal solid waste by producing 215 million tons per year and was expected to produce 480 million tons in 2030 (Cudjoe et al. 2020). This waste includes solid waste, such as papers, woods, meats, plastics, etc. China utilizes it as a substrate for biogas production. Therefore, China was the world's largest producer of biogas in 2014 with the total production of about 15 billion cubic meters. There is lots of ongoing research into the various upgrading and cleaning techniques in order to improve the quality of biogas (Xue et al. 2020). These improvements are classified on its behavior into two different groups – physiochemical and biological technologies. This technology includes cryogenic separation, hydrated separating technique, enriching membranes, anaerobic, high pressure, and multistage digestion (Theuerl et al. 2019; Passos et al. 2020). Still, there are lots of problems which is why biogas production cannot be implemented on a large scale in different countries. These problems are infrastructural challenges, due to the unavailability of feedstocks, improper segregation, and high installation cost. In spite of all these issues, biogas is one of the most important sources of energy, especially in rural areas, and it is also utilized as an alternative and renewable source of energy at a small scale whose flow sheet is explained in Figure 5 (Teferra and Wubu, 2018).

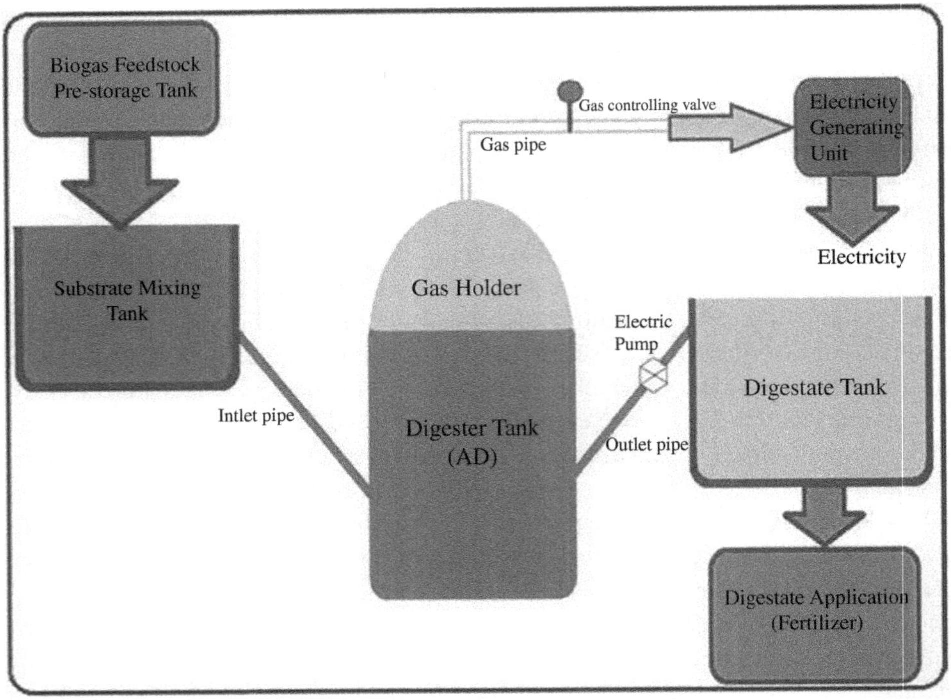

Figure 5 Main components and general process flow of biogas production [Teferra and Wubu (2018), CCBY 3.0].

6 Conclusion

In the present world, bioethanol and biodiesel are mostly produced and in demand as compared to other types of biofuel that exists in nature. Different developed countries, such as the USA, the UK, and Brazil, are the top producers of biodiesel and bioethanol. In 2019 the USA produced about 15.8 billion gallons of bioethanol and about 6.5 billion liters of biodiesel. Similar to this, Brazil produces about 5.9 billion liters of biodiesel and 8.6 billion gallons of bioethanol in the same working year. Lots of countries produce bioethanol because it can be produced by using large variety of carbohydrates as a substrate. It is easily possible to convert waste like woods, straws, and even daily use waste to bioethanol. Along with these, one of the most important reasons for producing bioethanol at this large of scale is that it can be used as a substituent or additive toward petrol. As fossil fuel decreases day by day, it is a major focus for every country to work on alternatives. Biodiesel is also an alternative of very important fossil fuel which can be easily used in any kind of diesel engines without even modifying it. Biodiesel is even better than diesel in terms of flashpoint, aromatic content, and using blend biodiesel compared to normal diesel gives even better fuel economy.

The basic difference between ethanol and biodiesel are that one is fuel and the other one is oil. Ethanol is an alcohol created through fermentation and can be used as a substitute or along with gasoline, whereas biodiesel is generated by extracting naturally occurring oils from differed plants and seeds by the process known as transesterification. The production of biofuel is focused on not only fulfilling the requirements of energy production at the decentralized level but also for fulfilling the requirements of transport. This generates interest from regional groups as well as that it involves the land of regional communities. This creates incentives for the local communities, especially if community lands are involved.

Acknowledgement

Author M.S. acknowledges the Science and Engineering Research Board for SERB Research Scientist award [SB/SRS/2018-19/48/PS] and also to DST for DST INSPIRE Faculty award [IFA-13-MS-02].

References

Abdullah, B., Muhammad, S.A.F.S., Shokravi, Z. et al. (2019). Fourth generation biofuel: a review on risks and mitigation strategies. *Renewable Sustainable Energy Rev.* 107: 37–50.

Algapani, D.E., Qiao, W., Min, S. et al. (2016). Bio-hydrolysis and bio-hydrogen production from food waste by thermophilic and hyperthermophilic anaerobic process. *Bioresour. Technol.* 216: 768–777.

Aparicio, E., Rodríguez-Jasso, R.M., Lara, A. et al. (2020). Biofuels production of third generation biorefinery from macroalgal biomass in the Mexican context: an overview. In: *Sustainable Seaweed Technologies*, 393–446. Elsevier.

Atabani, A.E., El-Sheekh, M.M., Kumar, G., and Shobana, S. (2017). Edible and nonedible biodiesel feedstocks: microalgae and future of biodiesel. In: *Clean Energy for Sustainable Development*, 507–556. Academic Press.

Basu, S., Lehman, S.J., and Miller, J.B. (2020). Constraints on fossil fuel CO2 emissions from measurements of 14C in atmospheric CO2. *Proc. Natl. Acad. Sci. U. S. A.* 117 (24), 13,300–13,307.

Bardhan, P., Gupta, K., Kishor, S. et al. (2020). Oleaginous yeasts isolated from traditional fermented foods and beverages of Manipur and Mizoram, India, as a potent source of microbial lipids for biodiesel production. *Ann. Microbiol.* 70 (1): 1–14.

Bhatia, S.K., Kim, S.-H., Yoon, J.-J., and Yang, Y.-H. (2017). Current status and strategies for second generation biofuel production using microbial systems. *Energy Convers. Manage.* 148: 1142–1156.

Bognar, J., Skogstad, G., and Mondou, M. (2020). Media coverage and public policy: reinforcing and undermining media images and advanced biofuels policies in Canada and the United States. *Environ. Commun.*: 1–18.

Boliko, M.C. (2019). FAO and the situation of food security and nutrition in the world. *J. Nutr. Sci. Vitaminol.* 65 (Supplement): S4–S8.

Brexó, R.P. and Sant'Ana, A.S. (2017). Impact and significance of microbial contamination during fermentation for bioethanol production. *Renewable Sustainable Energy Rev.* 73: 423–434.

Campanari, S., Augelletti, F., Rossetti, S. et al. (2017). Enhancing a multi-stage process for olive oil mill wastewater valorization towards polyhydroxyalkanoates and biogas production. *Chem. Eng. J.* 317: 280–289.

Chandrasekhar, K., Lee, Y.-J., and Lee, D.-W. (2015). Biohydrogen production: strategies to improve process efficiency through microbial routes. *Int. J. Mol. Sci.* 16 (4): 8266–8293.

Chen, H. and Xiaoguo, F. (2016). Industrial technologies for bioethanol production from lignocellulosic biomass. *Renewable Sustainable Energy Rev.* 57: 468–478.

Chen, J., Li, J., Dong, W. et al. (2018). The potential of microalgae in biodiesel production. *Renewable Sustainable Energy Rev.* 90: 336–346.

Cudjoe, D., Han, M.S., and Nandiwardhana, A.P. (2020). Electricity generation using biogas from organic fraction of municipal solid waste generated in provinces of China: techno-economic and environmental impact analysis. *Fuel Process. Technol.* 203: 106381.

Deepanraj, B., Sivasubramanian, V., and Jayaraj, S. (2017). Effect of substrate pretreatment on biogas production through anaerobic digestion of food waste. *Int. J. Hydrogen Energy.* 42 (42): 26522–26528.

Duong, V.T., Li, Y., Nowak, E., and Schenk, P.M. (2012). Microalgae isolation and selection for prospective biodiesel production. *Energies.* 5 (6): 1835–1849.

Eiben, C.B., Tian, T., Thompson, M.G. et al. (2020). Adenosine triphosphate and carbon efficient route to second generation biofuel isopentanol. *ACS Synth. Biol.* 9 (3): 468–474.

El-Dalatony, M.M., Salama, E.-S., Kurade, M.B. et al. (2017). Utilization of microalgalbiofractions for bioethanol, higher alcohols, and biodiesel production: a review. *Energies.* 10 (12): 2110.

Fernandez, M.S. (2018) A novel green biorefinery concept: protein refining by lactic acid fermentation and biogas production from green biomass. PhD diss., Aalborg University.

Gallagher, D., Parker, D., Allen, D.J., and Tsesmetzis, N. (2018). Dynamic bacterial and fungal microbiomes during sweet sorghum ensiling impact bioethanol production. *Bioresour. Technol.* 264: 163–173.

Hassan, G.K., Massanet-Nicolau, J., Dinsdale, R. et al. (2019). A novel method for increasing biohydrogen production from food waste using electrodialysis. *Int. J. Hydrogen Energy.* 44 (29): 14715–14720.

Hernández-Beltrán, J.U., Hernández-De Lira, I.O., Cruz-Santos, M.M., Saucedo-Luevanos, A., Hernández-Terán, F., and Balagurusamy, N. (2019). Insight into pretreatment methods of lignocellulosic biomass to increase biogas yield: current state, challenges, and opportunities. *Appl. Sci.* 9(18). doi: 10.3390/app9183721

Hegnar, O.A., Goodell, B., Felby, C. et al. (2019). Challenges and opportunities in mimicking non-enzymatic brown-rot decay mechanisms for pretreatment of Norway spruce. *Wood Sci. Technol.* 53 (2): 291–311.

Ishangulyyev, R., Kim, S., and Lee, S.H. (2019). Understanding food loss and waste—why are we losing and wasting food? *Foods.* 8 (8): 297.

Jia, T., Dai, Y., and Wang, R. (2018). Refining energy sources in winemaking industry by using solar energy as alternatives for fossil fuels: a review and perspective. *Renewable Sustainable Energy Rev.* 88: 278–296.

Karnauskas, K.B., Miller, S.L., and Schapiro, A.C. (2020). Fossil fuel combustion is driving indoor CO2 toward levels harmful to human cognition. *GeoHealth.* 4 (5): e2019GH000237.

Konur, O. (2020). The pioneering research on the bioethanol production from green macroalgae. In: *Handbook of Algal Science, Technology and Medicine* (ed. O. Konur). Elsevier, Amsterdam.

Karuppiah, T. and Azariah, V.E. (2019). Biomass pretreatment for enhancement of biogas production. In: *Anaerobic Digestion*. IntechOpen.

Kumar, S., Sharma, S., Thakur, S. et al. (2019). Bioprospecting of microbes for biohydrogen production: current status and future challenges. *Bioproc. Biomol. Prod.*: 443–471.

Kumarappan, S. and Joshi, S. (2011). Trading greenhouse gas emission benefits from biofuel use in US transportation: challenges and opportunities. *Biomass Bioenergy.* 35 (11): 4511–4518.

Leong, W.-H., Lim, J.-W., Lam, M.-K. et al. (2018). Third generation biofuels: a nutritional perspective in enhancing microbial lipid production. *Renewable Sustainable Energy Rev.* 91: 950–961.

Liu, D., Sun, Y., Li, Y., and Yuan, L. (2017). Perturbation of formate pathway and NADH pathway acting on the biohydrogen production. *Sci. Rep.* 7 (1): 1–8.

Mayumi, K.T. (2020). Credibility of scientific analysis, and assessment of PV systems and ethanol production. In: *Sustainable Energy and Economics in an Aging Population*, 51–77. Cham: Springer.

Milano, J., Ong, H.C., Masjuki, H.H. et al. (2016). Microalgae biofuels as an alternative to fossil fuel for power generation. *Renewable Sustainable Energy Rev.* 58: 180–197.

Mulat, D.G., Jacobi, H.F., Feilberg, A. et al. (2016). Changing feeding regimes to demonstrate flexible biogas production: effects on process performance, microbial community structure, and methanogenesis pathways. *Appl. Environ. Microbiol.* 82 (2): 438–449.

Nakamura, T. and Shima, J. (2018). Selection and development of stress-tolerant yeasts for bioethanol production. *Jpn Agric. Res. Q.* 52 (2): 137–142.

Passos, F., Bressani-Ribeiro, T., Rezende, S., and Chernicharo, C.A.L. (2020). Potential applications of biogas produced in small-scale UASB-based sewage treatment plants in Brazil. *Energies.* 13 (13): 3356.

Plugge, C.M. (2017). Biogas. *Microb. Biotechnol.* 10 (5): 1128–1130.

Prabakar, D., Manimudi, V.T., Sampath, S. et al. (2018). Advanced biohydrogen production using pretreated industrial waste: outlook and prospects. *Renewable Sustainable Energy Rev.* 96: 306–324.

Rodionova, M.V., Poudyal, R.S., Tiwari, I. et al. (2017). Biofuel production: challenges and opportunities. *Int. J. Hydrogen Energy.* 42 (12): 8450–8461.

Rudakiya, D.M. and Gupte, A. (2017). Degradation of hardwoods by treatment of white rot fungi and its pyrolysis kinetics studies. *Int. Biodeterior. Biodegrad.* 120: 21–35.

Sohedein, M.N.A., Wan, W.A.A.Q.I., Hui-Yin, Y. et al. (2020). Optimisation of biomass and lipid production of a tropical thraustochytrid *Aurantiochytrium* sp. UMACC-T023 in submerged-liquid fermentation for large-scale biodiesel production. *Biocatal. Agric. Biotechnol.* 23: 101496.

Teferra, D.M. and Wubu, W. (2018). Biogas for clean energy. In: *Anaerobic Digestion*. (ed. J. Rajesh Banu), IntechOpen. doi: 10.5772/intechopen.79534. Available from: https://www.intechopen.com/books/anaerobic-digestion/biogas-for-clean-energy

Theuerl, S., Herrmann, C., Heiermann, M. et al. (2019). The future agricultural biogas plant in Germany: a vision. *Energies.* 12 (3): 396.

Tongco, J.V., Kim, S., Oh, B.-R. et al. (2020). Enhancement of hydrolysis and biogas production of primary sludge by use of mixtures of protease and lipase. *Biotechnol. Bioprocess Eng.* 25 (1): 132–140.

Trinh, L.T.P., Lee, Y.-J., Park, C.S., and Bae, H.-J. (2019). Aqueous acidified ionic liquid pretreatment for bioethanol production and concentration of produced ethanol by pervaporation. *J. Ind. Eng. Chem.* 69: 57–65.

Vunnava, V.S.G. and Singh, S. (2020). Spatial life cycle analysis of soybean-based biodiesel production in Indiana, USA using process modeling. *Processes.* 8 (4): 392.

Wang, R., You, T., Yang, G., and Feng, X. (2017). Efficient short time white rot–brown rot fungal pretreatments for the enhancement of enzymatic saccharification of corn cobs. *ACS Sustainable Chem. Eng.* 5 (11): 10849–10857.

Wei, N., Oh, E.J., Million, G., Cate, J.H.D., and Jin, Y.-S. (2015). Simultaneous utilization of cellobiose, xylose, and acetic acid from lignocellulosic biomass for biofuel production by an engineered yeast platform. *ACS Synth. Biol.* 4(6): 707–713. doi: 10.1021/sb500364q

Xue, S., Song, J., Wang, X. et al. (2020). A systematic comparison of biogas development and related policies between China and Europe and corresponding insights. *Renewable Sustainable Energy Rev.* 117: 109474.

Yesilyurt, M.K., Cesur, C., Aslan, V., and Yilbasi, Z. (2020). The production of biodiesel from safflower (*Carthamus tinctorius* L.) oil as a potential feedstock and its usage in compression ignition engine: a comprehensive review. *Renewable Sustainable Energy Rev.* 119: 109574.

Yun, Y.-M., Lee, M.-K., Im, S.-W. et al. (2018). Biohydrogen production from food waste: current status, limitations, and future perspectives. *Bioresour. Technol.* 248: 79–87.

Zabed, H., Sahu, J.N., Suely, A. et al. (2017). Bioethanol production from renewable sources: current perspectives and technological progress. *Renewable Sustainable Energy Rev.* 71: 475–501.

1

Role of Enzymes in Biofuel Production

Ashok Kumar Yadav[1], *Surabhi Pandey*[2], *Abhishek Dutt Tripathi*[2] *and Veena Paul*[2]

[1] *Food Processing and Management, DDU Kaushal Kendra, Rajiv Gandhi South Campus, Banaras Hindu University, Barkachha, Mirzapur, UP, India*
[2] *Department of Dairy Science and Food Technology, Institute of Agricultural Sciences, Banaras Hindu University, Varanasi, UP, India*

1.1 Introduction

The International Energy Agency (2006) released a factsheet based on a survey stating that approximately 80.3% of fossil fuels utilized as a principal energy source, and 57.7% employed for transport purposes. The demand for energy will grow by about 37% by 2040 due to the significant change in the global energy system (International Energy Agency 2006). There is a need for sustainable development by producing renewable and alternative energy sources at reachable costs with minimum hindrances. Biofuel represents renewable energy sources that combust and emits a lower amount of CO_2 and SO_2. Hence, it can be used in the future to restore wholly insufficient petroleum oil. Biofuels can be obtained easily from the naturally available sources, such as oil and starch or carbohydrate.

Due to the relentless exhaustion of renewable energy resources, for their biodiversity, biofuels promise to be the most suitable and recommended oil for the future. They are ideal for replacing petroleum fuels that decrease the demand for imported fuels. Biomass conversion is a sustainable way to produce biofuels like biodiesel, biohydrogen, and bioethanol, which prevents the release of harmful gases and is also accountable for reducing gases like CO_2. This sustainable behavior of biofuel enhances the demand of enzymes for increased production of biofuel. If enzymes were going to be used more and more for the production of biofuel, then it may prove to give a positive impact on environment with improved biofuel quality (Christy et al. 2014). Enzymes play an essential role in biofuel production, as they reduce the use of synthetic methods and are advantageous over these methods. The enzymatic method of biofuel production is energy-saving and eco-friendly as compared to the synthetic process. The enzymes act as a biocatalyst that employs fatty acids and cellulose as feedstock for biofuel production. Enzymes are used for the hydrolysis of feedstock into simpler components for its fermentation. The enzymatic hydrolysis process is cheaper than other methods as it is done in controlled and natural conditions without the need for any treatment when used for further steps (Christy et al. 2014). Table 1.1 depicts the benefits and drawbacks of biofuels.

Bioenergy Research: Evaluating Strategies for Commercialization and Sustainability, First Edition.
Edited by Neha Srivastava and Manish Srivastava.

Table 1.1 Benefits and drawbacks of biofuel.

With respect to	Benefits	Drawbacks
Price	Energy security Reduces the import of crude oil	High cost of manufacturing
Vehicles	Safe Produce less noise pollution Increases lubrication Increases ignition	Causes corrosion of engine
Pollution	Reduces the emission of CO_2 and carbon particles Biodegradable Nontoxic in nature	Increases the emission of NO_2 Increases the emission of aldehydes

This chapter focusses on the application of enzymes for the production of biofuels in the crude oil industry. It also deals with advanced techniques related to biofuel production.

1.2 Biofuel Classification

Based on the raw material used, we can classify biofuels as the first, second, and third generation of biofuels. The sources of first-generation biofuels are agricultural crops and grains, second-generation biofuels are obtained from agricultural mass like feedstock (Table 1.2), and third-generation biofuels use microorganisms like bacteria, fungi, and cyanobacteria as their feedstock (Tabatabaei et al. 2011).

Table 1.2 Biofuel classification with examples based on the sources.

"First generation biofuels" (sources are starch, grain, and cereal seeds)	"Second-generation biofuels" (sources are cellulosic mass, like waste of crops or crop woods)
Substitutes of Gasoline:	
• Products of fermentation of sugar and starch like butanol and ethanol	• Products of enzymatic hydrolysis like butanol and ethanol
Substitutes of Petroleum Diesel	
• Products of transesterification of plant oils like biodiesel	• General alcohol • Gasoline • Methanol
Pure plant oils	
• Vegetable oil	• Ecofriendly diesel • Fischer-Tropsch (FT) diesel • Dimethyl ether

1.3 Enzymes Role in Biofuels

Due to consumer awareness toward biofuel, its market demand has increased tremendously. The feedstock used for the biofuel production process undergoes pretreatment, which helps to remove the impurities. The concept of employing enzymes in biofuel production process increases the kinetics of the process and leads to easier and simpler conversion process of free fatty acids (FFA) to fatty acid methyl esters (FAME). Enzymes increase the reaction kinetics for the transesterification. The conventional production process generates wastewater with the low recovery of glycerol, and the process is costly, whereas employing enzymes for the catalysis will subsequently reduce the wastewater generation (Kulkarni and Dalai 2006). With several advantages, the enzymatic process also has some disadvantages, such as a reduced rate of reaction, expense in industrial use, and generally low stability (Luković et al. 2011). However, enzymes act as a potent biocatalyst due to their product specificity, and they are less toxic. So, they are a suitable biocatalyst for biofuel production at the industrial level, and this will leads to a cheaper and economical production process. Various industries like food, beverages, medicine, textile, and cosmetics are using different enzymes for product development or product improvement because of their low waste generation and low energy demands. Except this, the catalytic activity of the enzyme for biofuel production is gaining the attention of researchers. However, employing enzymes as a catalyst in the chemical conversion process results in a reduced yield because of low enzyme stability. Enzyme stability during the production process is affected by the high-temperature treatment as well as the toxicity of the solvents. Hence, the research focus should be on improving the enzyme stability for quality yield (Chapman et al. 2018). Table 1.3 represents the widespread applications of industrial enzymes.

Table 1.3 Applications of different enzymes in different industries.

Industries	Name of the enzymes	Applications
Medicine	Lipase, Nitrile hydratase, Transaminase, Penicillin acylase, monoamine oxidase	Act as fillers
Food processing	Pectinase, Papain, Trypsin, Amylase, Glucose isomerase and many more	Enzymes convert starch to glucose, helpful in production of high fructose corn syrup, debittering of juice, tenderization of meat, clarification of beverages
Textile	Lipase, Protease, Amylase,	Removal of stain, fat, oil, and dirt marks, helpful in color retention.
Fuels	Lipase, Cellulase, Xylanase	Production of FAME, decomposition of cellulosic material for the production of bioethanol
Paper	Lipase, Cellulase, Xylanase	Improves bleaching, Improves the strength of the paper

1.4 Enzymatic Reaction

Enzymes convert FFA present in the feedstock in two stages: first, they hydrolyze glycerides into FFA, and then these FFA, along with methanol, produces FAME by esterification. Water is used for the activation of the enzyme. During the enzymatic reactions, various parameters, such as conversion rate, the reaction between enzyme, water, and methanol are essential and responsible for the quality yield and reusability of enzymes used.

1.5 Enzyme Recovery and Reuse

After completion of the reaction, the recovery of the enzyme is also essential. This recovery can be achieved in two steps. In the first step, the raw biofuels (light phase) are separated from the dense phase (enzymes, glycerol, methanol, and water). The light phase is then refined, and the dense phase is filtered. In the second step, the filtered dense phase is collected in the reaction vessel, and then due to flocculation or mechanical separation (membrane separation), the enzyme mitigates at the time of interaction between glycerin and the impure biofuel. The layer of the active enzyme along with leftover glycerin is utilized in the next catalytic reaction. This recovery method results in 8–10 times increased enzyme reusability. Vacuum processing can be used for the purification of glycerin. This method recovers 80% pure glycerin. However, using advanced techniques, such as employing resins will increases the purity of glycerin up to 97%. This glycerin can be used further for value addition.

1.6 Enzyme Immobilization

The enzyme can be immobilized, improving the reaction, enzyme stability, rate, and enzyme reuse. Immobilized enzymes are gaining attention in the biofuel industry because of their reusability (Nielsen et al. 2008). Enzyme immobilization means linkage of the particular enzymes over or within carrier binding, cross-linking, or entrapment in wall material, to enhance the enzyme activity and stability (Chapman et al. 2018). The advantage of the enzyme immobilization technique has enhanced the demand for an immobilized enzyme for the purpose of biofuels production. It has been observed that use of immobilized lipases results in a high recovery of desired fuel, high thermal stability, and reduced waste and effluent generation as compared to normal lipases (Li et al. 2012). Nevertheless, immobilized enzymes have some demerits, such as throughout the immobilization process, the enzymatic activity may lose expensive carrier material, and the reduced enzyme stability in the oil-water system (Luna et al. 2016). The commonly used immobilized enzyme is Novozyme 435. Enzyme immobilization can be achieved by employing various immobilization techniques, such as binding on specific carrier (adsorption over physical surface, formation of affinity, or covalent bonding), cross-linkage, and entanglement or envelopment (i.e. fiber-like entrapment or encapsulation) (Figure 1.1).

These different techniques of enzymes immobilization have been differentiated into two categories: one is irreversible process and the other is reversible process, which depends

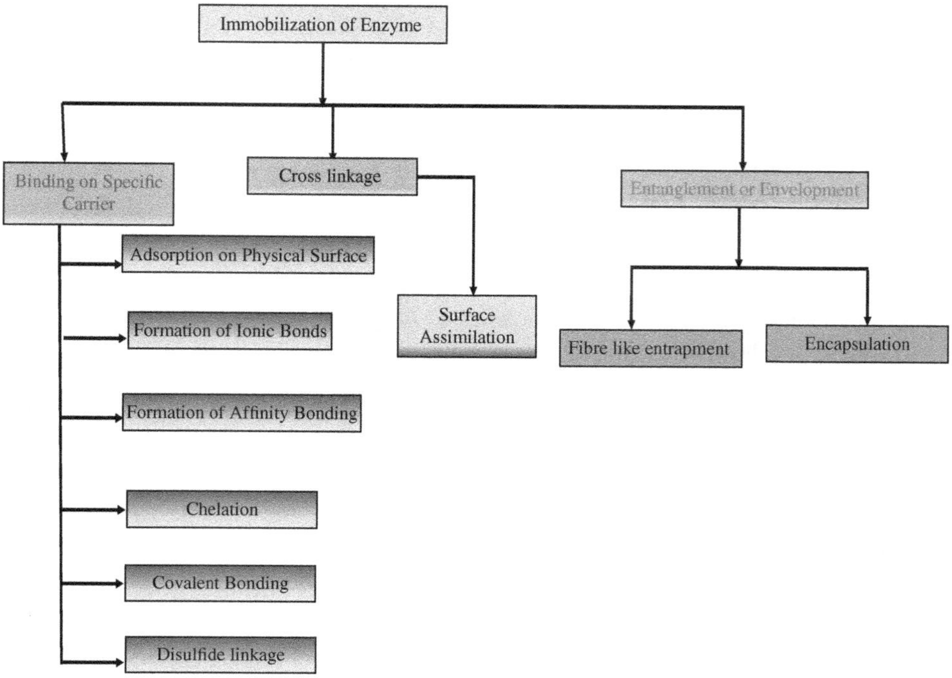

Figure 1.1 Enzyme immobilization techniques.

basically on nature of the bonding between particular enzymes with suitable carriers. Immobilization techniques, such as entrapment, cross-linking, and covalent bonding are irreversible processes, whereas techniques like affinity bonding are reversible (Zhao et al. 2015; Luna et al. 2016).

1.6.1 Adsorption on Physical Surface: Physical Adsorption

The most commonly used immobilization technique is physical adsorption. This technique requires no chemical linkages. Examples of physical adsorption are hydrophobic bonds, biospecific interaction, and electrostatic linkages. Immobilization using physical adsorption has several advantages, such as the low cost of carrier material used, the ease of the operation process, and no use of chemical additives and high recovery and reusability of the enzyme (Luna et al. 2016).

1.6.2 Ionic Bonding

This technique immobilizes the enzyme by ionic bonding, where ionic linkages immobilize the enzymes. In this technique, the ionic linkages are much more stable than that of the physical adsorption technique. Ionic bonding requires mild conditions for immobilization that result in modification of enzyme conformation and active site (Luna et al. 2016).

1.6.3 Entanglement or Envelopment

Entanglement or envelopment means the binding of desired enzyme in or into a polymeric wall material. This technique is more stable in comparison to other immobilization techniques, and the immobilization process is simple. However, the drawback is that during the biofuel manufacturing procedure, the life span is low and most important, the conversion rate of encapsulated enzyme is low (Luna et al. 2016).

1.6.4 Cross-Linkage

The cross-linkage technique for immobilizing enzymes involves the formation of cross-linkages or bonds. For this purpose, various cross-linking reagents (glutaraldehyde) are used (Zhao et al. 2015). The cross-linkage with enzyme forms a 3-D structure. The immobilized enzymes are present in the form of complex with cross-linking agents. The enzyme immobilized by this technique is stable for a broad range of pH and temperature. Despite these advantages, this technique requires a harsh environment in which the cross-linking reagents are used that are capable of altering the enzyme conformation, which results in a reduction of enzyme activity and low yield. These demerits can be overcome by using the cross-linkage immobilization technique in combination with other techniques, like physical adsorption (Luna et al. 2016).

1.7 Unique Techniques of Enzyme Immobilization

Nowadays, many advanced and unique technologies for immobilization of enzymes are used for the production of biofuel to enhance the stability of enzymes, and they can be further used, if necessary. Some of the novel immobilization techniques are electrospun nanofibers protein-coated microcrystals (PCMC), particles of magnetic carriers, and cross-linked PCMC. These novel techniques help in the easier separation of enzymes and fuel (Luna et al. 2016).

Enzymatic catalysis is a novel approach for biofuel production. These novel techniques will improve the sustainability of enzymes and increase the reactivity of the enzymes used. Finally, all this leads to a more sustainable and economical biofuel production process (Luna et al. 2016). Various novel approaches of enzyme immobilization for biofuel production are depicted in Table 1.4.

1.8 Application of Various Enzymes in Biofuel Production

The biofuel production process using enzyme catalysis is one of the best, unique, and efficient techniques to increase the production of biofuel. Some of the enzymes used in biofuel production are discussed in this chapter.

1.8.1 Amylases

The starch-rich feedstock comprises long polymeric chains of glucose that are needed to break down into simpler molecules. For the degradation of these long polymeric chains

Table 1.4 Unique methods for biodiesel production with their advantages and drawbacks.

Methods	Advantage	Drawbacks
Combination of lipase enzyme	Enhanced specificity of substrate, increase in yields, reaction time is less	Combination of enzymes is difficult and it can be done only by genetic engineering is sophisticated process
Use of ionic liquids as solvent	Specificity, action of enzyme and Life span gets increased	Not economical
Transesterification by the help of enzymes under the influence of supercritical fluid	Rate of reaction get increased, and improvement in rate of diffusion took place, oils can easily be removed in supercritical medium	Due to use sophisticated instruments it is also not economical
Transesterification in the presence of enzyme	Low cost of feedstock, produces less pollution	Transportation is difficult
process without solvent	Economical, produces less pollution and safe to use	Reaction has limitations due to mass transfer
Transesterification of microalgae by in situ method	No need of solvent or less required, less energy consumption	Economical only when the biomass has higher quantity of fat percentage

into simpler molecules, enzyme amylases are used. Amylase enzymes hydrolyze the α-1,4 linkage of starch molecules into dextrins. Dextrins are then further hydrolyzed into simpler sugar (maltose and isomaltose) for the easy conversion of feedstock in biofuels (Visioli et al. 2014), which is an advanced and unique mechanism process that polymerizes the starch using isomerases. This polymerization yields fructose that is further converted into 2,5-dimethylfuran. However, the use of these enzymes does not give promising results (Chheda et al. 2007).

1.8.2 Proteases

Protease enzymes are used in the biofuel industry as a nitrogen provider within the yeast fermentation. During the process of fermentation, the proteases alter the complex structure of feedstock by breaking the complex starch-gluten molecules into simpler molecules (Alvarez et al. 2010). Proteases are also used for hydrolyzing the oleosins, present in corn seed feedstock (Huang 1996). The utilization of this enzyme under optimal conditions will improve the biofuel yields (Harris et al. 2014).

1.8.3 Dehydrogenases

Dehydrogenases have emerged as a potent enzyme for converting carbon dioxide into methanol (Jiang et al. 2003). The conversion process takes place in three steps. First, formate dehydrogenase converts the carbon dioxide into formate, formaldehyde dehydrogenase converts the formate into formaldehyde, and lastly, alcohol dehydrogenase converts the formaldehyde into methanol. However, due to multiple stages of conversion, this method is not industrially acceptable.

1.8.4 Lipase

Lipases (Enzyme Council no. 3.1.1.3) are an important class of enzymes for biocatalysis for the biofuel production process. Lipases are generally enzymes of class hydrolases capable of hydrolyzing the complex reaction involved in the production process. Lipases are used mainly for biodiesels production (Luković et al. 2011). Mittelbach (1990) has studied the application of lipase for biodiesels production. The sources of this enzyme are many; we can extract them from plants, such as castor seed, stem of papaya, coarse cereal oat, animals (human pancreatic cells, pig), and microorganisms. For the industrial-scale production of lipases, microorganisms are preferred as the best source. Microorganisms are a potent source for the substrate conversion into the lipase enzyme. Bacteria and fungi can be used to isolate lipases for biodiesel production, like *Pseudomonas cepacia, Thermomyces lanuginose, Pseudomonas fluorescens, Rhizopus oryzae, Aspergillus niger, Streptomyces* sp., *Candida antarctica, Photobacterium lipolyticum, Chromobacterium viscosum, and Candida rugose* (Luković et al. 2011).

Lipases mainly hydrolyze the carboxylic ester bond of triacylglycerol to form FFA and glycerols. Due to its potential for esterification and transesterification, these are essential enzymes of the biotechnological industries and can be extra or intracellular. The extracellular lipases are recovered and purified from the fermentation broth, whereas intracellular lipases are produced from inside the cells (Luna et al. 2016).

Lipases are the biocatalyst that are used for transesterification reaction during the production of biofuel (Table 1.5). Simões et al. (2015) reported that SiO2 immobilized lipases produce 96% biodiesels using babassu oil as a feedstock. Similarly, Bergamasco et al. (2013) studied that lipases immobilized with poly vinyl alcohol (PVA) yielded 66.3% biodiesel. Employing lipases as a biocatalyst helps glycerol recovery and yields high purity biofuels.

Since yeast and fungal lipase are easily available and very economical, they are commonly used. Some of the bacterial lipases produced from *Pseudomonas* species, *Enterobacter aerogenes, and Bacillus subtilis* are used based on their methanol resistance. Besides, some commercial lipases are also used because of their suitability as a catalyst (Luković et al. 2011).

1.9 Biofuel Production Process

The biofuel production process is gaining more attention due to its various advantages such as reduced greenhouse gas emissions, energy-saving, renewable sources, and its easy applicability as a fuel in other sectors (Nielsen et al. 2008).

1.9.1 Bioethanol

Bioethanol is one of the vital biofuels due to its high demand for the market. Bioethanol is produced from starch, sugars, or lignocellulosic biomass as a feedstock. The production

Table 1.5 Biodiesel production with various lipases.

Lipases types	Oil as raw material	Acyl acceptor	Solvents	Time in hours/ Yield in %
Candida antarctica	Palm	Methanol	Tertiary butanol	4/79.1
Thermomyces lanuginosa	Soybean	Ethanol	Normal hexane/without solvent	10/70–100
T. lanuginosa *Pseudomonas fluorescens*	Sunflower	Methanol	solvent free	24/90
P. fluorescens *Burkholderia cepacia* *Candida rugosa* *Mucor javanicus*	Nettle spurge	Ethanol	No solvent is used	8/98
Candida spp	Cooking	Methanol	Normal hexane	10/91.08
C. antarctica	Cotton seed	Methanol	Tertiary butanol	24/97
Criptococcus spp	Rice bran	Methanol	solvent-free	120/80
Penicilium cyclopium *Rhizomucor miehei*	Soybean	Methanol	No solvent is used	12/68–95
C. antarctica	Soybean	Methyl acetate	No solvent is used	14/92
Bacillus subtilis	Cooking	Methanol	solvent-free	72/90
T. lanuginosa	Sunflower, Soybean	Methanol	No solvent is used	24/90–97
Penicillium expansum	Waste	Methanol	Tertiary amyl alcohol	24/92
C. antarctica	Sunflower	Methyl acetate	No solvent is used	12/95
Enterobacter aerogenes	Jatropha	Methanol	Tertiary butanol	48/94
C. antarctica	Cotton seed	Methanol Propanol Butanol Amyl alcohol	No solvent is used	7/91.5
C. antarctica	Rapeseed	Methanol	No solvent is used	24/91.1
T. lanuginosa *C. antarctica*	Rapeseed	Methanol	Tertiary butanol	12/95
B. cepacia	Soybean	Methanol	solvent-free	90/80
C. antarctica	Sunflower, Jatropa	Ethyl-acetate	No solvent is used	12/90
R. miehei	Sunflower	Methanol	Normal hexane	24/80

Source: Adapted from Luković et al. (2011).

process for bioethanol involves pretreatment, fermentation, and purification. The pretreatment of the feedstock varies based on the feedstock material used. Thus, milling, soaking, physical, chemical, biological, and enzymatic pretreatment can be used, depending on the raw material's composition. A very common feedstock in bioethanol production is cornstarch. Dry milling and wet milling are the pretreatment methods for cornstarch. In the former one, the corn is usually ground, whereas the corn is first soaked in water and then further ground for the removal of the nonstarch portion. The dry milling processed cornstarch is used for bioethanol production (Figure 1.2). The cornstarch slurry obtained from

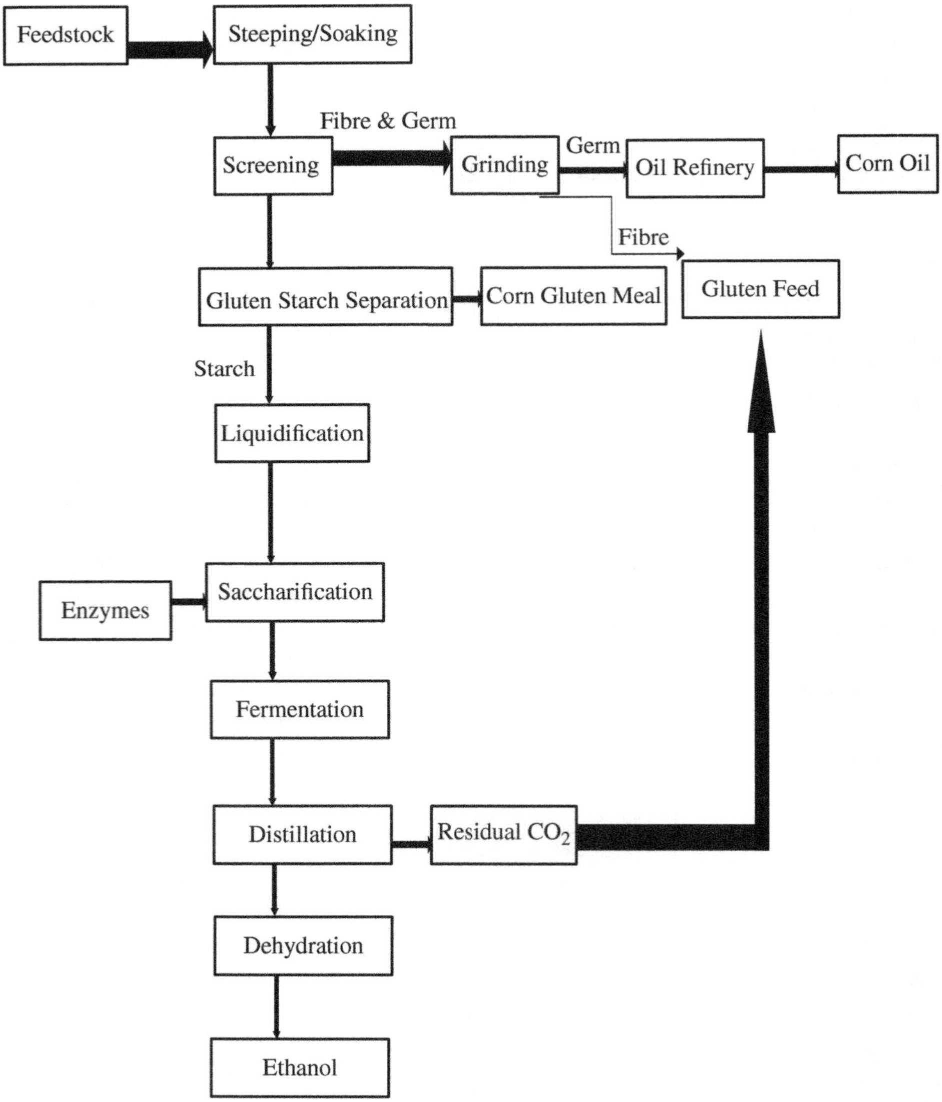

Figure 1.2 Ethanol production by wet milling method.

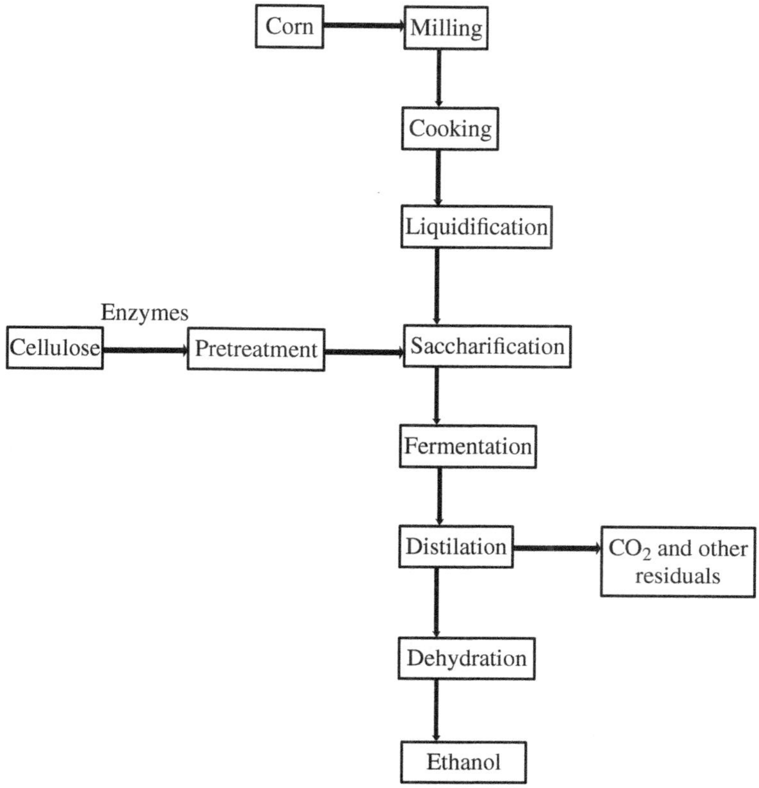

Figure 1.3 Ethanol production by dry milling method.

wet milling is further treated with amylase, which hydrolyzes the starch for bioethanol production (Figure 1.3) (Mollahoseini et al. 2015).

1.9.2 Biohydrogen

Biohydrogen is acknowledged as an alternative to petroleum fuels due to its high energy. For the production of biohydrogen, feedstock, and microorganisms, both are essential parameters, and this parameter depends on various factors like pH, temperature, feedstock composition, and age of inoculum used. A commonly used and preferred production process for this biofuel is fermentation, particularly in a dark process. The dark fermentation process involves two steps: in the first step, the substrate is hydrolyzed, followed by acidogenesis, which is the second step. Mesophilic or thermophilic microorganisms can produce biohydrogen. Among these, thermophilic microorganisms are preferred because of their high hydrolysis capability at 70–80 °C (Mollahoseini et al. 2015).

1.9.3 Biomethane

The biomethane production process involves hydrolysis, acidogenesis, dehydrogenation, and methanation steps. The hydrolysis of feedstock depends on its type and composition.

Microorganisms involved in the production of biomethane are *Clostridium and Bifidobacterium*. During biomethane production, propionates, hydrogen, butyrates, and acetates are produced as a by-product. In acidogenesis, the microbial culture produces alcohols, carbon dioxide, organic acids (acetic, butyric, and propionic acid), and hydrogen. In dehydrogenation steps, the products produced in the second step are considered as substrate and produce acetic acid and hydrogen as a by-product. In the last step, methanogens are used for the methanation. Microorganisms used for this conversion process are *Methanosarcina barkeri, Metanonococcus mazei*, and *Methanosaeta concilii*. This biomethane production process mainly depends upon factors like pH, temperature, substrate, and the enzyme used (Mollahoseini et al. 2015).

1.9.4 Biodiesel

The demand of biofuels has increased due to its environment friendly and renewable source of energy. It is a kind of biofuel that is used as an alternative source of petroleum. Biodiesel has a high combustion rate and cetane rating, and as a result, there is less sulphur and aromatic component discharge. Biodiesel also has some demerits, such as high visibility, less energy, high pour point and cloud point, high energy consumption, and expense (Luković et al. 2011). The flashpoint of biodiesel is much higher when compared with general diesel. A report by Chetri et al. (2008) suggested that it comprises 10–11% oxygen by mass and no sulphur and aromatic components. Table 1.6 describes the different compounds of biodiesel and diesel (Kulkarni et al. 2008).

Currently, the available transport fuels can be replaced by biodiesel with some modifications in engines and separation process. Generally, alkaline are used for reaction during the production of biodiesel because it is an energy-saving process. Lipases are preferred for biodiesel over other enzymes because they were more specific in their reaction, which enhances biodiesel efficiency and ensures purity. The drawback of this production process is the expensive purification process, and thus, the reaction rate is little bit slower.

Table 1.6 Compounds present in diesel and biodiesel (Kulkarni et al. 2008).

Compounds represented in %	Diesel	Biodiesel
Carbon (C)	86.4	79.6
Hydrogen (H)	13.6	10.5
C/H Ratio	6.35	7.58
Aliphatics	67.4	15.2
Oxygen	–	8.6
Aromatics	20.1	–
Nitrogen	–	1.3
Naphtens	9.1	–
Olephenics	3.4	84.7

Source: Kulkarni et al. (2008).

The reaction rate may be increased on adopting advanced molecular technologies (Luna et al. 2016).

Due to the presence of oxygen, biodiesels exhibit a negative polar structure, whereas traditionally available diesel shows positive polar structure, which results in a higher viscosity of biodiesels. Besides, there is a reduced biodiesel temperature due to the oxygen content (Kulkarni et al. 2008; Capehart 2007; Gerpen et al. 2004). Biodiesels are used either in pure form or in mixed form. The commonly used biodiesel is B2, in which 2% biodiesel are mixed with 98% diesel, another one having a mixture of 5% biodiesel and 95% diesel termed as B5, and the last one B20, which is having 20% biodiesel and 80% diesel, so the numbering is as per the percentage of biodiesel present in the mixture (Pandey 2008). The enzymes act as a biocatalyst during biodiesel manufacturing and this helps to enhance the reaction rate. These enzymatic reactions are more efficient than that of chemical reactions yielding a quality biodiesel by using a more straightforward separation process. The merit of using enzymes in the production process is that it can be reused in the future, which makes this process eco-friendly (Mittelbach and Remschmidt 2004; Akoh et al. 2007; Jegannathan et al. 2008; Ghaly et al. 2010). During biodiesel production, enzymes like lipase are helpful in both types: esterification reaction and transesterification reaction. Figure 1.4 depicts the enzymatic transesterification during the biodiesel production process (Fan 2012).

There are certain demerits of using chemical methods for biodiesel production. These are: (i) low glycerol content; (ii) expensive recycling; and (iii) high chemical waste generation. These problems can be controlled by employing the use of enzymes. These enzymes act as a selective catalyst for enhancing the overall yield. Lipase are used in production process, thus, the process is energy-saving and inhibits the formation of unwanted products. This enzyme also favors the use of free fatty acids as a reactant in the production of biodiesel, which is converted to FAME through esterification and favors the utilization of triacylglycerols for the production of FAME by transesterification.

In the production process of biodiesel, acyl acceptors are methanol. The various catalysts involved in the transesterification reaction are alkaline catalysts (NaOH, KOH, and CH_3NaO), acid catalyst (H_2SO_4, HCl, and H_3PO_4), and enzymatic catalysts (lipase). Among all these, alkaline catalysts are preferable because they are economical, highly efficient and specific for product formation. However, the drawback is the saponification that occur

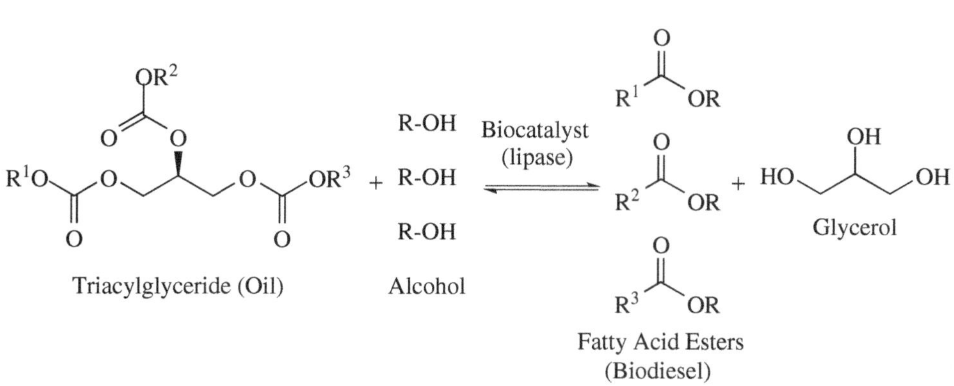

Figure 1.4 Transesterification reaction for biodiesel production using lipase as a catalyst.

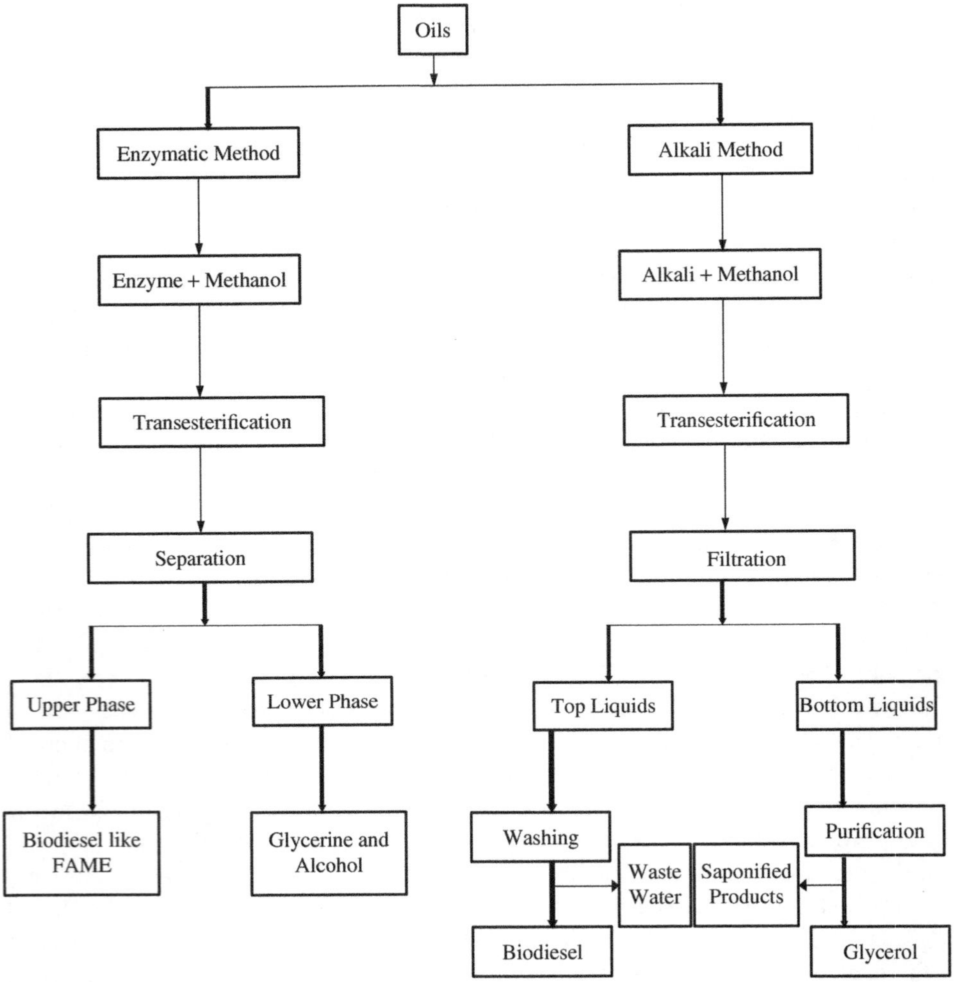

Figure 1.5 Comparative analysis of biodiesel production by enzymatic and alkali methods.

during the separation step. Figure 1.5 depicts biodiesel production process using alkaline and enzymatic catalyst. Because of several advantages, the enzymatic catalysis process is preferable.

1.10 Production of Biodiesel by Enzymatic Catalysis

For the production of biodiesel by enzymatic catalysis, bioreactor's design and process parameters are essential, as these parameters help improve the stability and recovery of enzymes. Three different bioreactors are used, namely, batch, continuous, and packed-bed reactors (Nielsen et al. 2008). Various steps that are involved in production of biodiesel has been shown in Figure 1.6.

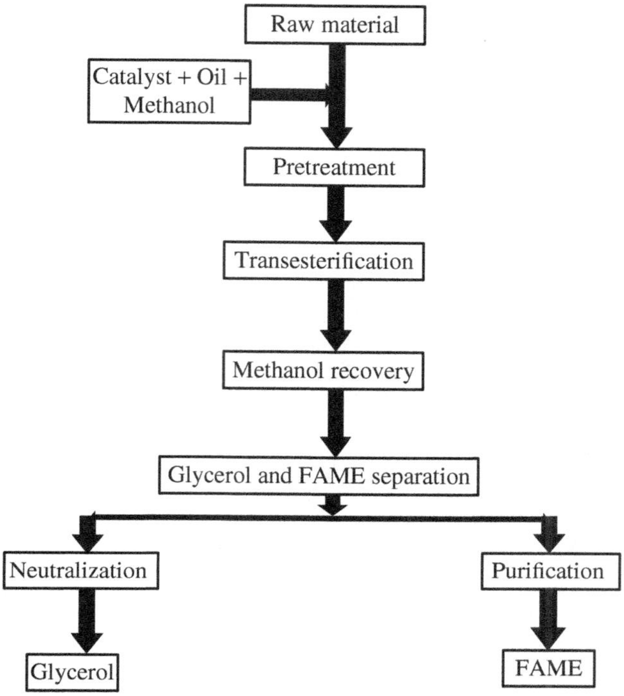

Figure 1.6 Flow chart for biodiesel production.

1.10.1 Batch Method

Batch method of fermentation is widely taken into account for the production of biodiesel because various factors can be monitored easily in this process. This process is best suited for the production of enzymes. However, this process has certain drawbacks, such as the need for large-volume tanks, a decline in enzymatic activity with time, and the lack of continuous process (Nielsen et al. 2008).

1.10.2 Continuous Stirred-Tank Method

There is a continuous supply of raw material and continuous removal of the finished product takes place in the continuous stirred-tank method. This process needs multiple tanks and ensures equal conversion rate in the equal reaction time, which indicates that the total volume of tanks is significant. This process has the merit of using a different class of enzymes at the same time (Nielsen et al. 2008).

1.10.3 Packed-Bed Columns

In this type of bioreactor, immobilized enzymes are used, which act as a solid catalyst. The advantage of using this process is that it increases the ratio of the enzyme to the substrate at a specific time. This process is commonly used for the interesterification of oils at the

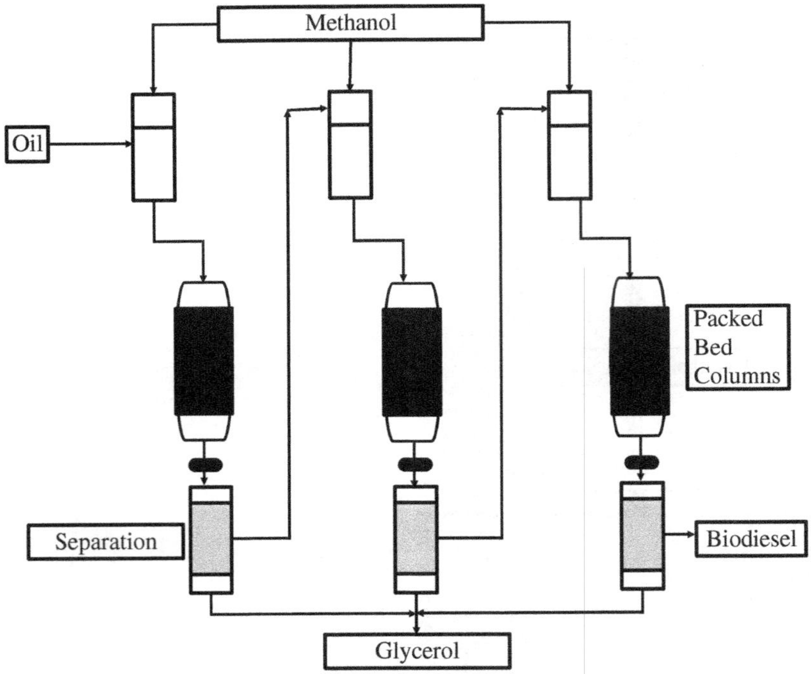

Figure 1.7 Packed-bed column method for biodiesel production.

industrial level. In biodiesel production, this process helps for the easy removal of glycerol (Figure 1.7) (Nielsen et al. 2008).

1.11 Future Prospects

The recent trend has shown that employing enzymes (free or immobilized) in biofuel production is gaining attention. The stability of the enzyme is one factor of concern. More research to improve enzymatic stability during the production process is needed. Metabolic engineering can also be employed to increase the efficacy, and to make the production method more economical. Gene cloning with recombinant cells of suitable microorganisms can be used for enhancing productivity. The novel immobilization technique will enhance enzymatic activity and stability with the availability of cost-effective immobilized enzymes.

1.12 Conclusion

The awareness of biofuel will lead to the rise of a sustainable and eco-friendly environment. Employing microbial culture and enzymes will help to make the biofuel production process sustainable. Replacing the biofuels with synthetic fuels will help minimize air pollution, as biofuels do not emit harmful gasses. However, the alkaline catalyst is preferred for biofuel production because of its low cost, which is needed to replace it with a biocatalyst, i.e.

enzymes. The advantages of employing enzymes will help in the development of a sustainable environment. Studies for the availability of low-cost immobilized enzymes are required for efficient biofuel production.

References

Akoh, C.C., Chang, S.W., Lee, G.C., and Shaw, J.F. (2007). Enzymatic approach to biodiesel production. *J. Agric. Food. Chem.* 55: 8995–9005.

Alvarez, M.M., Carrillio, P.E., and Saldivar, S.S. (2010). Effect of decortications and protease treatment on the kinetics of liquefaction, saccharification and ethanol production from sorghum. *J. Chem. Technol. Biotechnol.* 85: 1122–1130.

Bergamasco, J., de Araujo, M.V., de Vasconcellos, A. et al. (2013). Enzymatic transesterification of soybean oil with ethanol using lipases immobilized on highly crystalline PVA microspheres. *Biomass Bioenergy* 59: 218–233.

Capehart, B.L. (2007). *Encyclopedia of Energy Engineering and Technology*. CRC Press/Taylor & Francis LLC. ISBN-13: 978-0849336539.

Chapman, J., Ismail, A.E., and Dinu, C.Z. (2018). Industrial applications of enzymes: recent advances, techniques, and outlooks: review. *Catalyst* 8 (6): 238–258.

Chetri, A.B., Watts, K.C., and Islam, M.R. (2008). Waste cooking oil as an alternate feedstock for biodiesel production. *Energies* 1: 3–18.

Chheda, J.N., Roman-Leshkov, Y., and Dumesic, J.A. (2007). Production of 5-hydroxymethylfurfural and furfural by dehydration of biomass-derived mono- and polysaccharides. *Green Chem.* 9: 342–350.

Christy, P.M., Gopinath, L.R., and Divya, D. (2014). A review on anaerobic decomposition and enhancement of biogas production through enzymes and microorganisms. *Renewable Sustainable Energy Rev.* 34: 167–173.

Gerpen, J.V., Shanks, B., Pruszko, R. et al. (2004). *Biodiesel Production Technology*. CO, USA: National Renewable Energy Laboratory.

Ghaly, A.E., Dave, D., Brooks, M.S., and Budge, S. (2010). Production of biodiesel by enzymatic transesterification: review. *Am. J. Biochem. Biotechnol.* 6 (2): 54–76.

Harris, P.V., Xu, F., Kreel, N.E. et al. (2014). New enzyme insights drive advances in commercial ethanol production. *Curr. Opin. Chem. Biol.* 19: 162–170.

Huang, A.H. (1996). Oleosins and oil bodies in seeds and other organs. *Plant Physiol.* 110: 1055–1061.

International Energy Agency. Office of Energy Technology, R&D., & Group of Eight (Organization) (2006). *Energy Technology Perspectives*. International Energy Agency.

Jegannathan, K.R., Abang, S., Poncelet, D. et al. (2008). Production of biodiesel using immobilized lipase-a critical review. *Crit. Rev. Biotechnol.* 28: 253–264.

Jiang, Z., Wu, H., Xu, S., and Huang, S. (2003). Enzymatic conversion of carbon dioxide to methanol by dehydrogenases encapsulated in sol–gel matrix. In: *Utilization of Greenhouse Gases*, ACS Symposium Series, vol. 852 (eds. C.J. Lui, R.G. Mallinson and M. Aresta), 212–218. Washington, DC: American Chemical Society.

Kulkarni, M.G. and Dalai, A.K. (2006). Waste cooking oils an economical source for biodiesel : a review. *Ind. Eng. Chem. Res.* 45: 2901–2913.

Kulkarni, B.M., Bujar, B.G., and Shanmukhappa, S. (2008). Investigation of acid oil as a source of biodiesel. *Indian J. Chem. Tech.* 15: 467–471.

Li, J., Chen, S., Li, L. et al. (2012). Approach of directed evolution of microbial lipases and biodiesel catalysis. *Afr. J.Microbiol. Res.* 6: 637–642.

Luković, N., Knežević-Jugović, Z., and Bezbradica, D. (2011). Biodiesel fuel production by enzymatic transesterification of oils: recent trends, challenges and future perspectives. In: *Alternative Fuel* (ed. M. Manzanera), 47–72. Croatia: InTech Publishers.

Luna, C., Luna, D., Calero, J. et al. (2016). Biochemical catalytic production of biodiesel. In: *Handbook of Biofuels Production* (eds. R. Luque, C.S.K. Lin, K. Wilson and J. Clark), 165–199. Woodhead Publishing.

Mittelbach, M. (1990). Lipase catalyzed alcoholysis of sunflower oil. *J. Am. Oil Chem. Soc.* 67: 168–170.

Mittelbach, M. and Remschmidt, C. (2004). *Biodiesel the Comprehensive Handbook*, 69–80. Boersedruck Ges.m.b.H.

Mollahoseini, A., Tabatabaei, M., and Najafpour, G.D. (2015). Biofuel production. In: *Biochemical Engineering and Biotechnology*, 2e (ed. G. Najafpour), 597–630. Amsterdam: Elsevier.

Nielsen, P.M., Brask, J., and Fjerbaek, L. (2008). Enzymatic biodiesel production: technical and economical considerations. *Eur. J. Lipid Sci. Technol.* 110: 692–700.

Pandey, A. (2008). *Handbook of Plant-Based Biofuels*. CRC Press/Taylor & Francis LLC. ISBN 9781560221753.

Simões, A.S., Ramos, L., Freitas, L. et al. (2015). Performance of an enzymatic packed bed reactor running on babassu oil to yield fatty ethyl esters (FAEE) in a solvent-free system. *Biofuel Res. J.* 6: 242–247.

Tabatabaei, M., Sulaiman, A., Nikbakht, A.M. et al. (2011). Influential parameters on biomethane generation in anaerobic wastewater treatment plants. In: (ed. M. Manzanera), 227–262. Croatia: InTech Publishers https://doi.org/10.5772/24681.

Visioli, L.J., Enzweiler, H., Kuhn, R.C. et al. (2014). Recent advances on biobutanol production. *Sustain. Chem. Proc.* 2: 15–25.

Zhao, X., Qi, F., Yuan, C. et al. (2015). Lipase-catalyzed process for biodiesel production: enzyme immobilization, process simulation and optimization. *Renewable Sustainable Energy Rev.* 44: 182–197.

2

Microbial Technology for Biofuel Production

Spriha Raven[1], Sashita Bindu Ekka[2], Stephen Edward Chattree[3], Shivani Smita Sadanand[4], Lipi Rina[5] and Archana Tiwari[1]

[1] *Diatom Research Laboratory, Amity Institute of Biotechnology, Amity University, Noida, UP, India*
[2] *Department of Environmental Science, Indira Gandhi National Tribal University, Amarkantak, MP, India*
[3] *Galgotias University, Greater Noida, UP, India*
[4] *Department of Agriculture, SHUATS, Prayagraj, UP, India*
[5] *Department of Agronomy, SHUATS, Prayagraj, UP, India*

2.1 Introduction

The dwindling of fuels and increase in the deposition of greenhouse gasses has become the global concern of today, which is leading to climatic changes and thus food production concerns (Hallenbeck 2012). The production of biofuels (e.g. alcohols, biodiesels, bioethanol, biobutanol, hydrogen, and biogas) has simultaneously taken place to replace the conventional fuels to reduce the climatic concerns. Several technologies are used to produce various varieties of biofuels from various biomass and crude materials, such as microbes, animal and agricultural residues, seeds, vegetable oil, and lignocellulose (Makishah 2017).

As the primary concern about climatic changes, along with the rising of crude oil prices and the increased demand of energy has increased, the interest toward biofuel production is rising; the sustainable, renewable, efficient, and cost-effective energy that they produce contains much lower levels of carbon dioxide released into the atmosphere. Increased industrialization has increased the demand of petroleum-based fuels, particularly 80% fossil fuels have been used in the world, while 50% alone is used in the transportation sector (Escobar et al. 2009). The increased necessity of fuels in transportation has projected to be increased by 40% over the period 2010–2040. The demand of production of various biofuels made by microorganisms has attracted the interest of the global platform in recent years (Hallenbeck 2012).

There are two different microbial processes to convert biomass into biofuel: thermochemical and biochemical. Thermochemical conversion processes, such as direct combustion, gasification, and pyrolysis, converts organic components into biofuels thermochemically. Pyrolysis or gasification process is considered to be the best process to convert biomass in the form of liquid-fuel (bio-oil). Anaerobic digestion and alcoholic fermentation are the biochemical processes of energy conversion into biofuels. Currently, biodiesels and bioethanol are the commercially produced biofuels that are obtained by

Bioenergy Research: Evaluating Strategies for Commercialization and Sustainability, First Edition.
Edited by Neha Srivastava and Manish Srivastava.

the transesterification process (biochemical process) by processing of plants, animal fats, and vegetable oils. Biodiesels and bioethanol are the most important biofuels and can be used with petrol, diesels, and as transport fuels. The alcoholic-based biofuels can replace gasoline in spark ignition engines, and biodiesel, green diesels, and dimethyl ether (DME) can be used in compression ignition engines. The variety of hydrocarbons are produced by the Fischer-Tropsch liquid (FTL) process, which are used in compression engines (Nigam and Singh 2011). Biogas and biomethane are also used in some transport vehicles with some additional biofuels that are produced by lipid manures and other raw materials. Fermentation of sugars – converting plant sources into sugar – is the most strategically important and active part of the biofuel research, catalytic technology to convert ethanol to mixed hydrocarbon, and mixed acid fermentation by enteric bacteria are all processes used in microbial biofuel production. The establishment of improved technologies will accelerate the system efficiency and production of biofuels at a low cost.

2.2 Microbial Biofuel

The production of microbial biofuel has been commercialized, particularly biodiesel, bioethanol, biobutanol, microbial lipids, and biogas, and are the only alternative energy fuels to minimize environmental damages. Primarily microbial biofuels are already in use in the form of sugar fermentation. Production of biofuels from agricultural food materials like cereals, corn, maize, or soybean are known as first-generation biofuels. Lignocellulosic feedstocks are available abundantly all over the world, and due to the stability of their polymers and their cost-effectiveness they are considered as second-generation biofuels. The third-generation biofuels are produced from cyanobacteria, algae, and other microbes, and hold promising alternatives for future production as algae can produce 200 times more biomass per area than first- and second-generation biofuel sources (Balan et al. 2013).

In recent years, the microbial biofuel production has been recognized and well-studied. The studies show that biofuel production by a biofilm of cyanobacteria and microalgae is a promising pathway. It has been recognized as a new platform for biomass production pathways.

Biodiesel is produced from microorganism such as algae, fungi, yeast, and bacteria through the transesterification process as they biosynthesize and store a huge amount of intercellular lipids in their biomass. Microalgae oil provides 20–60% of dry weight biomass and is considered as the most favorable crude material to provide a high amount of lipids, carbohydrates, and proteins that are synthesized directly into biodiesel. Biodiesel is engineered by using *Escherichia coli* and *Saccharomyces cerevisiae,* which contain heterologous genes.

Bioethanol as a biofuel is the most environment-friendly product produced by using a renewable biomass of ethanol. Bioethanol can be produced by some crude material, which contains a good amount of carbohydrates, and which provides the fermentable sugars such as corn, wheat, sugar beet, sugarcane, and maize. About 50% of algae is composed of oil, hence it is considered to be used for biofuel production. *Spirulina* species of algae is used in biofuel production through gasification process.

Biobutanol is considered as promising biofuel because of its high energy level, high volatility, and hygroscopicity, and they are produced by the fermentation process of *Clostridium* species, which converts sugars into solvents like butanol, acetone, and ethanol.

2.3 Microbial Pathway for Biofuel Production

Currently, the most cost-effective renewable energy biofuels, on a large scale, are produced by engineering of microorganisms. In recent advancements, advanced biofuels that are produced by microbial synthetic pathways are esters, alcohols, and fatty acid pathways. Industrial processes of fermentation are used in advanced biofuels production in which engineering of microorganisms is included to improve the biofuel production. At last, lignocellulose biomass can also be used in biofuel production.

Microorganisms have different and specific metabolic pathway and catalytic enzymes to produce biofuels, e.g. in *S. cerevisiae*, the direct decarboxylation of pyruvate results in direction formation of ethanol. While in *E. coli*, during decarboxylation of pyruvate, CoA activates acyl group, and reduces to ethanol.

E. coli is considered as the best choice for the biofuel production, which is advanced biofuel. Many recent developments of metabolic pathways of biofuel productions are discussed, e.g. conversion of sugar to alcohols in the glycolytic pathway, production of butanol by acetone-butanol-ethanol (ABE) fermentation pathway, and conversion of protein to alcohols and alcohol production by keto acid pathway.

2.3.1 Sugar Conversion to Alcohols/Glycolytic Pathway

Fermentation of ethanol by *S. cerevisiae* is primarily done by glycolysis pathway. Regardless of the starting material, the starch degradation produces cellulose or hemicellulose, which gives hexose and pentose. The hexoses and pentoses later converts into ethanol. In many processes of fermentation, ethanol is the end product. However, in the large scale, ethanol has to be produced at the end. In this pathway, microorganisms like *S. cerevisiae* produces pyruvate (Embden-Meyeroff or glycolytic pathway) or *Zymomonas mobilis* converts to alcohol through pyruvate decarboxylase (Enter-Douoroff pathway) (Elshahed 2010). Ethanol production by using *S. cerevisiae* is considered as the best process under industrial production. This process can occur only at high levels of substrate and high industrial levels and the strains are produced by engineering, which inhibits sugar fermentation in the process (Liu et al. 2008).

2.3.2 Butanol Synthetic Pathway/ABE Pathway

1-butanol can be naturally produced by acetone-butanol-ethanol pathway. *Clostridia* has diverse group of microorganisms, like gram-positive bacteria, and anaerobic and spore-forming bacteria. *Clostridium acetobutylicum* produces the acetone, butanol, and ethanol at a 3 : 6 : 1 ratio as the end product. The current process is known as ABE fermentation. The ABE pathway is valuable at producing lacquer solvent butyl-acetate. Different species of *clostridium* (Gheshlaghi et al. 2009), such as *Clostridium Sccharoperbutylacetonicum*, *C. acetobutylicum*, and *Clostridium beijierinckii,* have very similar pathways. When fermentation process takes place in *C. acetobutylicum*, the three major groups of products are produced, which are solvent (1-butanol, acetone, and ethanol), organic acids (butyric acid, lactic acid, and acetic acid), and gases (hydrogen and carbon dioxide). The acid fermentation takes place in an exponential phase, and elements such as pyruvate and acetate are formed, and solvent fermentation is carried out at the end. Acids that are excreted in solventogenic phase are carried up again and converts into ethanol and butanol. In this stage, more n-butanol and butyrate are produced than ethanol and acetate.

Due to difficulty in engineering, butanol production through clostridia fermentation, and the high price of substrates, there are several limitations in traditional methods. These problems are solved by using biomass of wheat sugar, cassava, corn starch, and glucose.

2.3.3 2-Keto Acid Pathways for Alcohols

(Atsumi et al. 2008a,b) reported that several high-chain alcohols like isobutanol, 2-methyl-1-butanol, 2-phenyehanol, and 3-methyl-1-butanol can be used for biofuel production. The production process is carried out from glucose by the biosynthetic pathway in *E. coli* by engineering of amino acid. In this process, 2-keto acids convert into aldehyde, and then convert to alcohol.

2.3.4 2-Keto Acid Pathway for Iso-Butanol

In anaerobic conditions iso-butanol are produced on the large scale. Microaerobic conditions are also preferred to produce iso-butanol. Since two cofactors in this pathway are dependent on nicotinamide adenine dinucleotide phosphate hydrogen (NADPH), the oxygen revitalize these cofactors, and as a result, nicotinamide adenine di-nucleotide hydrogen (NADH) is produced. (Bastian et al. 2011) introduced the pathway (NADH-dependent), which was constructed by engineering of ketol-acid reductoisomerase I1vC (*E. coli*) and ADH AdhA *(Lactococcus lactis*).

2.3.5 Protein into Alcohol

(Huo et al. 2011) reported that protein biomasses are useful for biofuel production apart from glucose. *E. coli* can be engineered to produce protein hydrolysates that breaks protein into three transamination cycles to amino acids from bacteria, microalgae, and yeast. This converts into 2-keto intermediates and then ultimately into C2–C6 alcohols. Figure 2.1 shows the different biochemical pathways for biofuel production.

2.4 Algal Biofuel Production

Microalgae is found to be the most leading and superior alternative over any other conventional feedstock resources. Algae, i.e. micro- and macroalgae, used for the extraction of biofuel, plays a vital role as an important source in reserving alternative source of energy

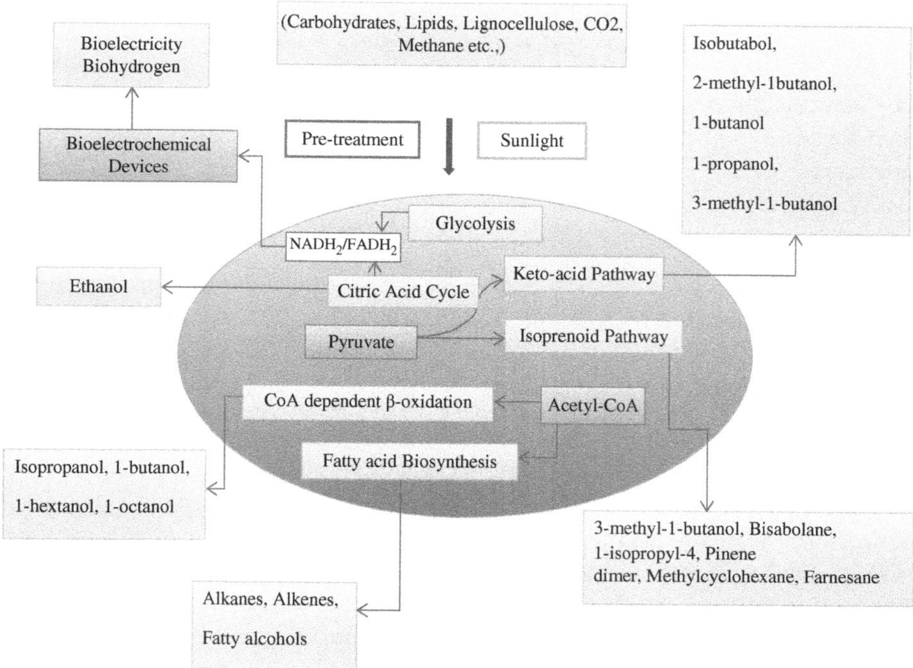

Figure 2.1 Biochemical pathway for biofuel production.

into the future. It also serves as an alternative biofuel feedstock. The following reasons make it an alternative biofuel feedstock (Ullah et al. 2015):

- The highest growth rate of the algae increases its viability, which makes it a good source of biomass.
- Another reason for its superiority and viability as a good source of biomass is that it mostly contains lipid oil content and it shows the highest yield of oil and its annual biodiesel productivity each year in comparison with other feedstock.
- The final reason for its superiority is its ability in greenhouse gas fixation.

Due to the higher lipid content of algae, it is quite simple to convert the algal biomass or the extracts obtained from algal biomass to be converted into several forms of fuels such as biogas, and a liquid form as kerosene, and a gaseous form as transportation fuels like ethanol, jet fuel, and hydrogen (Cuellar et al. 2014).

2.4.1 Microalgal Cultivation

Microalgal biomass cultivation is not very difficult because it does not require fertile land and can be grown anywhere, wherever there is access to reasonable sunlight and water, including deteriorated lands that were once used for intensive farming in the past. It also helps indirectly in decreasing habitat losses for certain native species and can be grown in wastewaters, thereby saving fresh water resources (Correa et al. 2017). Algal cultivation for biofuel production seems to be very simple as the growth requirements of algae are quite simple. There are several factors that effect the cultivation of the algae and its lipid content (Saad et al. 2019):

- micro- and macronutrients, their availability, and the concentration
- carbon dioxide
- temperature
- intensity of light
- pH

Temperature between 20 and 30 °C is considered to be the favorable temperature for the cultivation of the algae (VanWagenen et al. 2012; Alam et al. 2012). Microalgae cultivation is an important factor in biofuel production. There are three major types of microalgae cultivation based on their growth conditions:

1. **Photoautotrophic culture:** In this type of cultivation light energy (i.e. sun) is the ultimate energy source, which is converted into chemical energy by the photosynthetic reactions.
2. **Heterotrophic culture:** In this type of cultivation, organic carbon materials are used as CO_2 and root source of energy for the microalgal growth.
3. **Mixotrophic culture:** In this type of cultivation, organisms develop and grow either autotrophically or heterotrophically based on the amount of organic compound and the intensity of light.

Among all these three types of cultivation mentioned above, the most widely used type is photoautotrophic cultivation, because of its economic benefit or viability and suitability for the large-scale algal biomass production (Gambelli et al. 2017; Narala et al. 2016; Chye et al. 2018). This photoautotrophic cultivation is of two different systems:

- **Closed system (photo-bioreactors [PBRs]):** are the higher yielding systems that are highly controlled and are subjected to higher light intensity or accessibility and perfect stirring (Liao et al. 2018; Lee and Lee 2016). Photo-bioreactors are of different designs, which include flat plate and tubular or columns (Brennan et al. 2010). It is found that bubble columns and an airlift photo-bioreactor gives a higher production of microalgal biomass. Excess amount of oxygen causes negative effects on the growth of the algae, therefore an auxiliary tank is constructed, so as to separate it (Rawat et al. 2013). Even though the contamination is eliminated out, its cost plays a major role as a drawback (Huang et al. 2016; Brennan et al. 2010). Other drawbacks include overheating, scaling up, cell damage, oxygen accumulation, biocontamination (Adeniyi et al. 2018).
- **Open systems (also known as open ponds):** are less costly than closed systems, and are not highly controlled like closed systems, but are less controlled. Types of open system to be used frequently are the closed pond, circular pond tank, raceway pond, and shallow big pond. The most important feature of an open system is its ability to make use of the atmospheric CO_2. An open system depends on sunlight availability, and therefore, the location of the system is the main criterion (Brennan et al. 2010). The major challenge of an open system is the contamination caused by the bacteria and other microalgae (Suganya et al. 2016). Other challenges or the drawbacks faced by this system are temperature, evaporation, release of CO_2 into the air, and intensity of light (Adeniyi et al. 2018).

A **hybrid system** is the prominent and most appropriate system for large algal cultivation (Rawat et al. 2013). A hybrid system is made by combining both the open system and closed system together. It accomplishes higher productivity of biomass accompanied by high nutrient removal (Shaikh et al. 2017). In this system microalgae is first cultivated in a closed photo-bioreactor and then the microalgae is shifted into an open system, so as to increase the yield (Schenk et al. 2017).

2.4.2 Microalgae Harvesting

Algal harvesting is the collection of algal cells that does not affect the source and its water content (Saad et al. 2019; Butterfi and Jones 1996). Some techniques to harvest the microalgae are as follows: filtration, flotation, flocculation, centrifugation, precipitation, sonication, and sedimentation.

2.4.3 Conversion Techniques for Algal Biofuel Production

Microalgae are the third-generation category feedstock that contains higher potential in biofuel production, due to their rapid growth rate, greater yield of biomass, highest lipid content, and the carbohydrate content. After the harvesting of the microalgae biomass, it can be refined further to generate fuel through techniques such as (Raheem et al. 2015; Naik et al. 2010): thermochemical conversion, biochemical conversion, transesterification (or chemical) conversion, and photosynthetic microbial-fuel cell conversion.

The principle biofuels extracted by algal biomass are bioethanol, biodiesel, biohydrogen, biomethane, and biogas (Demirbas 2010; Cuellar et al. 2014). Algal carbohydrates are used to produce bioethanol and algal oils are useful to produce biodiesel and the rest of the biomass is used to produce methane or fuel oil. The residue of the biofuel production is used in the making of therapeutics, protein supplements, fertilizers, and animal feed (Hossain et al. 2008). Biogas or biomethane are obtained by the anaerobic digestion of organic matter. Biogas mostly contain methane (65–75%) and CO_2 (25–35%) (Ward et al. 2014). Figure 2.2 shows the techniques used in the algal biofuel production.

2.4.3.1 Thermochemical Conversion

This process is subjected to the thermal disintegration of algal biomass into organic chemical transformation, which produces biofuels as the final output (Raheem et al. 2018; Naik et al. 2010). Thermochemical conversion is composed of the following processes: gasification, pyrolysis, hydrothermal liquefaction (HTL).

These processes convert lipids or carbohydrates and complete microalgal biomass into biofuels (Hallenbeck et al. 2016; Chye et al. 2018).

- **Gasification:** The fractional oxidation of microalgal biomass within a reserved amount of air, oxygen, or steam at 700–1000 °C, which finally produces Syngas, this process of formation of Syngas is known as gasification (Chaiwong et al. 2013). The process of gasification is followed by other processes, like drying, pyrolysis, combustion, and reduction, which further produces hydrogen, methane, and carbon-dioxide, which collectively is known as Syngas, along with some solid by-products like tar and ash (Raheem et al. 2015;

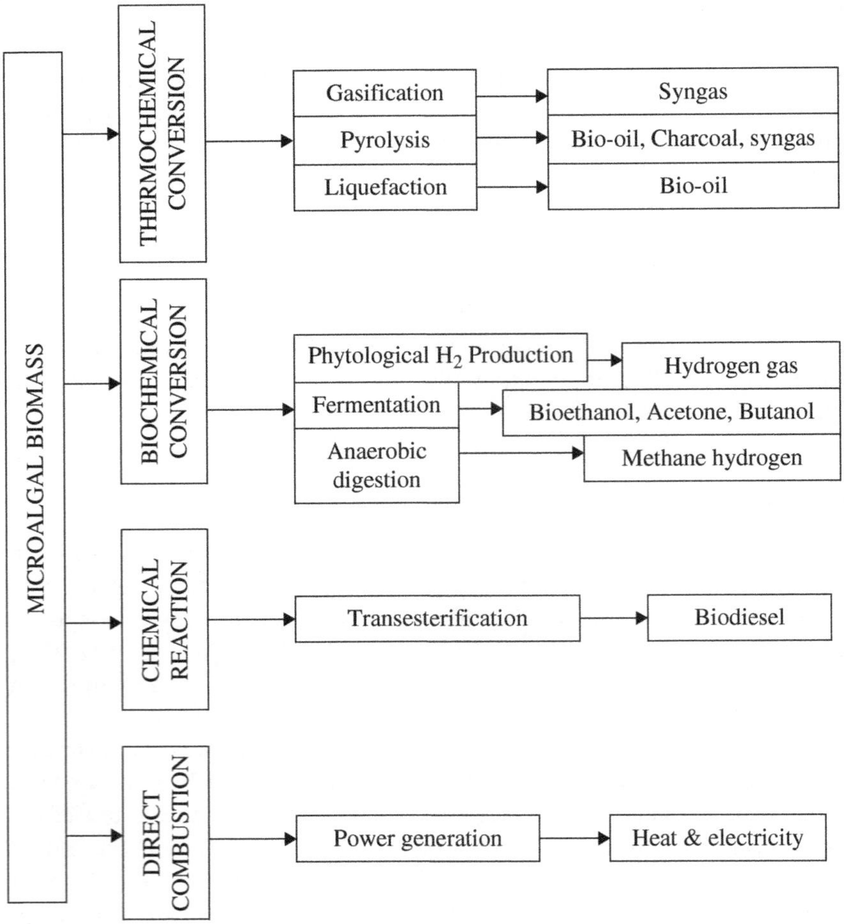

Figure 2.2 Techniques for Algal biofuel production.

Chye et al. 2018). The process of gasification comprises the two following main domains (Chen et al. 2015):

- **Conventional gasification:** Conventional gasification is performed at high temperature between 800 and 1000 °C, which either is fluidized bed or fixed bed.
- **Super-critical water gasification (SCWG):** SCWG is performed at a temperature of about 375–500 °C at 22.1–36 MPa with the help of a catalyst. SCWG does not require any energy-intensive drying process unlike conventional gasification, rather it directly converts microalgae into the gas product beyond the temperature (1.e1 375–500 °C at 22.1–36 MPa, known as water critical point).
- **Pyrolysis:** Pyrolysis is accomplished by the thermal decomposition of microalgae by heating it in the absence of oxygen\air. Operating conditions for different types of pyrolysis is at atmospheric pressure ranging with different temperatures (Chen et al. 2015): Temperature ranging between 400 and 600 °C for conventional pyrolysis, temperature ranging up to 800 °C for microwave pyrolysis, temperature ranging minimum of 300 °C for catalytic pyrolysis.

- **Hydrothermal liquefaction (HTL):** HTL of microalgae involves physical and chemical conversion, which is performed at a high temperature of about 250–500 °C of high autothermal water pressure of about 5–20 bar, either with or without using a catalyst (Raheem et al. 2015). Bio-oil and gaseous coproducts of methane are the products of liquefaction. HTL of microalgal biomass produces the yields only up to 54% without using a catalyst, which is less than the pyrolysis and gasification (Matsui et al. 1997).

2.4.3.2 Biochemical Conversion

Biochemical conversion is carried out by the hydrolysis of cell walls using bacteria to be transformed into fermented sugars. Fermentation is the process of anaerobic digestion of sugars into different biofuel production such as biogas, bioethanol, and biohydrogen (Raheem et al. 2018; Adeniyi et al. 2018). Also, according to Chye et al. (2015), biochemical conversion of microalgal biomass includes the use of microorganisms or enzymes in breaking down algae into liquid fuels. This conversion is relatively slower and less energy-efficient in comparison with the thermochemical conversion. The biochemical conversion involves three different transformation processes, and they are:

- **Anaerobic digestion:** Biochemical conversion of organic matter like microalgal biomass is transformed into biomethane and carbon dioxide with the help of enzymes and microorganisms in this process of anaerobic digestion (Brennan and Owende 2010).
- **Fermentation:** Fermentation is another type of biochemical conversion that involves the conversion of carbohydrate constituents such as cellulose, starch, or sugar into bioethanol. Fermentation is the enzymatic process carried out by the hydrolysis of starch into smaller forms of sugars, along with yeast fermentation leading to the formation of bioethanol (Milano et al. 2016). Extraction of ethanol through fermentation involves four stages (Suali and Sarbatly 2012)
  - Glycolysis: First step glycolysis is described by the formation of two molecules of pyruvate. Water and hydrogen ions (H^+) are released as the by-products after the breakdown of glucose.
  - Second step is the formation of acetaldehyde, CO_2, and H^+ ions from the pyruvate, catalyzed by the help of enzyme pyruvate decarboxylase as the catalyst.
  - Third step involves the conversion of acetaldehyde into ethanol anion, carried out by the help of co-enzyme called NADH, which is synthesized during the process of glycolysis.
  - Fourth step is carried out by the protonation of ethanol anion by the help of hydrogen ions, which produces ethanol.
- **Photobiological H_2 production:** In this process of biochemical conversion, hydrogen gas is produced through making use of microalgal biomass, specifically with the help of hydogenase enzymes, which converts electrons and protons into hydrogen gas.

$$2H^+ + 2e^- \leftrightarrow H_2 \uparrow$$

Microalgae biomass produces hydrogen gas in the dark and in the light by the process of fermentation and due to photosynthetic conditions respectively. Hydrogenase enzymes help in the formation of catalytic hydrogen in these microalgal biomasses. Abundance in solar energy and water helps in largescale production of hydrogen (H_2) photobiologically (Maness et al. 2009).

2.4.3.3 Transesterification (or Chemical Conversion)

This process involves reaction of triglycerides with an alcohol, most commonly methanol, and ethanol to produce biodiesel and glycerol using an acidic or a basic catalyst (Machado and Atsumi 2012). According to Milano et al. (2016), transesterification is described by the conversion of microalgal biomass to biodiesel and is also responsible in the chemical transesterification of algal biomass. Triglycerides or lipid reacts with alcohol with the help of a catalyst in the form of an acid, alkali, or enzyme, which forms fatty acid methyl ester (FAME) and glycerol. Transesterification is necessary in reducing algal oil consistency and in increasing its fluid capacity in response to be combined with petroleum diesel and used directly in engines (Adeniyi et al. 2018). Transesterification is of two types (Milano et al. 2016):

- Conventional transesterification: This type of transesterification requires drying and lipid extraction before the start of an actual process and also a consecutive purification.
- In situ or direct transesterification: It is simpler and cheap kind of microalgal biodiesel manufacturing in comparison to the conventional method. It does not require lipid extraction.

2.4.3.4 Photosynthetic Microbial Fuel Cell

Microbial fuel cells (MFCs) was invented as an outcome of forthcoming energy crisis (Mohan et al. 2014; Angenent et al. 2004). Past recent years showed the need of biological substitution in a sustainable manner to generate electricity, which led to the development of the MFCs (Logan and Regan 2006; Zou et al. 2009). The major process of MFCs works under the anaerobic conditions (Biffinger et al. 2008; Mohan et al. 2008). Sunlight is the main source of energy in the invention of self-sustaining MFCs (Cao et al. 2008; Malik et al. 2009; He et al. 2009). The activities in MFCs depend upon the photosynthetic oxygen, which is generated at the cathode, which increases the electron transfer from the anode (Saba et al. 2017).

The symbiosis between the bacterial fermentation at the anode and the oxygenic photosynthesis of microalgal biomass at the cathode generates the proper power output (Mohan et al. 2014).

The algal photosynthetic activity during the daytime increases the concentration of dissolved oxygen (DO). The concentration of DO is responsible for the reduction of reaction rate at the cathode, which helps in improving the bioelectrogenic activity. At nighttime, a lower concentration of the DO level decreases the power output (Mohan et al. 2014). Figure 2.3 shows the photosynthetic MFC.

2.5 Bioethanol

Biofuels have been used as a substitute for fossil fuels since the past decade, because of the crisis of oil resources, and due to the increasing demand of fuel and its consumption. Biomass can be described as any biological material, which can either be energy crops, waste material, or by-product or residue from agriculture and forestry background or any other materials related to biomass. The major biofuels are the following: (i) bioethanol,

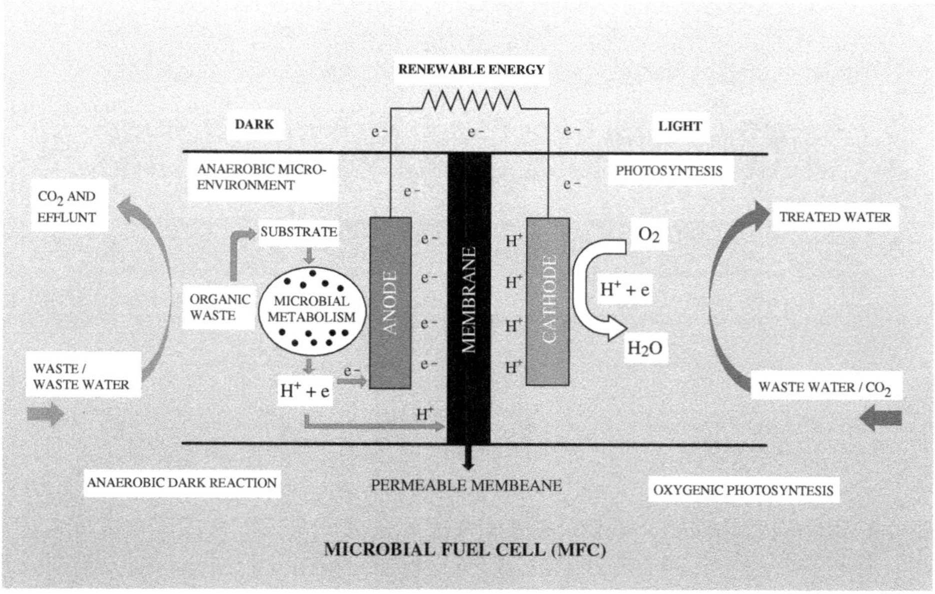

Figure 2.3 Photosynthetic microbial fuel cell (MFC). Source: Modified from Kotay and Debabrata (2008).

(ii) biodiesel, (iii) biohydrogen, and (iv) biogas. Some other biofuels in use are charcoal, fuelwood, and biomethane (Saladini et al. 2016; Chye et al. 2015; Cuellar et al. 2014; Demirbas 2010). Bioethanol is the principal biofuel used as a petrol for roadway transportation. Bioethanol is extracted mainly through the sugar fermentation, and can also be produced by the chemical process in which ethylene is reacted with steam to produce bioethanol. The main source of ethanol extraction are the sugars, which are extracted mostly from fuel or energy crops, that includes maize, wheat, corn, waste straw, sawdust, etc. Commercially, fermentation is used a leading process in the extraction of ethanol from the sugar crops and starch crops (Demirbas 2011). Some microalgal biomass are found to contain a heavy content of starch, cellulose, and glycogen as a raw material for the production of ethanol. Pretreatment of biomass is necessary for the ethanol production to release the carbohydrates found in the cells, and later ethanol is extracted by the fermentation of these carbohydrates (Cuellar et al. 2014; Demirbas 2011; Suali and Sarbatly 2012). Bioethanol has several advantages, making it better than the conventional fuels (Figure 2.4).

2.6 Biodiesel

Generally, biofuel can be developed using vegetable oil such as oil palm, coconut, jatropha, canola oil, which have a decent oil-yielding capacity per acre. In contrast to that, algal feedstock is a champion in the oil yield, having neutral lipid/oil (up to 50% dry cell weight) (Chisti 2007; Sheehan et al. 1998).

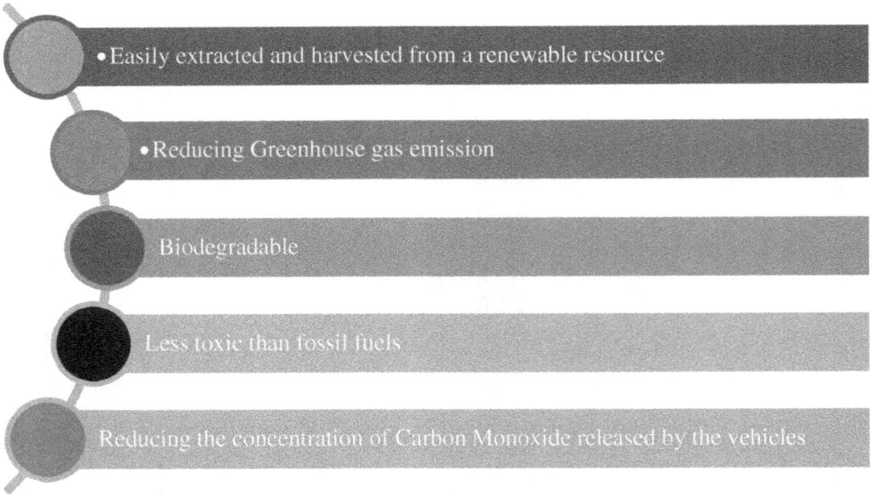

Figure 2.4 Advantages of bioethanol.

Types of algae feedstock for the creation of biofuel are macroalgae and microalgae. Macroalgae is a rapidly growing algae surpassing that of forest plants and growing up to 60 m in length. They can be helpful to produce methane using anaerobic fermentation and act as potential feedstock for ethanol production (Goh and Lee 2010). Microalgae are microorganisms synthesizing photosynthesis by taking in sunlight, water, and CO_2 and results in the formation of lipids and triacylglycerol (TAG). The microalgae can prove to be a possible feedstock for the expansion of biodiesel making, and can become one of the viable ways of biofuel production. Compared to the vegetable oils that are not reasonable to use in India (as India faces a food shortage and is one of the highest consumer of edible oil), hence, vegetable oil cannot be considered for the production of biofuel.

Microalgae is in abundance in nature, contains a high percentage of oil in a very small area, and has a large capacity of oil productivity, surpassing the best oil crops, like rapeseeds, soybeans, sunflower, safflower, corn, peanut, and coconut. Microalgae can be easily developed in brackish water or seawater, and does not require any resources, unlike conventional agriculture. They thrive in parched lands and desert lands, which do not support agriculture (Khan et al. 2009). Microalgae has a unicellular structure, which gives it an edge over the other bigger plants, as it helps in the efficient transformation of solar energy, and it being suspended in an aqueous solution makes CO_2, water, and other nutrients readily available, which makes it 30 times better in oil-yielding capacity. Microalgae are easy to cultivate, are renewable in nature, and can be grown on land, as well in water bodies offshore. If cultivated under water, it can enhance marine habitat and the aquaculture (Trent et al. 2012; Thomas et al. 2018, 2019).

Biodiesel is a type of fuel extracted from three main sources: plant oils, animal fat, and algal biomass. Biodiesel is generally a great alternative to conventional diesel, and one of its main advantages is that it can be implemented in the current diesel engine without bringing any alterations to it. While it has the same energy efficiency and fuel economy, it has the ability to mitigate pollution, thus plummeting the greenhouse gas release. It decreases the amount of carbon dioxide significantly. It lowers the particulate matter content by 86%, which is produced in straight diesel engines, even though it does not reduce the nitrogen oxide. It does, however, reduces the amount of hydrocarbons present, as well as 100% of the sulfur oxides, which are produced by biodiesel vehicles (McMillen et al. 2005). To make biodiesel, we remove impurities from the lipid or oil and change its viscosity by a chemical process known as transesterification, so that it can burn in a normal diesel engine without stuffing the fuel lines. Many countries are adopting biodiesel as an alternative fuel against the other fossil fuels, but have a long way to go as there are hurdles down the road of the production. One of the disadvantages of biodiesel is that it requires hard work to remove water from oil, and it is also a tedious job to cultivate the algae at a certain optimum temperature in order to maintain the productivity.

2.6.1 Stages of Biodiesel Production

2.6.1.1 Cultivation

i) **Open ponds/raceways:** Open ponds can be of various sizes and shape, but the most common one is the raceway pond. It has the shape of a race track, hence, the name. It is made of concrete with a shallow depth of 15–20 cm and is painted white and receives sunlight (Maity et al. 2014). The optimum temperature required is 25–30 °C. The continuous flow of water and mixing of nutrients are maintained by a paddlewheel situated at the start of the flow (Gong and Jiang 2011). The broth of algae is collected before the paddlewheel. Baffles are installed at the turns. One disadvantage of this is the water loss in the evaporation is the same as that of the land crops.
ii) **Closed bioreactors:** Scientists nowadays are opting for closed bioreactors as the open ponds can only be cultivated where there is a sufficient amount of sunlight throughout the year, such as in tropical developing countries (Ugwu et al. 2008). Water-saving features, less energy utilization, and less chemical usage makes these bioreactors more alluring to implement for the creation of algal feedstock, as the cost has been made so much cheaper (Schenk et al. 2008). They can be made of see-through objects like plastic, glass, etc., to help the sunlight pass through it, along with some amount of nutrients and CO_2 in the reactor. They can be various shapes like tubular, flat-plate, vertically column mounted, and bubble column (annular column) (Ugwu et al. 2008; Dillschneider and Posten 2013).
iii) **Hybrid system:** Open ponds are very competent and an economically viable solution to grow microalgae, but they get contaminated very quickly by unwanted species. Photo-bioreactors are excellent for maintaining uncontaminated cultures, but are expensive to set up, nearly 10 times the cost than that which is required to open ponds. A fusion of these two systems can prove to be a reasonable idea for the economical production of highly productive strains for biodiesel. The mixing of the open pond is done with the desired strain from a closed bioreactor. To maintain the desired species in the open system, the inoculum should be larger than the undesirable species. The open system, however, will become contaminated, and eventually we have to get it clean and re-inoculate it as a part of aquaculture routine; in this way, the open ponds become batch cultures. One of the microalgae is cultivated in a nutrient-rich

environment in the photo-bioreactor and then it is transferred into a limited nutrient open system. We let the culture to grow for 24 hours and after 72 hours during the peak, we harvest it, clean it, and then re-inoculate it (Schenk et al. 2008).

2.6.1.2 Harvesting/Dewatering

Since algae have a high content of water, in order to harvest the lipid content, the removal of water is required. There is no one best solution for it. Some of the processes involved are flocculation, micro-screening, centrifugation, filtration, sedimentation, coagulation, electrophoresis, air flotation, and reverse osmosis. Making the harvesting method cost-effective makes the whole biofuel production economical. Some species can be harvested more easily than others, hence the choice of strain determines the cost of the operation. Cyanobacterium *Spirulina*, which are spiral in shape, can be harvested easily by a cost-effective method of micro-screening. While some filamentous and large colonial microalgae can be treated with the filtration process, it poses many clogging problems when produced on a mass scale (Schenk et al. 2008). Keeping in mind the cost factor and efficiency while selecting the harvesting method is essential. These factors majorly depend on the physical features of the microalgal strain, culture method (open Pond or bioreactor), end products, and downstream processes (Gong and Jiang 2011).

Centrifugation is a secondary method for harvesting, but is a costly and energy-consuming method. A two-step process of coagulation and flocculation can prove to be an affordable pretreatment process used to accumulate negative charged microalgal cells, which make larger clumps, enhancing the processes of filtration, centrifugation, and sedimentation (Elmaleh et al. 1991; Jiang et al. 1993). Coagulation can be done by adding some of the metallic salts, like ferric chloride ($FeCl_{3)}$, ferric sulphate (Fe_2SO_4), aluminum sulfate (Al_2SO_4), and alum{$KAl(SO_4)_2 \cdot 12H_2O$}, which are commonly used agents for coagulation. The process should be agitated at 300 rounds per minute for about 20 minutes. Flocculation process requires a thorough agitation at 300 rpm for about 10 minutes, but for a large-scale production, this is not feasible, as they are too expensive and cannot be used for downstream applications, such as anaerobic digestion and animal feedstock production. Some organic cationic chitosan, such as poly (D) glucosamine are better substitutes than metal salts, as quantities required are less and they can also be inculcated in the downstream processing (Faried et al. 2017).

This process helps to reduce the additional chemicals remaining in the harvesting process. The next process is to dry the harvested biomass by keeping it under the sun on a polyethylene sheet for an estimated three to five hours. The biomass needs to be kept in mild sunlight and then it goes to the oven for heating at a temperature of 55 °C (Ashok Kumar et al. 2014).

2.6.1.3 Oil Extraction

There are two types of methods for oil extraction:

i) **Physical methods:** Presses, which can be of three types: screw, expeller, and piston. The screw type of press is generally used to remove oil from oil seed crops. However, application of this method has not been reported in any literature.
ii) **Chemical methods:** Pyrolysis, conventional solvent extraction, and enzymatic extraction are some chemical ways for extraction of oil, and some that can only be

implemented in a laboratories are osmotic shock, ultrasonic-enabled extraction, and supercritical carbon dioxide extraction.

In 1986, Germany was the first country reported to extract liquid fuel from pyrolyzing the microalgae. They knew that the technique of catalytic pyrolysis could produce petrol with the increased amount of aromatic hydrocarbon and high octane numbers (Milne et al. 1990). Pyrolysis is a process of breaking down the biomass at an increased temperature with the absence of oxygen. It has been used for the formation of biogases and bio-oil from a complex known as lignocellulose. However, this type of technology works well with a microalgae as high-quality oils are obtained with a lower temperature (Maggi and Delmon 1994).

Solvents, such as chloroform and hexane, are fast and efficient for the pulling out of oil from the whole cell. Press and solvent extraction are infrequently used together to achieve the purpose. The disadvantage of using this method is the amount of energy it consumes, and there is a potential danger of fire and explosion (Greenwell et al. 2010).

Supercritical CO_2 extraction is a promising process for the thorough withdrawal of oils. Its low viscosity and high diffusivity of supercritical CO_2 can give a high efficiency of extraction as well as CO_2 in this process gets recycled (Eller and King 2000).

2.6.1.4 Conversion

Since the viscosity of the microalgal oil is ample for the diesel engines, a method of conversion known as transesterification is applied to convert microalgal oil into biodiesel in its usable state. Transesterification is a process that involves mixing of alcohol and triglycerides using an acid catalyst for a faster reaction rate, and yield, which in turn gives methyl ester (biodiesel) and glycerol as a leftover product (Figure 2.5). Types of alcohols for usability are methanol, ethanol, propanol, butanol, and amyl alcohol. Out of these, only methanol is used, as it is economical to use it commercially on a bigger scale and due to its physical and chemical advantages.

2.7 Biohydrogen

The word biohydrogen is a Greek word; *bio* implies life and *hydro* implies water and biohydrogen refers to nonfossil and biodegradable natural substance extractant from plants, animals, and microorganisms acquired from organic sources. Biohydrogen from gas production from inexhaustible assets is termed as "green technology" (Kapdan and

Figure 2.5 Process of transesterification.

Kargi 2006). Hydrogen is a clean, recyclable, and dynamic vitality, and an endless supply of H_2 water is the main major side effect. In ongoing decades, the alluring and inexhaustible attributes of H_2 prompted the buildup of an assortment of organic courses for the creation of H_2. The chance of changing over hydrogen into power through power modules makes the utilization of hydrogen vitality promising. Broad endeavors have been held to create from fossil fuels, which are nondirtying and a sustainable wellspring of vitality. Lately, hydrogen gas is mostly delivered from fossil fuels, and elective courses of hydrogen production, which are savvy and contamination-free are still requested; biohydrogenation has been centered around the photolysis of water utilizing green growth and cyanobacteria to produce hydrogen (Chang et al. 2002). The major natural procedure used for hydrogen gas production is biophotolysis of water by dark algae and photo-fermentation of organic matter (Kapdan and Kargi 2006). Hydrogen (H_2) has the highest energy density (143 GJ ton^{-1}) and is really a carbon impartial fuel (Das and Veziroglu 2008; Tiwari and Pandey 2012).

Hydrogen is counted as the energy for the future, for it is a perfect vitality wellspring with a high vitality involved. Hydrogen (H_2) is not promptly accessible in nature. In nature, an assortment of life forms, together with the archaea, anaerobic, and facultative high-impact microscopic organisms, blue green algae, and unicellular organisms (that is green growth and single-celled organisms) produce H_2. The major hydrogen-generating biocatalysts are run in the typical heterotrophs in the fermentation procedure. Some photoheterotrophic microorganisms use natural acids, for example, acidic, lactic, and butyric acids to create hydrogen and carbon dioxide. In the process of anaerobic conditions, H_2 has been created as the residuum during transformation of organic waste into organic acids, which are then transformed for the methane gas. Photosynthetic procedure incorporates algae, which uses carbon oxide and hydrogen for hydrogen gas generation. Nonetheless, the pace of hydrogen generation is quite lower and the innovation for this particular procedure needs an imminent turn of occurrence (Levin et al. 2004).

2.7.1 Stages of Biohydrogen Production

Biological production of biohydrogen, utilizing life forms, is an energizing new zone of innovation advancement that provides the potential creation of utilizable hydrogen (H_2) from an assortment of sustainable assets. Biological frameworks give a wide scope of ways to deal with the production of hydrogen. Procedures for biological hydrogen creation, for the most part, run at ambient pressure and temperature, which are required to be less vitality concentrated than the thermochemical strategies for hydrogen (H_2) creation. The mentioned procedures can utilize an assortment of carbon (C) sources as a feedstock.

In view of the frameworks that develop H_2, an enormous number of various characteristic natural procedures are classified into four categories as shown in Figure 2.6.

2.7.1.1 Biophotolysis

At present, the most alluring and appealing H_2 creation process is biophotolysis, known as water-scattering photosynthesis. The oxygenic photosynthetic microorganisms, for example, green microalgae (e.g. *Chlamydomonas reinhardtii, Scenedesmus obliquus, Scenedesmus,* and *Chlorella*, among others) and cyanobacteria (e.g. *Nostoc punctiforme, Anabaena variabilis,* and *Synechocystis* sp., among others) utilize these procedures that

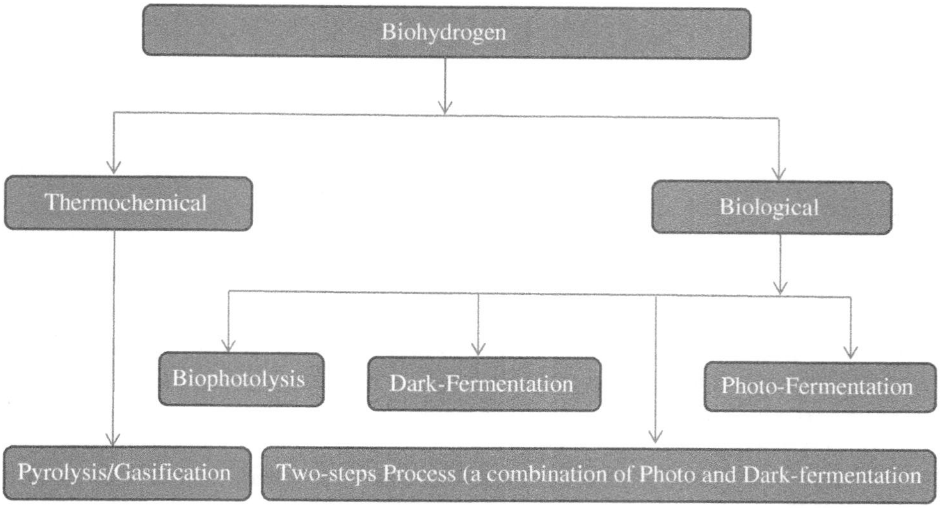

Figure 2.6 Hydrogen production routes from fungi.

Table 2.1 Challenges in photolysis.

Challenges faced in Direct biophotolysis	Challenges faced in Indirect biophotolysis
Need for customized photo bioreactors	Need of an external light source.
Lower in hydrogen production by lower light transformation technique	Lower hydrogen production convicted by hydrogenase (s)
Total light conversion efficiency is better	Total light conversion efficiency is very low

requires just water and daylight. The biophotolysis is additionally isolated into direct and indirect forms (Chandrasekhar et al. 2015). In direct biophotolysis, there is a generation of oxygen from water mediated by solar energy (Levin et al. 2004).

The sun energy is utilized to change over a promptly accessible substratum, H_2O, to O_2 and H_2. Some anaerobic thermophilic H_2 delivering microscopic organisms, for example, *Clostridium thermocellum*, can create cellulolytic enzymes for successful hydrolysis, and accordingly, can produce hydrogen gas in thermophilic acidogenic culture (Kanai et al. 2005; Liu et al. 2008). Table 2.1 show the challenges in direct and indirect photolysis.

In indirect biophotolysis, issues effect the hydrogen-developing process capacity to develop spatially oxygen advancement and hydrogen advancement. The indirect biophotolysis procedure includes division of the hydrogen and oxygen responses into isolated phases, coupled through CO_2 fixation development. Blue green bacteria have a one of a kind attribute of utilizing CO_2 noticeable in the air as a carbon wellspring and solar energy as a vitality source. The cells consume CO_2 first to deliver cell substances, which are thus utilized for hydrogen creation. The general instrument of hydrogen creation in cyanobacteria can be spoken to by the accompanying responses:

As a result of the higher paces of H_2 creation by *Anabaena* species and strains, these species are dependent upon serious examination (Levin et al. 2004). As indirect biophotolysis, freak strains of the species *Anabaena variabilis* have exhibited hydrogen productivity at the pace of 0.355 mmol/h/l (Sveshnikov et al. 1997).

2.7.1.2 Photo Fermentation

Photo fermentation additionally includes the transformation of daylight vitality to biomass with the creation of carbon dioxide (CO_2) and hydrogen (H_2). Some photoheterotrophic microbes are equipped for changing over natural acids (acidic, butyric, and lactic) under the anaerobic conditions to carbon dioxide (CO_2) and hydrogen (H_2) within the sight of the daylight. Instances of microscopic organisms are *Chloroflexus aurantiacus, Chlorobium vibrioforme, R. sphaeroides, Thiocapsa roseopersicina, Desulfuromonas acetoxidans,* and *Allochromatium vinosum* (Chandrasekhar et al. 2015). Carbon dioxide (CO_2) is necessary for certain cultures in the duration of the hydrogen (H_2) advancement stage (Pinto et al. 2002) in spite of the fact that hindrance impacts of CO_2 on the photo-production creation of hydrogen (H_2) (Tsygankov et al. 1998).

Purple nonsulfur microscopic organisms advance atomic H_2 enhanced by nitrogenase under nitrogen-lacking parameters utilizing sun light vitality and diminished mixes (Pinto et al. 2002).

In writing hydrogen creation paces of the request for 145–160 mmol/h/l, these photoheterotrophic microbes have been examined for their capability to change over light vitality into H_2 utilizing waste natural mixes.

Challenges Faced in Photo fermentation

1. Lower in hydrogen (H_2) yield brought about by incredibly low light transformation effectiveness.
2. The procedure is constrained by day and night cycles, with daylight as the light source.
3. Extremely in need of an external light source.

2.7.1.3 Dark Fermentation

Dark fermentation is characterized as the transformation of a natural substrate to biohydrogen without light, e.g. *Enterobacter cloacae, Enterobacter aerogenes, Citrobacter intermedius,* and *E. coli,* and anaerobic microorganisms, e.g. *Clostridium paraputrificum, Ruminococcus albus,* and *Clostridium beijerinckii.* Dark fermentative microscopic organisms don't require sunlight-based vitality as a vital resource ,and can endure oxygen (O_2) insufficient conditions; these living microorganisms are committed anaerobes, which are additionally arranged independently with their affectability to oxygen (O_2) and with required development temperature (Chandrasekhar et al. 2015).

An assortment of organisms anaerobic microscopic organisms (*Clostridium* sp.), facultative anaerobes (*Enterobacter* and *Baccilus* sp.) just as bacterial consortium from natural squanders, (anaerobic digester slime, soil, creature dung, and so forth) can be utilized for dark H_2 fermentation. Anaerobic microorganisms use natural compounds as a sole wellspring of electrons (e−) and vitality, changing them over into H_2. The responses engaged with H_2 creation are quick and these procedures do not require a lot of sunlight-based radiation, which makes them helpful for utilizing huge amounts of natural waste by utilizing a suitable fermenting agent.

Table 2.2 Advantages and limitations of dark fermentation.

Advantage in Dark fermentation	Disadvantage in Dark fermentation
Inexpensive	Lower in H_2 yield
Lower energy requirement	Lower substrate conversion efficiency.
Process enhanced with the help of micro-organisms at ambient pressure and temperature in an aqueous environment.	Limitations in thermodynamic parameters. In the mixture of CO_2 and H_2 gases as products, which require separation

Hypothetically maximal hydrogen (H_2) generated from dark fermentation has 4 H_2 mol/glucose mol. Moreover, since the process of fermentation is just a fragmented corruption of natural substrates, the creation of H_2 gas is joined by an arrangement of acetic acid derivation or potentially butyrate stoichiometrical proportion of 2 mol H_2 per mol acetic acid derivation or butyrate.

Different natural carbon sources as feedstock incorporate agrarian harvests and their loss by items, wood squander, food handling waste, and amphibian plants, green growth, and effluents delivered in creature living spaces can be appropriately utilized by oleaginous microorganisms, for example, microalgae, yeasts, organisms, and so on. The procedure of dark fermentation happens at a bit of a higher rate than the procedures of photolysis and photo fermentation (Saratale et al. 2008). Table 2.2 highlights the benefits and limitations of dark fermentation.

2.7.1.4 Two-Step Process (a Combination of Photo and Dark Fermentation)

In the process of fermentation, the whole oxidation of 1 glucose mole yields 12 hydrogen moles. Be that as it may, the complete oxidizing process of glucose ($C_6H_{12}O_6$) into carbon dioxide (CO_2) and hydrogen (H_2) is beyond the realm of imagination, as the relating response isn't achievable thermodynamically.

With the outside vital energy (photon vitality in the photo fermentation), hypothetically, 12 hydrogen moles for every glucose mole can be delivered. Anyway, this procedure can't be worked without light. Then again, without outside vitality (on account of dark fermentation), oxidizing of glucose with the help of fermentative microscopic organisms brings about other side effects additionally, and at the most extreme, four hydrogen moles are created for every glucose mole utilization with acetic acid derivation as the sole result (Bala and Murugesan 2011).

Dark fermentation stages can be oxidized to produce acetate with the help of photosynthetic bacteria to produce hydrogen.

Henceforth, continuous generation of hydrogen at its most extreme yield can be accomplished by incorporating dim and photograph-aging techniques. In consequence, a portion of the research facility concentrated on this two-phase process (Kotay and Debabrata 2008) in Table 2.3.

Table 2.3 Comparison of important biological hydrogen production processes.

Process	Advantages	Microorganism	Reference
Direct Biophotolysis	It can create H_2 legitimately from water and daylight Solar transformation vitality expanded by 10 times when contrasted with trees, crops.	*Chlamydomonas reinhardtii*	Pinto et al. (2002)
Photo Fermentation	It can utilize diverse waste materials like refinery effluents, squander, and so on.	*Rhodobacter spheriods*	Levin et al. (2004)
Dark Fermentation	It produces valuable metabolites such as butyric, lactic, and acetic acids as by products.	*Enterobacter cloacae*	Saratale et al. (2008)
Two stage Fermentation (Dark + photo)	Stoichiometric yield of 12 mol H_2 per mol hexose represents the ultimate target for biohydrogen.	*Enterobacter cloa-cae + Rhodobacter spheriods*	

2.8 Applications of Biofuel Production

Biofuels give more benefits compared to petroleum-based fuels. The advantages of biofuels are that they provide better lubricants than petrol fuels, mitigate CO_2 emission, and they are biodegradable, which helps in greenhouse gas emission, which make the environment pollutant-free. Biofuels are good options for future use and can be produced by natural processes in several countries. In developing countries like the United States and Brazil, ethanol is used in pure form or with gasoline in Otto cycle engines. Several uses are presented in Figure 2.7. However, their production strategies require little modification to get their highest benefits. In any industry, biofuel requires a high production capacity and the transformation from discrete batch process to continuous flow process (Rajagopal et al. 2007).

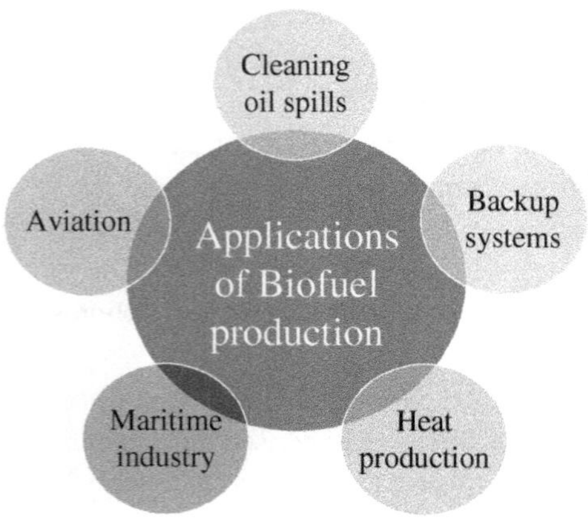

Figure 2.7 Applications of biofuel production.

2.8.1 In Aviation

Biofuel utilization in aviation and the maritime industries is taking place worldwide. In 2011, United Continental Holdings used the first blend of advanced biofuel with petroleum-based fuel in airline for a commercial flight, while Green Flight International was the first airline to use 100% biofuel in their jet aircraft. The aviation industries are responsible for producing 2–3% of carbon dioxide, where biofuel use in aviation can reduce greenhouse gas emission by 60–80%. The blending of biofuels from algae with existing jet fuels are beneficial in aviation.

2.8.2 Maritime Industry

The high emission and sulfur produced by the marine industry causes the pollution in a marine environment. The emission of CO_2 is lowest in the marine industry but due to high demand of shipping can triple by 2050 if left untreated. The need for alternative fuel, which contains the low amount of sulfur, drives the marine industry attention to biofuel. Biofules are cleaner than natural fuels and degrades other contaminations forming in the processing systems.

2.8.3 Heat

In many countries the heat is obtained by combustion process, which includes direct burning of biomass, such as woods, dung, peat, etc. Producing heat by combustion releases CO_2 and SO_2, which pollutes the environment and is also not good for health (World Health Organization 2006). Heat production by biofuel has grown in some countries and is already in use over the past few years. The substantial materials used in heat producing reduces the emission of nitrogen and sulfur both.

2.8.4 Backup Systems

Biofuels can be used as energy-generating applications in backup systems where emission rates are high as it causes release of carbon monoxide. The use of biofuels as generators in hospitals, schools, and other public areas results in the reduction of poisonous gases (UC Riverside 2001).

2.8.5 Cleaning Oil Spills

Biofuels have methyl ester, which is an effective solvent to crude oils. The seashores are 80–90% affected by oil spills, which causes a high cost investment in order to clear it from the seashore. Biofuels like biodiesel have the capacity to dissolve crude oil, and lowers its viscosity, which later helps in cleaning oil spills. The results show that about 80% of oil is removed from sand.

2.8.6 Microalgae Applications

Microalgae such as green algae and diatoms are available all year and can also be produced throughout the year. They contain about 20–50% oil of dry biomass, and after oil extraction, they also give protein, lipids, polysaccharides, pigments, vitamins, and residual biomass, which are coproducts of microalgae. These coproducts can also be used for the production of methane and ethanol. As they are biodegradable, they are cultivated for algal biofuel production. Many species of microalgae are beneficial for producing lipids and hydrocarbons in large amounts, such as *Chlorella* sp., *Botryococcus braunii,* and *Crypthecodinium cohnii. B braunii* has the highest amount of oil content, which is about 80%, while *Chorella* has about 50% lipids. The green jet fuel from algae is used in commercial traveling.

2.9 Conclusion

In a fossil-fuel dependent economy, biofuel production helps to form a balance between renewable and fossil-fuel consumption. The need for biofuel production has been a hot topic of debate in various international forums and organizations. Researchers are devising the kind of material that can become a potential feedstock for biofuel production as the cost involved in the production is a big factor that requires much attention. There is a continuous effort to reduce the energy consumption while producing the energy. The prime focus is to maintain the balance at the agriculture level, economic development, and the environment for a better tomorrow.

References

Adeniyi, O.M., Azimov, U., and Burluka, A. (2018). Algae biofuel: current status and future applications. *Renewable Sustainable Energy Rev.* 90: 316–335.

Alam, F., Date, A., Rasjidin, R. et al. (2012). Biofuel from algae- is it a viable alternative? *Procedia Eng.* 49: 221–227.

Almodares, A. and Hadi, M.R. (2009). Production of bioethanol from sweet sorghum: a review. *Afr. J. Agric. Res.* 4 (9): 772–780.

Angenent, L.T., Karim, K., Al-Dahhan, M.H. et al. (2004). Production of bioenergy and biochemicals from industrial and agricultural wastewater. *Trends Biotechnol.* 22: 477–485.

Ashok Kumar, V., Agila, E., Sivakumar, P. et al. (2014). Optimization and characterization of biodiesel production from microalgae *Botryococcus* grown at semi-continuous system. *Energy Convers. Manag.* 88: 936–946.

Atsumi, S., Cann, A.F., Connor, M.R. et al. (2008a). Metabolic engineering of *Escherichia coli* for 1- butanol production. *Metab. Eng.* 10: 305–311.

Atsumi, S., Hanai, T., and Liao, J.C. (2008b). Non-fermentative pathways for synthesis of branched-chain higher alcohols as biofuels. *Nature* 451 (7174): 86–89.

Bala, A.K. and Murugesan, A.G. (2011). Biological hydrogen production by the algal biomass *Chlorella vulgaris* MSU 01 strain isolated from pond sediment. *Bioresour. Technol.* 102: 194–199.

Balan, V., Chiaramonti, D., and Kumar, S. (2013). Review of US and EU initiatives toward development, demonstration, and commercialization of lignocellulosic biofuels. *Biofuels Bioprod. Biorefin.* 7: 732–759.

Balat, M., Balat, H., and Öz, C. (2008). Progress in bioethanol processing. *Prog. Energy Combust. Sci.* 34 (5): 551–573.

Bastian, S., Liu, X., Meyerowitz, J.T. et al. (2011). Engineered ketol-acid reductoiso-merase and alcohol dehydrogenase enable anaerobic2-methylpropan-1- oil production at the oretical yield in *Escherichia coli. Metab. Eng.* 13: 345–352.

Biffinger, J.C., Byrd, J.N., Dudley, B.L., and Ringeisen, B.R. (2008). Oxygen exposure promotes fuel diversity for *Shewanella oneidensis* microbial fuel cells. *Biosens. Bioelectron.* 23: 820–826.

Brennan, L. and Owende, P. (2010). Biofuels from microalgae—a review of technologies for production, processing, andextractions of biofuels and co-products. *Renewable Sustainable Energy Rev.* 14: 557–577.

Butterfi, B.A. and Jones, J. (1996). Harvesting of algae grown in agricultural wastewaters. *Trans. Geophys. Union* 50: 612.

Cao, X., Huang, N., Boon, N. et al. (2008). Electricity generation by an enriched phototrophic consortium in a microbial fuel cell. *Electrochem. Commun.* 10: 1392–1395.

Chaiwong, K., Kiatsiriroat, T., Vorayos, N., and Thararax, C. (2013). Study of bio-oil and bio-char production from algae by slow pyrolysis. *Biomass Bioenergy* 56: 600–606.

Chandrasekhar, K., Lee, Y.J., and Lee, D.W. (2015). Biohydrogen production strategies to improve process efficiency through microbial routes. *Int. J. Mol. Sci.* 16: 8266–8293.

Chang, J.S., Lee, K.S., and Lin, P.J. (2002). Biohydrogen production with fixed-bed bioreactors. *Int. J. Hydrog. Energy* 27: 1167–1174.

Chen, W.H., Lin, B.J., Huang, M.Y., and Chang, J.S. (2015). Thermochemical conversion of microalgal biomass into biofuels: a review. *Bioresour. Technol.* 184: 314–327.

Chisti, Y. (2007). Biodiesel from microalgae. *Biotechnol. Adv.* 25 (3): 294–306.

Chye, J.T.T., Jun, L.Y., Yon, L.S. et al. (2018). Biofuel production from algal biomass. doi: https://doi.org/10.1201/9781351228138-3

Correa, D.F., Beyer, H.L., Possingham, H.P. et al. (2017). Biodiversity impacts of bioenergy production: microalgae vs. first generation biofuels. *Renewable Sustainable Energy Rev.* 74: 1131–1146.

Cuellar, S.P., Romero, M.A., Rittmannb, B.E., and Parra, R. (2014). Algae biofuels production processes, carbon dioxide fixation and biorefinery concept. *J. Pet. Environ. Biotechnol.* 4 https://doi.org/10.4172/2157-7463.1000185.

Das, D. and Veziroglu, T.N. (2008). Advances in biological hydrogen production processes. *Int. J. Hydrog. Energy* 33: 6046–6057.

Demirbas, M.F. (2011). Biofuels from algae for sustainable development. *Appl. Energy* 88: 3473–3480.

Demirbas, A. and Demirbas, M.F. (2010). Algae energy: algae as a new source of biodiesel. *Green Energy.* 2010.

Dillschneider, R. and Posten, C. (2013). Closed bioreactors as tools for microalgae production. *Adv. Biofuels Bioprod.*: 629–649.

Eller, F.J. and King, J.W. (2000). Supercritical carbon dioxide extraction of cedarwood oil: a study of extraction parameters and oil characteristics. *Phytochem. Anal.* 11 (4): 226–231.

Elmaleh, S., Coma, J., Grasmick, A., and Bourgade, L. (1991). Magnesium induced algal flocculation in a fluidized bed. *Water Sci. Technol.* 23 (7–9): 1695–1702.

Elshahed, M.S. (2010). Microbiological aspects of biofuel production: current status and future directions. *J. Adv. Res.* 1: 103–111.

Escobar, J.C., Lora, E.S., Venturini, O.J. et al. (2009). Biofuels: environment, technology and food security. *Renewable Sustainable Energy Rev.* 13: 1275–1287.

Faried, M., Samer, M., Abdelsalam, E. et al. (2017). Biodiesel production from microalgae: processes, technologies and recent advancements. *Renewable Sustainable Energy Rev.* 79: 893–913.

Gambelli, D., Alberti, F., Solfanelli, F. et al. (2017). Third generation algae biofuels in Italy by 2030: a scenario analysis using Bayesian networks. *Energy Policy* 103: 165–178.

Gheshlaghi, R., Scharer, J.M., Moo-Young, M., and Chou, C.P. (2009). Metabolic pathways of clostridia for producing butanol. *Biotechnol. Adv.* 27: 764–781.

Goh, C.S. and Lee, K.T. (2010). A visionary and conceptual macroalgae-based third generation (TGB) biorefinery in Sabah, Malaysia as an underlay for renewable and sustainable development. *Renewable Sustainable Energy Rev.* 14 (2): 842–848.

Gong, Y. and Jiang, M. (2011). Biodiesel production with microalgae as feedstock: from strains to biodiesel. *Biotechnol. Lett.* 33 (7): 1269–1284.

Greenwell, H.C., Laurens, L.M., Shields, R.J. et al. (2010). Placing microalgae on the biofuels priority list: a review of the technological challenges. *J. R. Soc. Interface* 7 (46): 703–726.

Hallenbeck, P.C. (2012). *Microbial Technologies in Advanced Biofuels Production*. Boston: Springer https://dio.org/10.1007/978-1-4614-1208-3.

Hallenbeck, P.C. and Benemann, J.R. (2002). Biological hydrogen production: fundamentals and limiting processes. *Int. J. Hydrog. Energy* 27: 1185–1193.

Hallenbeck, P.C., Grogger, M., Mraz, M., and Veverka, D. (2016). Solar biofuels production with microalgae. *Appl. Energy* 179: 136–145.

He, Z., Kan, J., Mansfeld, F. et al. (2009). Self-sustained phototrophic microbial fuel cells based on the synergistic cooperation between photosynthetic microorganisms and heterotrophic bacteria. *Environ. Sci. Technol.* 43: 1648–1654.

Hossain, A., Salleh, A., Boyce, A. et al. (2008). Biodiesel fuel production from microalgae as renewable energy. *Am. J. Biochem. Biotechnol.* 4: 250–254.

Huang, G., Chen, F., Wei, D. et al. (2016). Biodiesel production by microalgal biotechnology. *Appl. Energy* 87: 38–46.

Huo, Y.X., Cho, K.M., Rivera, J.G. et al. (2011). Conversion of proteins into biofuels by engineering nitrogenflux. *Nat. Biotechnol.* 29: 346–351.

Jiang, J.Q., Graham, N.J., and Harward, C. (1993). Comparison of polyferric sulphate with other coagulants for the removal of algae and algae-derived organic matter. *Water Sci. Technol.* 27 (11): 221–230.

Kanai, T., Imanaka, H., Nakajima, A. et al. (2005). Continuous hydrogen production by the hyperthermophilicarchaeon, Thermococcus kodakaraensis KOD1. *J. Biotechnol.* 116: 271–282.

Kapdan, I.K. and Kargi, F. (2006). Bio-hydrogen production from waste materials. *Enzym. Microb. Technol.* 38: 568–582.

Khan, S.A., Hussain, M.Z., Prasad, S., and Banerjee, U.C. (2009). Prospects of biodiesel production from microalgae in India. *Renewable Sustainable Energy Rev.* 13 (9): 2360–2372.

Kosourov, S., Tsygankov, A., Seibert, M., and Ghirardi, M.L. (2002). Sustained hydrogen photoproduction by *Chlamydomonas reinhardtii*: effects of culture parameters. *Biotechnol. Bioeng.* 78: 731–740.

Kotay, S.M. and Debabrata, D. (2008). Biohydrogen as a renewable energy resource — Prospects and potentials. *Int. J. Hydrog. Energy* 33: 258–263.

Lee, O.K. and Lee, E.Y. (2016). Sustainable production of bioethanol from renewable brown algae biomass. *Biomass Bioenergy* 92: 70–75.

Levin, D.B., Pitt, L., and Love, M. (2004). Biohydrogen production: prospects and limitations to practical application. *Int. J. Hydrog. Energy* 29: 1179–1186.

Liao, Q., Chang, J.S., Herrmann, C., and Xia, A. (2018). *Bioreactors for Microbial Biomass and Energy Conversion*. Springer.

Logan, B.E. and Regan, J.M. (2006). Electricity-producing bacterial communities in microbial fuel cells. *Trends Microbiol.* 14: 512–518.

Machado, I.M. and Atsumi, S. (2012). Cyanobacterial biofuel production. *J. Biotechnol.* 162: 50–56.

Maggi, R. and Delmon, B. (1994). Comparison between 'slow' and 'flash' pyrolysis oils from biomass. *Fuel* 73 (5): 671–677.

Maity, J.P., Bundschuh, J., Chen, C.Y., and Bhattacharya, P. (2014). Microalgae for third generation biofuel production, mitigation of greenhouse gas emissions and wastewater treatment: present and future perspectives–a mini review. *Energy* 78: 104–113.

Makishah, N.H. (2017). Bioenergy: microbial biofuel production advancement. *Int. J. Pharm. Res. Alli. Sci.* 6 (3): 93–106.

Malik, S., Drott, E., Grisdela, P. et al. (2009). A self-assembling self-repairing microbial photoelectrochemical solar cell. *Energy Environ. Sci.* 2: 292–298.

Maness, P.C., Yu, J., Eckert, C. and Ghirardi, M.L. (2009). Photobiological hydrogen production – prospects and challenges. Efforts to scale up the capacity of green algae and cyanobacteriato use sunlight to convert water into hydrogen gas for energy use. https://doi.org/10.1128/microbe.4.275.1.

Matsui, T.O., Nishihara, A., Ueda, C. et al. (1997). Liquefaction of micro-algae with iron catalyst. *Fuel* 76: 1043–1048.

McMillen, S., Shaw, P., Jolly, N., Goulding, B., and Finkle, V. (2005). Biodiesel: Fuel for Thought, Fuel for Connecticut's Future. Connecticut Centre for Economic Analysis Report.

Melis, A., Zhang, L., Forestier, M. et al. (2000). Sustained photobiological hydrogen gas production upon reversible inactivation of oxygen evolution in the green alga *Chlamydomonas reinhardtii*. *Plant Physiol.* 122: 127–135.

Milano, J., Ong, H.C., Masjuki, H.H. et al. (2016). Microalgae biofuels as an alternative to fossil fuel for power generation. *Renewable Sustainable Energy Rev.* 58: 180–197.

Milne, T.A., Evans, R.J., and Nagle, N. (1990). Catalytic conversion of microalgae and vegetable oils to premium gasoline, with shape-selective zeolites. *Biomass* 21 (3): 219–232.

Mohan, S.V., Mohanakrishna, G., and Sarma, P.N. (2008). Effect of anodic metabolic function on bioelectricity generation and substrate degradation in single chambered microbial fuel cell. *Environ. Sci. Technol.* 42: 8088–8094.

Mohan, S.V., Srikanth, S., Chiranjeevi, P. et al. (2014). Algal biocathode for in situ terminal electron acceptor (TEA) production: synergetic association of bacteria—microalgae metabolism for the functioning of biofuel cell. *Bioresour. Technol.* 166: 566–574.

Naik, S.N., Goud, V.V., Rout, P.K., and Dalai, A.K. (2010). Production of first- and second-generation biofuels: a comprehensive review. *Renewable Sustainable Energy Rev.* 14: 578–597.

Narala, R.R., Garg, S., Sharma, K., and Schenk, P.M. (2016). Comparison of microalgae cultivation in photobioreactor, open raceway pond, and a two-stage hybrid system. *Front. Energy Res.* 4 (29): 1–10.

Nigam, P.S. and Singh, A. (2011). Production of liquid biofuels from renewable resources. *Prog. Energy Combust. Sci.* 37: 52–68.

Pinto, F., Troshina, O., and Lindbald, P. (2002). A brief look at three decades of research on cyanobaceterial hydrogen evolution. *Int. J. Hydrog. Energy* 27: 1209–1215.

Raheem, A., Azlina, W.W., Yap, Y.T. et al. (2015). Thermochemical conversion of microalgal biomass for biofuel production. *Renewable Sustainable Energy Rev.* 49: 990–999.

Raheem, A., Prinsen, P., Vuppaladadiyam, A.K. et al. (2018). A review on sustainable microalgae-based biofuel and bioenergy production: recent developments. *J. Clean. Prod.* 181: 42–59.

Rajagopal, D., Sexton, S.E., Roland-Holst, D., and Zilberman, D. (2007). Challenge of biofuel: filling the tank without emptying the stomach? *Environ. Res. Lett.* 2 (4).

Raven, S., Francis, A., Srivastava, C. et al. (2019a). Fungal biofuels: innovative approaches. In: *Recent Advancement in White Biotechnology Through Fungi*, 385–405. Cham: Springer.

Raven, S., Srivastava, C., Kaushik, H. et al. (2019b). Fungal cellulases: new avenues in biofuel production. In: *Approaches to Enhance Industrial Production of Fungal Cellulases*, 1–18. Cham.: Springer.

Rawat, I., Ranjith, K.R., Mutanda, T., and Bux, F. (2013). Biodiesel from microalgae: A critical evaluation from laboratory to large scale production. *Appl. Energy* 103: 444–467.

Saad, M.G., Dosoky, N.S., Zoromba, M.S., and Shafik, H.M. (2019). Algal biofuels: current status and key challenges. *Energies* 12 (10): 1920.

Saba, B., Christy, A.D., Yu, Z., and Co, A.C. (2017). Sustainable power generation from bacterio-algal microbial fuel cells (MFCs): an overview. *Renewable Sustainable Energy Rev.* 73: 75–84.

Saladini, F., Patrizi, N., Pulselli, F.M. et al. (2016). Guidelines for emergy evaluation of first, second and third generation biofuels. *Renewable Sustainable Energy Rev.* 66: 221–227.

Saratale, G.D., Chen, S.D., Lo, Y.C. et al. (2008). Outlook of bio-hydrogen production from linocellulosic feedstock using dark fermentation – a review. *J. Sci. Ind. Res.* 67: 962–979.

Sarkar, N., Ghosh, S.K., Bannerjee, S., and Aikat, K. (2012). Bioethanol production from agricultural wastes: an overview. *Renew. Energy* 37 (1): 19–27.

Schenk, P.M., Thomas-Hall, S.R., Stephens, E. et al. (2008). Second generation biofuels: high-efficiency microalgae for biodiesel production. *Bioenergy Res.* 1: 20–43.

Shaikh, A.R., Saad Aldin, M.A., Mohammad, M.H., and Hugo, L. (2017). Biological CO_2 fixation with production of microalgae in wastewater: a review. *Renewable Sustainable Energy Rev.* 76: 379–390.

Shenk, P.M., Thomas-Hall, S.R., Stephens, E. et al. (2008). Second generation biofuels: high-efficiency microalgae for biodiesel production. *Bioenergy Res.* 1 (1): 20–43.

Sheehan, J., Dunahay, T., Benemann, J., and Roessler, P. (1998). *A look back at the US Department of Energy's aquatic species program: biodiesel from algae*, vol. 328, 1–294. National Renewable Energy Laboratory.

Suali, E. and Sarbatly, V. (2012). Conversion of microalgae to biofuel. *Renewable Sustainable Energy Rev.* 16 (6): 4316–4342.

Suganya, T., Varman, M., Masjuki, H.H., and Renganathan, S. (2016). Macroalgae and microalgae as a potential source for commercial applications along with biofuels production: a biorefinery approach. *Renewable Sustainable Energy Rev.* 55: 909–941.

Sveshnikov, D.A., Sveshnikov, N.V., Rao, K.K., and Hall, D.O. (1997). Hydrogen metabolism of Anabaena variabilis in continuous cultures and under nutritional stress. *FEBS Lett.* 147: 297–301.

Tabil, L., Adapa, P., Kashaninejad M. (2011). Biomass feedstock pre-processing–part 1: pretreatment, Biofuel's engineering process technology.

Thomas Kiran, M., Parine, N.R., and Tiwari, A. (2018). Potential of diatom consortium developed by nutrient enrichment for biodiesel production and simultaneous nutrient removal from wastewater. *Saudi J. Biol. Sci.* 25: 704–709.

Thomas Kiran, M., Datta, A., Patil, M.D. et al. (2019). Biodiesel production through algal cultivation in urban wastewater using algal floway. *Bioresour. Technol.* 280: 222–228.

Tiwari, A. and Pandey, A. (2012). Cyanobacterial hydrogen production – a step towards clean environment. *Int. J. Hydrog. Energy* 37: 139–150.

Trent, J., Wiley, P. et al. (2012). Research spotlight: the future of biofuels: is it in the bag? *Biofuels*: 7:521–7:524.

Tsygankov, A.A., Serebryakova, L.T., Rao, K.K., and Hall, D.O. (1998). Acetylene reduction and hydrogen photoproduction by wild type and mutant strains of anabaena at different CO_2 and O_2 concentrations. *FEMS Microbiol. Lett.* 167: 13–17.

Ugwu, C.U., Aoyagi, H., and Uchiyama, H. (2008). Photobioreactors for mass cultivation of algae. *Bioresour. Technol.* 99 (10): 4021–4028.

Ullah, K., Ahmad, A., Sharma, V.K. et al. (2015). Assessing the potential of algal biomass opportunities for bioenergy industry: a review. *Fuel* 143: 414–423.

VanWagenen, J., Miller, T.W., Hobbs, S. et al. (2012). Effects of light and temperature on fatty acid production in *Nannochloropsis salina*. *Energies* 5: 731–740.

Ward, A.J., Lewis, D.M., and Green, F.B. (2014). Anaerobic digestion of algae biomass: a review. *Algal Res.* 5: 204–214.

World Health Organization (2006). *Fuel for Life*. Geneva.: Household Energy and Health.

Zou, Y., Pisciotta, J., Billmyre, R.B., and Baskakov, I.V. (2009). *Photosynthetic Microbial Fuel Cells with Positive Light*. Wiley InterScience www.interscience.wiley.com, https://doi.org/10.1002/bit.22466.

3

Biohydrogen Production from Cellulosic Waste Biomass

Enosh Phillips

Department of Biotechnology, St. Aloysius College (Autonomous), Jabalpur, Madhya Pradesh, India

3.1 Introduction

Hydrogen is the most abundant element present on earth. Apart from being a part of life-sustaining molecules like H_2O, carbohydrate, and many more, hydrogen has been extensively used as fuel. It is a promising fuel for the future, as it can replace fossil fuel, which raises environmental concerns since its combustion produces greenhouse gases. It is known that the combustion of hydrocarbons produces greenhouse gases like CO_x, NO_x, and SO_x, which are a constant threat to the environment. Hydrogen (H_2), on the other hand, upon consumption, has a cyclic nature, that is, changes from water to hydrogen and then from hydrogen to water. The utilization of H_2 may help in the mitigation of CO_2 to reduce the risk of global climate change (Momirlan and Veziroglu 2005). Although there could be present many obstacles in the path of switching from a fossil-based economy to a hydrogen-based economy, like the introduction of new energy infrastructure and production cost. At present, combustible energy production infrastructure is for fossil fuel, and much investment is needed for the development of a state-of-the-art facility for hydrogen production. Lowering of hydrogen production cost by biological assays will encourage the commercial production of hydrogen (Jacobson 2005). Now, why is hydrogen-based fuel promising for future energy demands? As energy controls a lot of aspects of our everyday life like living standard, geopolitical tension, and climatic change. Fossil-based fuels tend to increase geopolitical tension and cause an imbalance in living standards across the border, since fossil deposition is unequal across the earth. Also, the level of deposition of fossil fuel is different in landmasses. Economical disbalance is seen due to this. To provide equal chances for efficient growth and development in all the regions of the world there is a need to switch from fossil-based fuel to alternative sources, which can be created by oneself without being dependent on anybody else, especially in the case of developing nation. The establishment of the relationship between hydrogen and electricity production makes the hydrogen-based fuels an eligible substitute for fossil fuels. Figure 3.1 shows the importance of hydrogen if used as fuel for our environment as on combustion it produces water and heat and no greenhouse gases ("Hydrogen explained: Use of hydrogen" 2020).

Presently, the primary method for H_2 production is by water splitting. But employed techniques for water splitting are energy expensive and power is sourced from nonrenewable

Bioenergy Research: Evaluating Strategies for Commercialization and Sustainability, First Edition.
Edited by Neha Srivastava and Manish Srivastava.

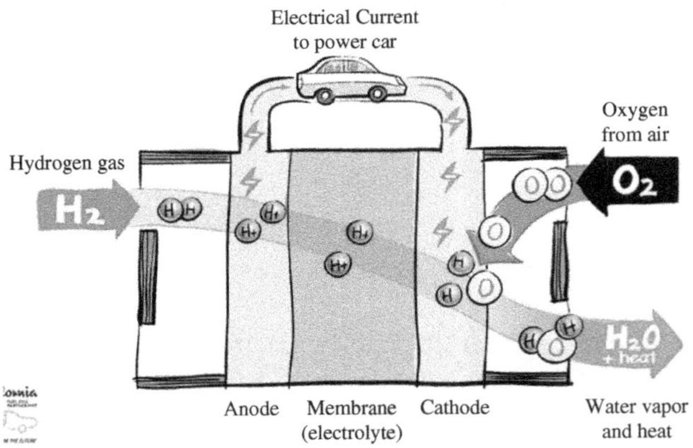

Figure 3.1 A representation of the significance of the hydrogen fuel. If hydrogen is used to fuel the automobiles to energize them, then on combustion it produces water vapor and heat instead of greenhouse gases. Source: Modified from "Hydrogen explained: Use of hydrogen," 2020.

energy. However, novel strategies are developing that are focused on mounting methods that utilize renewable energy for splitting water. The present conventional method requires more energy to split water than what is recovered on recombination. At present, water is split with 80% efficiency, and about 60% of electricity is generated by the process. This in itself calls for thorough input for the alternative method that could increase output efficiency (Crabtree and Dresselhaus 2008).

3.2 History of Hydrogen Fuel

Sir William Robert Groove of Wales is credited for the invention of the first hydrogen-based fuel cell. In 1839, he developed a fuel cell by mixing hydrogen and oxygen with electrolyte inserted in the system. In this attempt, he tried to produce water and electricity from this fuel cell. The experiment was a success, but the hydrogen fuel cell did not produce enough electricity. His effort was the first stone laid in the path for the development of hydrogen-based fuel for energy generation. The term fuel cell was later coined by Ludwig Mond and Charles Langer in 1889. They tried to develop fuel cells from the air and from industrial coal. The present-day solid oxide fuel cells were synthesized due to the research and development undertaken by Germany in the 1920s. Table 3.1 describes the historical events that undertook for fuel cell development. However, much research is still undertaken for an efficient process.

3.3 Biohydrogen Fuel Cell

The efficiency of the fuel cell is much more than that of internal combustion engines. The working of hydrogen fuel cell is depicted in Figure 3.2. It is reported that fuel cells could achieve 40–50% efficiency in terms of transportation when fuel cells like PEMFC

Table 3.1 Historical events in the development of the hydrogen fuel cell.

S.No.	Developer Name	Year	Work	Reference
1.	Sir William Robert Groove	1839	Developed the first fuel cell by mixing hydrogen and oxygen with a platinum electrode	Serov et al. (2018)
2.	Charles Langer and Ludwig Mond	1889	Used monoporus electrode to stabilize electrolyte and gave the term fuel cell	Guaitolini and Fardin (2018)
3.	Walther Hermann Nernst	1899	Discovery of solid oxide fuel cell	Behling (2013)
4.	Francis Thomas Bacon	1939	Bacon cell – Used nickel electrode and alkaline electrolyte for producing energy by combining H_2 and O_2 to produce water	Guaitolini and Fardin (2018)
5.	Harry Karl Ihrig	1959	Used fuel cell to drive a 20Hp tractor	Panayiotou et al. (2008)
6.	International Fuel Cell Company	1960s and 1970s	Developed fuel cells for NASA space shuttle Apollo and Orbiter	Panayiotou et al. (2008)
7.	USA Senate	1996	Hydrogen Future Act	Sopian and Wan Daud (2006)

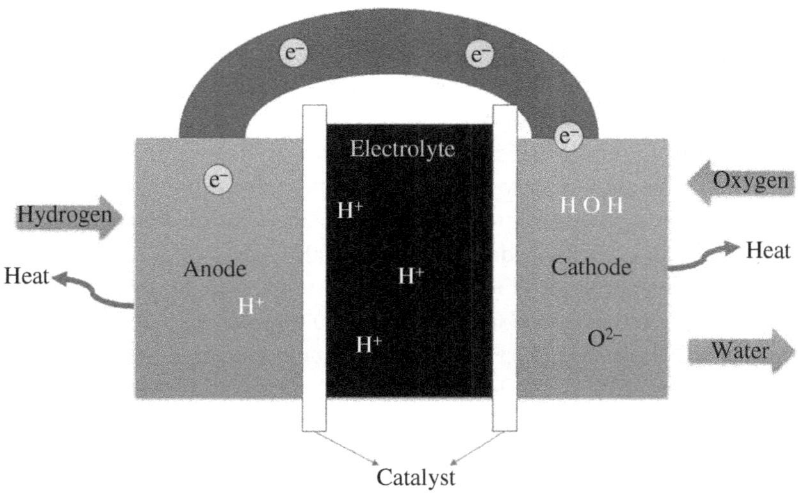

Figure 3.2 In the hydrogen fuel cell, hydrogen and oxygen are fed, and electrons flow between the two electrodes. Hydrogen fuel cell exhausts are water and heat. This makes it favorable toward the environment.

(proton exchange membrane fuel cells) and SOFC (solid oxide fuel cells) were used. In the future, the use of biohydrogen- and biomethane-based fuel cell will be preferred for electricity and transportation as they are CO_2-neutral and environmentally friendly. Due to the fact of being CO_2-neutral, it has attracted researchers and governments to get involved thorough work for an efficient process of H_2 production (Yusuf et al. 2015). One critical issue in H_2 production is storage and transportation. Its storage is difficult; on the other hand, the transportation of anaerobically produced H_2 is costly. These two factors serve the major concerns restricting the use of H_2 for commercial production and use. In recent times, H_2 produced anaerobically is directly transferred to PEMFC for electricity uses. This, however, reduces the cost. Wei et al. (2010) in an attempt to work on this issue, produced H_2 anaerobically and tried it to combine it with PEMFC. The results were encouraging in lowering the cost (Wei et al. 2010).

In terms of energy content, H_2 has the maximum energy value when compared to methane, ethanol, methanol, and other biofuels. Therefore, it would be a legitimate decision to use hydrogen as a biofuel in the future, and to develop engines that could be powered up by hydrogen fuel. The energy produced by the combustion of H_2 is 142 KJ g^{-1}. Thus it can be used for direct combustion in internal combustion engines and as fuel in fuel cells. At present, Toyota, Mercedes, Honda, and Lexus have developed vehicles that run on fuel cells. Currently, hydrogen is produced from natural gas, heavy oil, coal, Naptha, electrolysis of water, and less from biomass. In this case, the former is limited and nonrenewable. Thus, constraints in production can be sensed. Hydrogen production from natural gas, coal, and heavy oil generates a high amount of heat and greenhouse gases, therefore the production of biohydrogen for fuel cell and energy from biomass is very useful in comparison. Bioconversion of biomass for biohydrogen production is both energy-efficient and cost-effective. Since a large amount of biomass is produced from the agriculture sector and industries that depend on raw material from agriculture, there is an abundance of biomass, much of which is considered as waste for hydrogen production. Microorganisms are used in the conversion of various biomass from all such sectors for hydrogen production. Hydrogen produced from biomass-utilizing microorganisms or their enzyme is called biohydrogen, abbreviated as bio-H_2. Although there are some issues related to biohydrogen production, viz., low yield, scale-up challenges, hydrodynamics problem in the bioreactor, and process optimization. Not many commercial applications have been developed so far in using biohydrogen and only lab-scale or pilot level work is going on. Charder et al. (2009) produced biohydrogen from microalgae and with syringe transferred to PEMFC. Lin and his coworkers (Lin et al. 2007) produced biohydrogen by dark fermentation and fed it directly to PEMFC. Yet so much work is required for the development of such an integrated system that connects production and conversion to fuel cells so that the use of biohydrogen becomes commercial (Lin et al. 2007; Chader et al. 2009; Rahman et al. 2016).

3.4 Cellulosic Biohydrogen Production from Waste Biomass

A natural polymer-like cellulose is widely used for biofuel production and work continues for understanding the appropriate mechanism underlying, for a better and efficient way of biofuel production. Cellulose is abundantly found in nature and is synthesized by a wide range of living organisms, from bacteria to trees. It is a polymer, composed of D-anhydroglucose rings joined together by β-1,4 glycosidic oxygen linkages as depicted in Figure 3.3. The protein responsible for cellulose synthesis is cellulose synthase and its gene is identified in bacteria as well as in plants. It can be extracted from plants or biosynthesized by microbes in a bioreactor. It is biodegradable and thermostable. These and many other such properties make it a suitable polymer for fermentative breakdown for product formation (Saxena and Brown 2001; Ansell and Mwaikambo 2009; Rudin and Choi 2013).

A major issue concerning the commercial level production of biohydrogen is cost-effectiveness, as it is produced from nonrenewable resources. To lower down the production cost that will attract industries and researchers for production is the use of renewable material. The lignocellulosic polymer is one such material that is abundant and renewable. It is one of the major waste from agriculture houses with low or no usage so far. As it is regarded as waste, therefore, valorization of lignocellulosic material is needed to commercialize its usage and establishing it as a major feedstock for biofuel industries. Since its use in hydrogen production is challenging, and hurdles of inadequate enzymatic action due to the presence of lignin, and optimized pretreatment is required to enhance the availability of free cellulose to be converted to glucose and would increase the yield of H_2. Figure 3.4 shows a simple mechanism involved in hydrogen production from free cellulose. The crystalline nature of cellulose is also a hurdle. This allows novel platforms to work upon ways to develop new methodologies that could bring down the processing cost of cellulosic material (Cheng et al. 2011).

3.4.1 Biohydrogen Production from Wheat Straw and Wheat Bran

The biological approach for hydrogen production is much more acceptable than conventional methods like thermochemical and electrochemical, as the former requires less energy

Figure 3.3 Structure of cellulose.

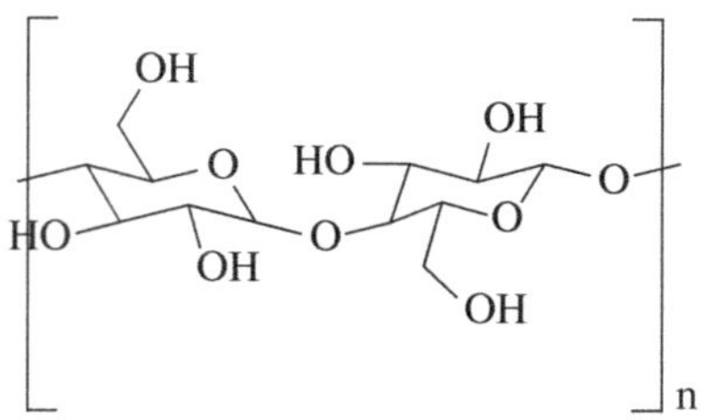

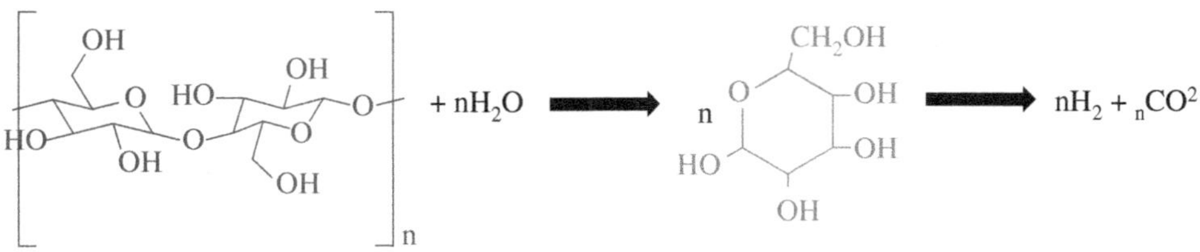

Figure 3.4 Free cellulose on hydrolysis produces glucose, which is enzymatically converted to hydrogen by microbes present in the innoculum.

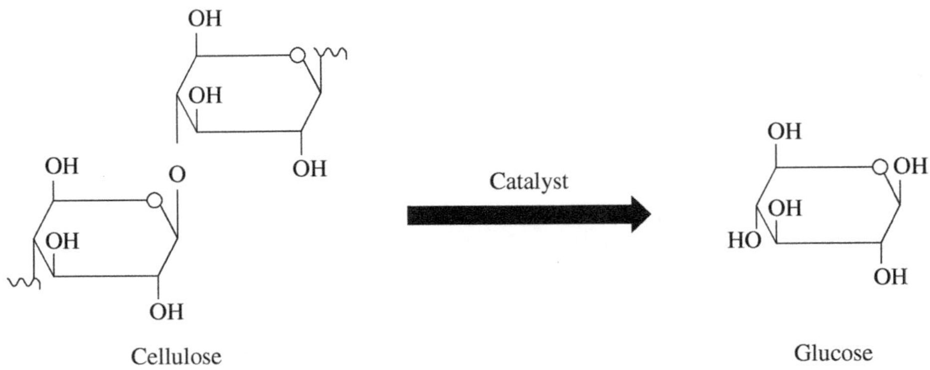

$$C_6H_{12}O_6 + H_2O \longrightarrow \text{Pyruvate} + \text{NADH} \longrightarrow 2CH_3COOH + 2CO_2 + 4H_2$$

Figure 3.5 During dark fermentation, cellulose is converted into glucose through acid, alkali, or enzymatic means, or by catalyst pretreatment. This glucose via glycolysis produces pyruvate. This then, through microbial enzymes, is converted to acetic acid and hydrogen.

input and is environmentally-friendly. Fermentation is, therefore, a useful way for biohydrogen production. Figure 3.5 shows the complete mechanism of dark fermentation ouccuring for hydrogen production from cellulosic material. From the Figure 3.5 it is evident that cellulose is catalytically converted to glucose. Glucose is then metabolized to pyruvate via glycolysis. Microbes through acetic acid synthesis process produces hydrogen. It is a useful way of converting the cellulosic waste into biohydrogen. The waste, like cow dung, is generally pretreated to reduce the natural microflora and give space to spore-forming organisms that are required for cellulose degradation. However, this pretreatment affects the population of spore-forming organisms also, decreasing their availability, and hence, lowering production efficiency. Cellulose-degrading microbes, their enzymes, and hydrogen-producing microbial strains play a vital role in increasing the hydrogen yield (Ren et al. 2011).

Wheat straw is produced in huge amounts all around the world as a by-product of wheat cultivation. Although being used as feedstock for animals, much of the wheat straw is burnt and is regarded as environment pollutant due to the emission of greenhouse gases. Wheat straw contains lignocellulosic material and can be utilized for the production of hydrogen by the dark fermentative process (Chu et al. 2011). Mixed microbial cultures (MMCs) have shown enhanced H_2 production during the fermentative process. From the various studies conducted for the use of MMCs in the dark fermentative method of H_2 production, few combinations were found to be beneficial are *Rhodobium-Vibrio/lactobacillus, Enterobacter-Rhodobacter/Clostridium,* and *Clostridium-Rhodopseudomonas/S. cerevisiae.* Patel et al. (2010) isolated certain strains of *Bacillus* sp. to develop the consortia for H_2 production. Different combinations of consortia and individual strains having hydrogen-producing activity are determined. For fed digestion, lignocellulosic waste materials like coconut choir, groundnut shells, and pea shells are packed in PVC (polyvinylchloride) tubes. Aspirator bottles are used for mixing the inoculum and fed. Different MMCs consortia are utilized. Only two MMCs combinations consisting of *B. cereus* strain, *B. thuringiensis* strain, *Bacillus sp.* strain, *B. pumilus* strain, and *E. aerogenes* strain and *P. mirabilis* strain in one and the other combination of *B. cereus* strain, *B. megaterium* strain, *B. pumilus* strain, *E. aerogenes* strain, and *P. mirabilis* strains showed better results than the other. However, the yield is low as predicted based on individual strain performance. Studies indicate that carbohydrate-rich food processing waste is a valuable substrate for hydrogen production by the dark fermentative method (Patel et al. 2010). In recent times focus is on carrying out biohydrogen production (through fermentation) in extreme thermophilic conditions. It is suggested because studies have indicated that under thermophilic conditions, the high yield of hydrogen production can be achieved. At higher temperatures, there is an increase in hydrolysis activity, pathogenic destruction, and a decrease in the risk of contamination from methanogens. These attributes are industrial friendly and hence are favorable for better and enhanced hydrogen production. Thermodynamic favourability is also a factor. From a study, it is known that after hydrothermal pretreatment of the wheat straw, two types of the substrate are produced – solid and liquid fraction. The solid fraction contains cellulose, and liquid fraction contains hexose and pentose. These fractions are not eligible for ethanol production but can be used for hydrogen production. A study is conducted on fermenting wheat straw hydrolysate with the mixed culture at a thermophilic temperature at about 70 °C. Results show that in both batch and continuous culture system under extreme thermophilic temperature, hydrogen is successfully produced even when there is a drawback of low cell count due to increased temperature and about 184 ml-H_2/day/$L_{reactor}$ is produced. Such a high production is because at high temperatures only extreme thermophilic microbes capable of hydrogen fermentation could withstand high-temperature fermentative conditions, and such organisms were *T. subteraneus, C. subterraneus,* and *T. thermosaccharolyticum*. Under such conditions, microbes like *Lactobacillus* sp. have retarded growth due to high temperatures. This ensures a high hydrogen yield. Thermophilic conditions are stringent to hydrogen production inhibiting or controlling the inhibitory activities and increasing the yield (Kongjan et al. 2010).

Wheat bran is about 14–19% of a wheat grain weight and is produced as a by-product during wheat milling and is regarded as a waste product. More than 20 million tons of wheat bran is produced annually. It is composed of hemicellulose, cellulose, and starchy residuals, which could be used for hydrogen production. Due to the complexity of cellulosic material, pretreatment is necessary before feeding it to microbial culture for hydrogen production. Physical, chemical, and enzymatic methods are employed for pretreatment. However, only a few show the essential features profitable to industrial processes, such as low cost, enhanced reactive carbohydrates, negligible microbial inhibitors, and low environmental damage. The selection of an appropriate pretreatment method is required before feeding the hydrosylate into fermenting tanks. Anaerobic fermentation is found to be profitable for hydrogen production, as it is a sustainable, green, and clean method. For anaerobic fermentative hydrogen production dominant hydrogen-producing bacteria are isolated and tested for species identification before adding into the inoculum media for future analysis. Wheat bran is subjected to acid hydrolysis. Apart from this conventional acid hydrolysis, pressure acid hydrolysis, and microwave-assisted acid pretreatment is also found to be beneficial. Anaerobic bacterias have a significant effect on hydrogen production from wheat bran. Along with its different pretreatment conditions, substrate concentration and inoculum concentration shows a positive effect on hydrogen production. The study indicates the production of volatile fatty acids like butyrate and acetate were created as valuable by-products during the process. Higher values of volatile fatty acids are indicative of enhanced fermentation and good yield. Wheat bran is a promising material for hydrogen production (Pan et al. 2008).

3.4.2 Biohydrogen Production from Corn Stalk

Corn stalk is a thick and strong structure comprising leaves and stems. They have good nutritional value. It contains cellulose, hemicellulose, and lignin. These three components are cross-linked to each other, and therefore, require pretreatment for industrial processes (Chen 2015). Many studies are being conducted for the use of corn stalk for biohydrogen production, which is evident that corn is recalcitrant due to cross-linking and needs pretreatment. In a study by Fan and coworkers, corn stalk is used for bio-H_2 production by a lesser panda manure method. In this, corn stalk is subjected to lactic acid pretreatment and neutralized to pH 7. Also, another batch is pretreated with microbial additives. These pretreatment methods were optimized for increasing glucose concentration in the medium by the breakdown of cellulose. H_2-producing bacteria are used for the preparation of inoculum. In such experiments, gas chromatography is used to analyze $H_{2,}$ volatile fatty acids, and alcohol production. Such a study is indicative of the fact that biotreatment and process optimization leads to higher production of bio-H_2 from recalcitrant corn stalk. They were able to produce 176 ml H_2/g-TS (total solid) (Fan et al. 2008). The use of microbes for corn stalk pretreatment is seen as an ecofriendly method with better results. However, there is a need for the acid pretreatment before microbial degradation of corn stalk for bio-H_2 production. Substrate concentration and pH at which the reaction is carried out plays an important role. At optimum level and value respectively bio-H_2 production peaks (Zhang et al. 2007).

Owing to the complex nature of corn stalk, the yield of cellulose from conventional acid hydrolysis and enzymatic treatment produces about 20% cellulose, and much of the cellulose is disrupted and lost and is not available for H_2 production. These methods are also energy expensive. Given this, the white-rot fungus has attracted researchers for treating corn stalk with this organism as it is energy efficient. Zhao et al. (2012) pretreated corn stalk with white-rot fungus, isolated, and cultured over potato dextrose agar (PDA). *Phanerochaete chrysosporium* is used for the bio-pretreatment of the feedstock. Treated cornstalk is then can be used for enzymatic pretreatment (*Trichoderma viride* is used for enzymatic hydrolysis). Such work indicates that after bio-pretreatment lignin concentration decreases substantially, but holocellulose loss is also observed under such condition. Studies suggest that such pretreatment is beneficial, ecofriendly, and energy inexpensive giving a high yield of bio-H_2 (Zhao et al. 2012). Fermentative hydrogen production is being impacted by experimental design. A study on orthogonal experimental design for bio-H_2 production from corn stalk indicates that substrate concentration has an overall role in high yield. Enzymatic temperature and time, substrate concentration, and initial pH are the four factors taken into consideration in the orthogonal design. Such a study indicates the production of bio-H_2 about 141.29 ml g1 –CS (Ma et al. 2011). Bio-H_2 production from corn stalk can also be enhanced by the co-culture method. A combination of cellulose hydrolyzing bacteria and hydrogen-producing bacteria is suitable for high yield and may also reduce the need for pretreatment. Such an approach of direct microbial conversion of cellulosic biomass to H_2 is beneficial and completely ecofriendly. It is evident from it that the co-culture method of bio-H_2 production from corn stalk yields more when compared to anaerobic methods. These studies open new paths for finding ways that lead to high yield (Li and Liu 2012).

An estimation suggests that 1017 million tons of corn is produced in China. Corn strover is produced on gathering of corn grains. About 0.94 tons of it is produced for every tons of grain developed from corn crop. Corn stalk rind and pith are major components of corn strover. Corn stalk is widely used in paper industry. The corn stalk pith introduces obstacles like high chemical consumption, reduced sheet opacity, and dewatering, because of which it is removed prior to paper development. This corn stalk pith is regarded as waste, but due to high carbohydrate content can be used for hydrogen production. Corn stalk pith is enzyme digestable. The enzyme hydrosylates from corn stalk pith is subjected to photo fermentation using photosynthetic consortium. Such an attempt produces $0.18\,g\,l^{-1}$ of hydrogen. Corn stalk is ignored so far for hydrogen production. Looking at its production annually it can become a vital feedstock for hydrogen production (Jiang et al. 2016). Hydrogen production is a discontimous process due to cell wash out. This could be overcomed by cell immobilization. It may enhance production rate and also provide stability to continuous hydrogen production. In 2016, Kirli and coworkers used different types of immobilizing material. They used sponge pad, black porous sponge, plastic nylon sponge, and others. They used wheat hydrosylate as feed stock. Immobilization showed two times higher production of hydrogen from hydrosylate when compared to nonimmobilized system; however, consumption of xylose in cellulosic hydrosylate does not occur and remains ignored. This can tackled by the choice of immobilizing material. Mycelia pellets formed from *Aspergillus sp.* is shown to be efficient immobilizing material and 2.45 mol H_2/glucose molecule is produced under such system. Mycelial pellets have shown higher hydrogen production as compared to other immobilizing materials. Clostridium mycelia pellets are also being used and about 14.2 mmol $H_2\ l^{-1}\ h^{-1}$ is produced. The efficiency of energy conversion is found to be 17.8% as reported (Zhao et al. 2017; Kirli and Kapdan 2016; Kumar et al. 2016a).

3.4.3 Biohydrogen from Rice Straw and Rice Bran

Rice straw is produced in millions of tons and is mostly burnt or buried for remedial purposes, and causes environmental issues. Rice straw has a high cellulosic content and can be used for bio-H_2 production. As is evident, anaerobic digestion is used widely for the conversion of rice straw to bio-$H_{2.}$ Rice straw is dried for several days and is grounded and stored before being used. Controlling the moisture content, dried rice straw is added with acid for its hydrolysis. Rice straw hydrosylate has 5-hydroxymethylfurfural as a result of acid hydrolysis, which may be poisonous for anaerobic microbes and may inhibit their growth. This could hamper hydrogen production. Overliming is done with the help of $Ca(OH)_2$ to overcome this obstacle. Dark fermentation is employed for H_2 production. Rice straw contains 33.1% cellulose and on hydrolysis produces $99.3\,g\,l^{-1}$ of hexose and $80.1\,g\,l^{-1}$ of pentose theoretically. Good yields of hydrogen are produced under different fermentation conditions. Rice straw under acid hydrolyzing conditions produces a good amount of reducing sugars. These reducing sugars are then worked on by enzymes for bio-H_2 production. It is suggested from studies that increasing reducing sugar concentration enhances the production of hydrogen gas. More studies are required for enhanced production (Chang et al. 2011). Rice straw is a lignocellulosic material and requires pretreatment before being fed to enzymes. Much stress is given in developing strategies for treating rice straw to produce unhampered cellulose. Organoslov pretreatment can be an effective way of producing cellulose, and also lignin, which can be used for the production of other chemicals. In this method, dried rice straw is sieved to attain a particular size. This is then mixed and blended with an ethanol-water solution with sulfuric acid acting as a catalyst. This mixture is maintained for a specific time duration at a particular temperature. This mixture is then washed with the water-ethanol solution. This mixture is now known as organosolv liquor, which is mixed with water to precipitate lignin. On centrifuging for 5 minutes at 6000 rpm, solid hydrosylate are separated, which are subjected to enzyme hydrolysis and then fed to microbes for bio-H_2 production. The result of such a study indicates a high concentration of glucose and therefore has high yields of H_2 gas. If such a method is employed by the top 10 rice-producing nations, then about 355.78 kilotons of hydrogen gas can be produced annually. China alone can produce 32% of global H_2 gas (Asadi and Zilouei 2017). In a study conducted by Liu et al. (2013) suggests that continuous culture is better than batch culture for bio-H_2 production. Through their study, they established that continuous culture produced 1.5 times more bio-H_2 from rice straw when compared with the same in batch culture. Moreover, substrate utilization is found to decrease in the case of batch culture. This suggests that continuous culture produces more hydrogen gas from rice straw with comparatively fewer substrate requirements (Liu et al. 2013). Co-digestion of rice straw and the activated sludge has shown enhanced bio-H_2 production. Heat treatment of activated sludge is proved to be beneficial in the co-digestion method as it increases the yield of hydrogen gas. The study indicates that the reduction of rice straw size to powdered form, which is mostly used in the lab-scale and harms H_2 production. Larger size particles are better feeds for hydrogen production (Alemahdi et al. 2015).

In southern Asian countries like China, India, Indonesia, Bangladesh, Vietnam, and Thailand, rice is a major part of everyday food and agriculture products. An annual increase in rice production can be seen since the advent. Every year about 480 million tons of rice are produced globally. After the rice is harvested from the field, the product is known as paddy rice. Paddy rice has three parts – the outer layer is known as husk, a middle layer known as rice bran, and at the center is the white rice. The rice bran is mainly used for edible oil production. After the rice bran is used for oil production, the leftover residue is known as de-oiled rice bran. It is a solid residue. De-oiled rice bran is a lignocellulosic waste that contains 24% lignin, 39% cellulose, 28% hemicellulose, and about 9% protein (Muthayya et al. 2014; Sereewatthanawut et al. 2008). Since de-oiled rice bran is produced in a huge amount every year, which is low in cost biomass and can be used for bio-H_2 production. Using such a feedstock for hydrogen production will reduce bio-H_2 production costs and encourage industrial production. Dark fermentation is one of the most preferred methods for hydrogen production. In this process, microorganisms are made to grow on carbohydrate-based feedstock in oxygen-derived conditions in the absence of light. Dark fermentation gives several advantages like less processing time, low energy input, low cost, and less adversity to the environment. Very few studies are made for bio-H_2 production from deoiled rice bran. In one of the studies, *Clostridium acetobutyricum* is used for the production of bio-H_2 from deoiled rice bran (DRB). DRB is stored at 4–8 °C before use. This is used for the production of oil later with the help of a soxhlet extractor. After the oil is extracted, the DRB produced is dried and then acid hydrolyzed. Acid hydrolysis is carried out with sulfuric acid. This mixture is shaken vigorously for efficient mixing and maximum interaction between DRB and sulfuric acid. This acid-treated DRB is then heated to bring about complete hydrolysis. The mixture is cooled and then filtered using Whatman filter paper to obtain hydrosylates. Hydrosylates are then mixed with *C. aectobutyricum* in the calculated amount and are incubated for hydrogen production. The result shows that DRB is an efficient feedstock for bio-H_2 production. A maximum of 574.6 ml of bio-H_2 production is reported in this study. Production can be enhanced by process optimization as the study suggests. The temperature during incubation has a significant effect on bio-H_2 production (Azman et al. 2016). Rice bran de-oiled wastewater, which is rich in carbohydrate is also used for bio-H_2 production. This wastewater is anaerobically treated for bio-H_2 production. The study shows about 330 ml of bio-H_2 production from rice bran. However, various parameters affect bio-H_2 production. Thermophilic temperature is found to be favorable for hydrogen production from rice bran de-oiled wastewater. At thermophilic temperature, the nonhydrogen-producing microbes are eliminated, decreasing the contamination chances (Sivaramakrishna et al. 2010). Thus, rice bran is an efficient feedstock for hydrogen production and is available in usable form even after it is being used for other purposes.

3.4.4 Biohydrogen Production from Food Waste

Millions of tons of food waste are produced every day globally and are a burden on the environment. It is a major source of decay, odor, and leachate and its transportation to dumping areas is a costly affair. Food waste has a high energy content. It is suggested through various studies that food waste can be utilized by generating energy from it. Hydrogen production food waste could be one of the strategies for energy generation. Bio-H_2 production from food waste depends on the fact that food waste must contain a high amount of carbohydrate, which could be used for hydrogen metabolism by microbes. Different components like carbohydrates, proteins, and lipids are present in the food waste and are degraded at their optimum environmental condition. Study shows that food waste from the dining room can be used for hydrogen production. The collected food waste contains vegetables, grains, and meat. All of them are fed to a leaching bed reactor mixed with anaerobic sporing bacteria (*Clostridium* sp.). Fermentation nature is again anaerobic, commonly employed for hydrogen production. Results from such an experiment predict that optimum dilution rate is beneficial for hydrogen production under anaerobic fermentation reaction. Since the dilution rate helps in maintaining the continuous addition of media, the microbial culture is always high at metabolizing the substrate into the product. However, one needs to determine the optimal dilution rate before proceeding at an industrial scale. Prediction of optimal dilution rate becomes essential as it affects the value of volatile fatty acids, decreasing it, and has an impact on hydrogen production. A decrease in volatile fatty acids increases pH and decreases hydrogen production. Determining the dilution rate increases hydrogen production efficiency from food waste by 70% (Han 2004). In a country like Korea, sewage waste and organic waste is problematic. Both are produced in huge amount and needs a scientific solution for cost-effective disposal. Co-digestion of sewage waste and food waste for hydrogen production could be an approach for this issue. Kim (2004) reports of blending the two before taking for anaerobic digestion. The two are mixed in different ratios and mixed with anaerobic bacteria. Results indicate that in co-digestion, food wastes contribute more toward hydrogen production as compared to sewage waste. Sewage waste has only an enhancing effect on hydrogen production by 13–19%. Volatile fatty acid concentration affects the hydrogen production (Kim 2004). Another study shows that the co-digestion of these two is positively affected by the addition of AR (age refuse). Refuse is left over in landfills, kept for several years, and then get stabilized there. The refuse obtained from such a place is known as aged refuse. Aged refuse is used as seed material for inoculum preparation. It contains a broad spectrum of microbes that have strong degradation property on both biodegradable organic matter and also on refractory organic matter found in sewage waste. The alkaline nature of AR prompts for bacterial metabolism for hydrogen production. When AR is added in a 50% amount, it is an encouraging enhancement in the bio-H_2 production in the co-digestion process. AR in the future may become a vital source of microorganisms, which can be utilized for various industrial production from wastes (Li et al. 2008).

Pretreatment of organic waste plays an important role in hydrogen production. Substrate pretreatment burst open the microbe for efficient digestion. Different pretreatment (thermal, ultrasonic, acidification, and alkaline) have different effects on hydrogen production from food waste. A study conducted on the effect of pretreatment on bio-H_2 production from food waste indicates that ultrasonication with base has the enhancing effect on substrate solubilization. Treated substrate always showed higher bio-H_2 production as compared to the untreated substrate. Alkaline treatment shows lesser hydrogen production compared to other pretreatment methods. The highest production is seen in ultrasonication with acid. Therefore, the choice of pretreatment becomes an essential criterion for enhanced bio-H_2 production from food waste since different pretreatment has a positive and negative impact on production. A suitable pretreatment is required for efficient production. Pretreatment increases the availability of glucose in the medium to be used for hydrogen production (Elbeshbishy et al. 2011).

3.4.5 Biohydrogen from Bagasse

Industrially sugarcane is an important crop used in sugar production. Billions of tons of sugarcane are produced worldwide. Sugarcane bagasse is the leftover residue after juice extraction. Bagasse alone accounts for 25% of the dry mass of sugarcane. In this manner, several million tons of bagasse are produced every year. It is mainly used for energy production by the combustion process. Combustion of bagasse has serious environmental issues. Apart from combustion, it is also used for ethanol production. It contains cellulose, hemicellulose, and lignin. On acid hydrolysis, it produces glucose and xylose. This can be used for the production of hydrogen by microbes like *Clostridium sp.* Bagasse hemicellulose through acid hydrolysis can be used for the production of hydrogen by microorganisms. Bagasse is subjected to acid hydrolysis by sulfuric acid at various concentrations that produce hydrosylates from hemicellulose. *Clostridium sp.* is used in batch fermentation to ferment hydrosylate into hydrogen. Gas chromatography is employed for analysis of various product concentration. Sugar fermentation generally produces volatile fatty acids, ethanol, butanol, propionic acid, butyric acid, and acetic acid apart from hydrogen. From such work, it is suggested that hydrosylates from bagasse hemicellulose are potent for hydrogen production due to high sugar concentration. It has a very low effect of inhibitors. Hydrogen is produced in a good amount. *Clostridium butyricum* is suggested to be used for hydrogen production. Since during the process no product is formed that has inhibition, neither toxic compound is synthesized, therefore optimizing the process means a great amount of hydrogen can be produced in the future from sugarcane bagasse (Pattra et al. 2008).

So far, much of the study is focused on single-step protocols for hydrogen production by fermentation. The two-step process for hydrogen production has not attracted much. Also, hydrosylate from bagasse is used alone for the production of hydrogen. The use of hydrosylate and bagasse cellulose together is still not understood well. So Rai et al. (2014) and coworkers studied the two-step process (dark and photo fermentation) using bagasse hydrosylate and bagasse cellulose together. Bagasse is chopped into 2–5 mm pieces to increase surface area and then subjected to acid hydrolysis. *C. fimi* isolated is cultured on suitable media, and the hydrosylate from bagasse is subjected to dark fermentation by the use of *E. aerogens*. Simultaneously, bagasse cellulose is hydrolyzed by *C. fimi* in another reactor. The filtrate from these two is gathered in separate vessels and is photo fermented using *Rhodopseudomonas*. Analysis by gas chromatography, HPLC, and spectrophotometry reveals that the two-step process enhances hydrogen production. During dark fermentation, about 1600 ml/2 l of bio-H_2 is produced and during photofermentation about 900 ml/2 l of bio-H_2 is produced. Such a study indicates that the two-step process gives more yield and better substrate utilization for hydrogen production. Using hydrosylates from hemicellulose and cellulose increases sugar concentration. The use of different microbes (both aerobic and anaerobic) is beneficial for the process. It reduces the toxic components and inhibitors, giving more chances to enzymes for substrate processing. More work is required, as it is in the infancy stage. The two-step process uses both cellulose and hemicellulose for the microbial breakdown (Rai et al. 2014).

Pretreatment is always found to be beneficial, and understanding process optimization by various parameters could be beneficial in hydrogen production from bagasse. The alkali treatment of baggage has shown to be advantageous in hydrogen production. Since bagasse has a high concentration of lignin, recalcitrant for the process, the use of ammonium hydroxide is found to be useful, due to its specific reactivity toward the lignin. However, NaOH is recommended for the removal of hemicellulose. Alkali pretreatment increases the cellulose-only concentration and also decreases its crystallization. The co-culture adopted could be a futuristic aspect as it gives a novel insight toward the cost-effectiveness of the process. It also proposes that ethanol can be produced with hydrogen during the process, which could be a constructive process for the industries in terms of income generation. Since two important bio-based fuels are generated from the same substrate, such work should be promoted at an industrial scale to decrease the dependency on fossil-based fuels (Cheng and Zhu 2013).

3.4.6 Biohydrogen Production from Mushroom Cultivation Waste

Among the various types of cellulosic waste considered for hydrogen production, mushroom cultivation waste (MCW) has attracted much attention recently due to easy availability and abundance. Mushroom is cultivated in polypropylene bags, which is stuffed with wood flour and other nutrients. After mushroom is cultivated, the bag becomes useless as it cannot support the growth of mushrooms any longer; it becomes deficient in the nutrient required for its growth. This bag is now known as MCW. It is estimated that about 150 million bags containing 80 000 tons of MCW are produced annually. This MCW is usually burnt and is a serious threat to the environment. MCW requires quick and serious waste management to avoid any such issues. Production of biohydrogen from it could be an effective method for its management. Studies have indicated that MCW can be used for bio-H_2 production. It could be a cost-effective method of bio-H_2 production as it solves the environmental problem. Many reaches are undertaken in this aspect. MCW has $0.46\,g\,l^{-1}$ of carbohydrate. It has 31% cellulose, 6.9% hemicellulose, and 14% lignin. The dried and powdered MCW undergoes anaerobic fermentation for bio-H_2 production. Such a study indicates that MCW is an efficient bio-H_2 producer. About 0.073 mmolH_2/g is produced from it, which is encouraging. It is also found from studies that MCW fermentation for bio-H_2 production can be enhanced by adding sewage waste. The addition of sewage waste nullifies the requirement of nutrients in the fermentation medium, as it contains all the required nutrients for culture growth. It is also evident from the study that sludge waste like cow dung could enhance bio-H_2 production. Using cow dung with MCW avoids the use of any extra nutrient and also degrades MCW efficiently as it contains cellulose-degrading bacteria. Using cow dung as inoculum with MCW has shown to produce hydrogen at 10.11 mmol H_2/L/d rate. Such work is encouraging for extending the type of cellulosic feedstock that could be used for hydrogen production simultaneously managing wastes produced in tons (Lay et al. 2012). However, the production of hydrogen from wastes like MCW is still in its infancy state and requires more attention to utilizing this inexpensive feedstock. The use of such feedstock could in future decrease the production cost of hydrogen and may help us to move toward the hydrogen-based economy, a need of the hour looking at the effects of fossil-based fuel on the environment. And it is not only MCW, but there are many wastes available freely and dumped due to uselessness in industrial product synthesis, which can be used for hydrogen production. One of the studies conducted in Taiwan has suggested that using substrates like MCW could reduce their carbon footprint in high amounts. As estimated that millions of tons of MCW are produced annually, the use of it in bio-H_2 production can produce about 1842 kWh of bioenergy every year and reducing about 114–178 kg of CO_2 per year. Heat pretreatment can be key in increasing bio-H_2 production as it inhibits heat-resistant methanogens (Lin et al. 2015).

Spent mushroom compost (SMC) is an budding cellulosic component. An estimation suggests that 30 million tons of SMC is produced annually by the solid-state fermentation industry; a 1 kg mushroom produces 5 kg of SMC. Although SMC is used for many value-added purposes like bioremediation, fertilizer, animal feed, and others. It contains 18–62% cellulose. *Clostridium thermocellum* has been widely used for biofuel production from cellulosic biomass. The same is applied for hydrogen production from SMC. The study shows *C. thermocellum* effectively fermented SMC for hydrogen production. The need for pretrearmnet of SMC is diminished over the use of *C. thermocellum*. Under various conditions, about 30 mmol l^{-1} of hydrogen is produced. This establishes that SMC is a vital source for hydrogen production. *C. thermocellum* on the other hand can be used for lowering the cost of ferementation and encouraging market for hydrogen fuel. Utilization of *C. thermocellum* eradicates pretreatment and sterilization process. This could reduce the cost. However, more supportive work is required (Lin et al. 2017).

3.4.7 Biohydrogen Production from Sweet Potato Starch Residue

Sweet potato is produced in millions of tons and is used for the production of starch. In a study, it is reported that in south Japan alone, about 8×10^4 tons of sweet potato starch residue (SPSR) is produced. This residue is produced as a by-product during starch production from sweet potato. SPSR is known to contain about 80% water. Apart from water, it has fibers, cellulose, and hemicellulose. Mostly SPSR is used for citric acid production. Most of the studies have indicated efficient hydrogen production from starchy material using *C. butyricum*. Hence, a study is being conducted to understand how starchy residue can be used for hydrogen production. Yokoi and coworkers (Yokoi et al. 2001) attempted to used SPSR for hydrogen production. *C. butyricum* and *E. aerogenes* were used for the production of bio-H_2 anaerobically using SPSR. SPSR is dried. Dried SPSR contains about 49% starch along with moisture, ash, and other components (proteins, fats, and fibers). The fibers contain cellulose and hemicellulose. Since *C. butyricum* activity is inhibited by L-cysteine, the use of *E. aerogenes* is suggested. The co-culture method has revealed that about 2 mol of H_2 is produced per mole of glucose. The amount of H_2 produce by this technique from SPSR is about 76 ml/batch culture. Organic acids were produced as by-products. These organic acids (acetic, butyric, and lactic acid) can be used for increasing hydrogen production by using *Rhodobacter*. By this continuous method, about 7 mol of H_2/mol of glucose can be produced (Yokoi et al. 2001). Pretreatment of SPSR could enhance the overall process, as it will breakdown cellulose and hemicellulose, availing more amounts of glucose for hydrogen production. Work on SPSR indicates that the by-product formed during the hydrogen production from organic wastes can be further utilized for hydrogen production. In doing so, the number of hydrogen molecules produced per glucose molecule can be increased, valorizing the waste, and used fully.

3.4.8 Biohydrogen from De-Oiled Jatropha

Jatropha is a plant species used for biodiesel production. It belongs to the family *Euphorbiaceae*. It is cultivated in many different regions of the world. The dried seeds of jatropha are used for the extraction of oil and studies have indicated that about 75–85% of oil can be extracted from jatropha. During the process, kernel cake is left over, which is an important by-product (Misra and Murthy 2011). Nonfood cellulosic biomass is preferred for bio-H_2 production. De-oiled jatropha waste is a solid residue after oil extraction. This contains certain toxic compounds, and therefore, cannot be used for animal feed. However, it contains cellulose and hemicellulose. An estimated amount of 2.5–3 tons of de-oiled jatropha waste is produced per ton oil extracted from jatropha. Mostly this waste is used for methane production under anaerobic conditions. Variation in pretreatment if done, can be used for hydrogen production. In one such work, de-oiled jatropha waste is mixed with municipal sewage sludge. The two are mixed in varying concentrations to determine the optimum value for maximum production. Under anaerobic conditions, hydrogen is produced along with methane. Results indicate that about 200 g l^{-1} of hydrogen is produced at 55 ° C. Temperature and pH affect the process significantly. pH between 6.5–7.5 sees a peak in production (Kumar and Lin 2013). Pretreatment is key in hydrogen production. In one such work, de-oiled jatropha waste is subjected to acid hydrolysis. The hydrosylate is used for anaerobic fermentation. Pretreatment increases sugar concentration when pretreating agent concentration is increased. However, on increasing pretreatment, agent fermenting inhibitors are produced, decreasing hydrogen production. Comparing HCL and H_2SO_4 in the case of acid hydrolysis, HCl is seen as less harsh. After pretreatment about 786 ml, H_2/L-d is produced, which shows an enhanced production rate as compared to the other works where pretreatment is not done (Kumar et al. 2015).

In one of the methods, de-oiled jatropha waste is mixed with sludge. The mixture is then mixed with hybrid immobilized cells. For immobilization heat-treated inoculum is mixed with sodium alginate, chitosan, silicon oxide, and activated carbon. The mixture is then dropped in $CaCl_2$ for bead formation. Also, the substrate is acid and enzymatically hydrolyzed. The results were encouraging, about 3.65 L H_2/Ld was produced. Immobilization is yet another way, other than pretreatment for hydrogen production. It could decrease the cost, since the immobilized cell or enzyme can be reused again and again. Immobilization has shown to increase activity and stability (Kumar et al. 2016a,b).

3.4.9 Biohydrogen Production Banyan Leaves and Maize Leaves

Banyan leaves are also used for hydrogen production by microwave pyrolysis and plasma. A 35% conversion rate is obtained (Lin et al. 2014). About 700 million tons of maize leftover biomass are produced per year. It can be used for hydrogen production. Hydrogen from maize leaves is produced through batch fermentation in 48 hours. In several attempts of bio-H_2 production, it is found that the yield is low by many folds as compared to other biomass. To increase the yield, *B. amyloliquifaciens* is used. This bacteria produces cellulase enzyme, which is used for treating the substrate. The treated substrate is air-dried and used for hydrogen production using *C. sacchrolyticus*. This effort proved to work as it increased yield by 2.1 times as compared to sweet sorghum plant biomass. The untreated maize leaves are found to produce 1.8 mol of H_2/glucose molecule. On the other hand, treated maize leaf is found to produce 3.6 mol of hydrogen per molecule of glucose consumed, which is 31% more as compared to untreated maize leave. Untreated maize leaf produced 16 ml of hydrogen per gram of dry matter, whereas treated maize leaf produces 64.5 ml of H_2 per gram of dry matter. An increase of fourfold from untreated indicates that enzymatic pretreatment of maize leaves holds significance in bio-H_2 production from maize leaves (Ivanova et al. 2009).

3.5 Conclusion

Our present world economy and infrastructure development depend on fossil fuel. Fossil fuels have energized our system and sustained society to date all while fulfilling its

Table 3.2 Various cellulosic biomass produces different amount of hydrogen.

S.No.	Cellulosic waste	Bio-H_2 produced		Reference
		In mol-H_2/sugar mol	In ml/l	
1.	Wheat bran	2.61 mol-H_2/ sugar mol	—	Patel et al. (2015)
2.	Wheat straw	—	212 ml/g sugar	Kongjan and Angelidaki (2010)
3.	Corn stalk	—	429.72 ml/l/d	Wang et al. (2018)
4.	Rice straw	—	72.5 ml/g sugar	Dong et al. (2019)
5.	Rice bran	—	545 ml/l	Tandon et al. (2018)
6.	Food waste	1.65–1.79 mol-H_2/ sugar mol	—	Jarunglumlert et al. (2018)
7.	Baggase	1.24 mmol-H_2/ sugar mol	—	Saratale et al. (2018)
8.	Mushroom	0.29 mmol-H_2/ sugar mol	—	Lin et al. (2015)
9.	Sweet potato starch residue	2.4 mol-H_2/ sugar molecule	—	Vendruscolo (2014)
10.	De-oiled jatropa	—	0.9 l/l/d	Kumar et al. (2017)
11.	Banyan leaves	—	0.06 ml/g	Lin et al. (2014)
12.	Maize leaves	3.6 mol-H_2/ sugar molecule	—	Ivanova et al. (2009)

It becomes evident from the table above that cellulosic waste possesses the potential for harnessing energy demands.

demands. But the use of fossil-based fuels is toxic to our environment. The combustion of fossil fuels produces greenhouse gases that are responsible for global warming. Thus there is an urgency of steps need to be taken to reduce greenhouse gas emissions. Among various strategies employed for the reduction of greenhouse gas, biohydrogen is an efficient tool for replacing fossil fuel. At present, hydrogen is produced from fossil fuel via steam reforming. Natural gas, oil, coal, and water splitting are the major substrate for hydrogen production (Chandrasekhar et al. 2015). But these are energy expensive, cost-ineffective, and have constrained in process. Looking at these issues, there is a need to look for the alternative substrate for hydrogen production. Cellulosic biomass is one of them. Cellulosic biomass is currently employed for biofuel production like bioethanol, biobutanol, and others. But there are other sources of cellulosic biomass, which are considered as waste. Billions of tons of cellulosic biowaste are produced annually, like rice bran, rice straw, wheat bran, wheat straw, de-oiled rice bran and jatropha, maize leaves, food waste, sewage from municipal entities, and many others. These wastes are high in cellulosic content ever after being used for the different major processes. Recently, these have attracted many researchers to use them as a substrate for bio-H_2 production. Table 3.2 shows the amount of hydrogen produced from cellulosic waste biomass.

Using cellulosic biowaste establishes a consent that it should not be considered useless, since it still contains vital components, which, if used properly, can be efficient enough to solve our energy demands. Cellulosic biowaste is proved to be an efficient substrate for biohydrogen production. Studies have shown that dark fermentation is a much-preferred fermentation condition for biohydrogen production. As most of the hydrogen-producing microbes are anaerobic, therefore dark fermentation is useful for the process. *Clostridium sp.* is favored for dark fermentation. Many other *Bacillus* species are also used. From the studies on bio-H_2 production, it is evident now that pretreatment plays a vital role in enhancing bio-H_2 production from cellulosic waste. Pretreatment via acid, alkali, or enzymatic has shown to increase sugar concentration and volatile fatty acids, which enhances hydrogen production. Enzymatic treatment will always be preferred as it is ecofriendly and doesn't produce toxic hydrosylate. Temperature and pH also has a role to play, as hydrogen-producing microbes can metabolize at optimum temperature and pH. Co-culture is found to be beneficial for biohydrogen production. Various studies have indicated that cellulosic waste yields 3–4.5 mol H_2/glucose mol. However, the optimizing process can enhance yield value.

There is a need to work more on these cellulosic waste, optimizing pretreatment, and other factors to increase hydrogen production. Bio-H_2 is the future fuel and much of its production depends on finding such cellulosic waste, as they are easy to collect and are inexpensive, decreasing the production cost and supporting its use to replace fossil fuel.

References

Alemahdi, N., Man, H.C., Rahman, N.A. et al. (2015). Enhanced mesophilic bio-hydrogen production of raw rice straw and activated sewage sludge by co-digestion. *Int. J. Hydrog. Energy* 40 (46): 16033–16044. https://doi.org/10.1016/j.ijhydene.2015.08.106.

Ansell, M.P. and Mwaikambo, L.Y. (2009). The structure of cotton and other plant fibers. *Handbook of Textile Fibre Structure*: 62–94. https://doi.org/10.1533/9781845697310.1.62.

Asadi, N. and Zilouei, H. (2017). Optimization of organosolv pretreatment of rice straw for enhanced biohydrogen production using Enterobacter aerogenes. *Bioresour. Technol.* 227: 335–344. https://doi.org/10.1016/j.biortech.2016.12.073.

Azman, N.F., Abdeshahian, P., Kadier, A. et al. (2016). Biohydrogen production from de-oiled rice bran as sustainable feedstock in fermentative process. *Int. J. Hydrog. Energy* 41 (1): 145–156. https://doi.org/10.1016/j.ijhydene.2015.10.018.

Behling, N.H. (2013). History of solid oxide fuel cells. In: *Fuel Cells: Current Technology Challenges and Future Research Needs* (ed. N.H. Behling), 226–421. Elsevier https://doi.org/10.1016/B978-0-444-56325-5.00006-5.

Chader, S., Hacene, H., and Agathos, S.N. (2009). Study of hydrogen production by three strains of Chlorella isolated from the soil in the Algerian Sahara. *Int. J. Hydrog. Energy* 34: 4941–4946. https://doi.org/10.1016/j.ijhydene.2008.10.058.

Chandrasekhar, K., Lee, Y., and Lee, D. (2015). Biohydrogen production: strategies to improve process efficiency through microbial routes. *Int. J. Mol. Sci.* 16 (12): 8266–8293. https://doi.org/10.3390/ijms16048266.

Chang, A., Tu, Y., Huang, M. et al. (2011). Hydrogen production by the anaerobic fermentation from acid hydrolyzed rice straw hydrolysate. *Int. J. Hydrog. Energy* 36 (21): 14280–14288. https://doi.org/10.1016/j.ijhydene.2011.04.142.

Chen, H. (2015). Integrated industrial lignocellulose biorefinery chains. In: *Lignocellulose Biorefinery Engineering*, 219–245. Elsevier https://doi.org/10.1016/b978-0-08-100135-6.00007-7.

Cheng, J. and Zhu, M. (2013). A novel anaerobic co-culture system for bio-hydrogen production from sugarcane bagasse. *Bioresour. Technol.* 144: 623–631. https://doi.org/10.1016/j.biortech.2013.07.018.

Cheng, C.L., Lo, Y.C., Lee, K.S. et al. (2011). Biohydrogen production from lignocellulosic feedstock. *Bioresour. Technol.* 102 (18): 8514–8523. https://doi.org/10.1016/j.biortech.2011.04.059.

Chu, Y., Wei, Y., Yuan, X., and Shi, X. (2011). Bioconversion of wheat stalk to hydrogen by dark fermentation: effect of different mixed microflora on hydrogen yield and cellulose solubilization. *Bioresour. Technol.* 102 (4): 3805–3809. https://doi.org/10.1016/j.biortech.2010.11.092.

Crabtree, G.W. and Dresselhaus, M.S. (2008). The hydrogen fuel alternative. *MRS Bull.* 33 (04): 421–428. https://doi.org/10.1557/mrs2008.84.

Dong, L., Cao, G., Wu, J. et al. (2019). High-solid pretreatment of rice straw at cold temperature using NaOH/urea for enhanced enzymatic conversion and hydrogen production. *Bioresour. Technol.* 287: 121399. https://doi.org/10.1016/j.biortech.2019.121399.

Elbeshbishy, E., Hafez, H., Dhar, B., and Nakhla, G. (2011). Single and combined effect of various pretreatment methods for biohydrogen production from food waste. *Int. J. Hydrog. Energy* 36 (17): 11379–11387. https://doi.org/10.1016/j.ijhydene.2011.02.067.

Fan, Y., Xing, Y., Ma, H. et al. (2008). Enhanced cellulose-hydrogen production from corn stalk by lesser panda manure. *Int. J. Hydrog. Energy* 33 (21): 6058–6065. https://doi.org/10.1016/j.ijhydene.2008.08.005.

Guaitolini, S.V.M. and Fardin, J.F. (2018). Fuel cells: history (short remind), principles of operation, Main features, and applications. *Advances in Renewable Energies and Power Technologies*: 123–150. https://doi.org/10.1016/b978-0-12-813185-5.00013-9.

Han, S. (2004). Biohydrogen production by anaerobic fermentation of food waste. *Int. J. Hydrog. Energy* 29 (6): 569–577. https://doi.org/10.1016/j.ijhydene.2003.09.001.

Hydrogen explained: Use of hydrogen. (2020). Retrieved 16 June 2020, from www.eia.gov/energyexplained/hydrogen/use-of-hydrogen.php

Ivanova, G., Rákhely, G., and Kovács, K.L. (2009). Thermophilic biohydrogen production from energy plants by Caldicellulosiruptor saccharolyticus and comparison with related studies. *Int. J. Hydrog. Energy* 34 (9): 3659–3670. https://doi.org/10.1016/j.ijhydene.2009.02.082.

Jacobson, M.Z. (2005). Cleaning the air and improving health with hydrogen fuel-cell vehicles. *Science* 308 (5730): 1901–1905. https://doi.org/10.1126/science.1109157.

Jarunglumlert, T., Prommuak, C., Putmai, N., and Pavasant, P. (2018). Scaling-up bio-hydrogen production from food waste: feasibilities and challenges. *Int. J. Hydrog. Energy* 43 (2): 634–648. https://doi.org/10.1016/j.ijhydene.2017.10.013.

Jiang, D., Ge, X., Zhang, T. et al. (2016). Photo-fermentative hydrogen production from enzymatic hydrolysate of corn stalk pith with a photosynthetic consortium. *Int. J. Hydrog. Energy* 41 (38): 16778–16785. https://doi.org/10.1016/j.ijhydene.2016.07.129.

Kim, S. (2004). Feasibility of biohydrogen production by anaerobic co-digestion of food waste and sewage sludge. *Int. J. Hydrog. Energy* 29 (15): 1607–1616. https://doi.org/10.1016/j.ijhydene.2004.02.018.

Kirli, B. and Kapdan, I. (2016). Selection of microorganism immobilization particle for dark fermentative biohydrogen production by repeated batch operation. *Renew. Energy* 87: 697–702. https://doi.org/10.1016/j.renene.2015.11.003.

Kongjan, P. and Angelidaki, I. (2010). Extreme thermophilic biohydrogen production from wheat straw hydrolysate using mixed culture fermentation: effect of reactor configuration. *Bioresour. Technol.* 101 (20): 7789–7796. https://doi.org/10.1016/j.biortech.2010.05.024.

Kongjan, P., O-Thong, S., Kotay, M. et al. (2010). Biohydrogen production from wheat straw hydrolysate by dark fermentation using extreme thermophilic mixed culture. *Biotechnol. Bioeng.* 105 (5): 899–908. https://doi.org/10.1002/bit.22616.

Kumar, G. and Lin, C.-Y. (2013). Bioconversion of de-oiled Jatropha Waste (DJW) to hydrogen and methane gas by anaerobic fermentation: influence of substrate concentration, temperature and pH. *Int. J. Hydrog. Energy* 38 (1): 63–72. https://doi.org/10.1016/j.ijhydene.2012.10.053.

Kumar, G., Sen, B., Sivagurunathan, P., and Lin, C.-Y. (2015). Comparative evaluation of hydrogen fermentation of de-oiled Jatropha waste hydrolyzates. *Int. J. Hydrog. Energy* 40 (34): 10766–10774. https://doi.org/10.1016/j.ijhydene.2015.06.118.

Kumar, G., Mudhoo, A., Sivagurunathan, P. et al. (2016a). Recent insights into the cell immobilization technology applied for dark fermentative hydrogen production. *Bioresour. Technol.* 219: 725–737. https://doi.org/10.1016/j.biortech.2016.08.065.

Kumar, G., Sen, B., Sivagurunathan, P., and Lin, C.-Y. (2016b). High rate hydrogen fermentation of cello-lignin fraction in de-oiled jatropha waste using hybrid immobilized cell system. *Fuel* 182: 131–140. https://doi.org/10.1016/j.fuel.2016.05.088.

Kumar, G., Sivagurunathan, P., Sen, B. et al. (2017). Mesophilic continuous fermentative hydrogen production from acid pretreated de-oiled jatropha waste hydrolysate using immobilized microorganisms. *Bioresour. Technol.* 240: 137–143. https://doi.org/10.1016/j.biortech.2017.03.059.

Lay, C.-H., Sung, I.-Y., Kumar, G. et al. (2012). Optimizing biohydrogen production from mushroom cultivation waste using anaerobic mixed cultures. *Int. J. Hydrog. Energy* 37 (21): 16473–16478.

Li, Q. and Liu, C. (2012). Co-culture of *Clostridium thermocellum* and *Clostridium thermosaccharolyticum* for enhancing hydrogen production via thermophilic fermentation of cornstalk waste. *Int. J. Hydrog. Energy* 37 (14): 10648–10654. https://doi.org/10.1016/j.ijhydene.2012.04.115.

Li, M., Zhao, Y., Guo, Q. et al. (2008). Bio-hydrogen production from food waste and sewage sludge in the presence of aged refuse excavated from refuse landfill. *Renew. Energy* 33 (12): 2573–2579. https://doi.org/10.1016/j.renene.2008.02.018.

Lin, C., Wu, S., Lee, K. et al. (2007). Integration of fermentative hydrogen process and fuel cell for on-line electricity generation. *Int. J. Hydrog. Energy* 32 (7): 802–808. https://doi.org/10.1016/j.ijhydene.2006.09.047.

Lin, Y.-C., Wu, T.-Y., Jhang, S.-R. et al. (2014). Hydrogen production from banyan leaves using an atmospheric-pressure microwave plasma reactor. *Bioresour. Technol.* 161: 304–309. https://doi.org/10.1016/j.biortech.2014.03.067.

Lin, C.-Y., Lay, C.-H., Chen, C.-C. et al. (2015). Biohydrogen production from mushroom cultivation waste by anaerobic solid-state fermentation. *J. Chin. Chem. Soc.* 63 (2): 199–204.

Lin, H., Wang, Y., and Zhu, M. (2017). Evaluation of spent mushroom compost as a lignocellulosic substrate for hydrogen production by clostridium thermocellum. *Int. J. Hydrog. Energy* 42 (43): 26687–26694. https://doi.org/10.1016/j.ijhydene.2017.09.04.

Liu, C., Chu, C., Lee, W. et al. (2013). Biohydrogen production evaluation from rice straw hydrolysate by concentrated acid pre-treatment in both batch and continuous systems. *Int. J. Hydrog. Energy* 38 (35): 15823–15829. https://doi.org/10.1016/j.ijhydene.2013.07.055.

Ma, S., Wang, H., Wang, Y. et al. (2011). Bio-hydrogen production from cornstalk wastes by orthogonal design method. *Renew. Energy* 36 (2): 709–713. https://doi.org/10.1016/j.renene.2010.08.019.

Misra, R.D. and Murthy, M.S. (2011). Jatropha—the future fuel of India. *Renew. Sust. Energ. Rev.* 15 (2): 1350–1359. https://doi.org/10.1016/j.rser.2010.10.011.

Momirlan, M. and Veziroglu, T. (2005). The properties of hydrogen as fuel tomorrow in sustainable energy system for a cleaner planetEds. (Cheng 2011). *Int. J. Hydrog. Energy* 30 (7): 795–802. https://doi.org/10.1016/j.ijhydene.2004.10.011.

Muthayya, S., Sugimoto, J.D., Montgomery, S., and Maberly, G.F. (2014). An overview of global rice production, supply, trade, and consumption. *Ann. N. Y. Acad. Sci.* 1324 (1): 7–14. https://doi.org/10.1111/nyas.12540.

Pan, C., Fan, Y., and Hou, H. (2008). Fermentative production of hydrogen from wheat bran by mixed anaerobic cultures. *Ind. Eng. Chem. Res.* 47 (16): 5812–5818. https://doi.org/10.1021/ie701789c.

Panayiotou, G., Axaopoulos, P., and Fyrippis, I. (2008). Evaluation and simulation of a commercial 100 W polymer electrolyte membrane fuel cell (PEMFC) stack. *Wseas Trans. Power Syst.* 10 (3): 633–342.

Patel, S.K.S., Purohit, H.J., and Kalia, V.C. (2010). Dark fermentative hydrogen production by defined mixed microbial cultures immobilized on ligno-cellulosic waste materials. *Int. J. Hydrog. Energy* 35 (19): 10674–10681. https://doi.org/10.1016/j.ijhydene.2010.03.025.

Patel, A., Debroy, A., Sharma, S. et al. (2015). Biohydrogen production from a novel alkalophilic isolate *Clostridium* sp. IODB-O3. *Bioresour. Technol.* 175: 291–297. https://doi.org/10.1016/j.biortech.2014.10.110.

Pattra, S., Sangyoka, S., Boonmee, M., and Reungsang, A. (2008). Bio-hydrogen production from the fermentation of sugarcane bagasse hydrolysate by *Clostridium butyricum*. *Int. J. Hydrog. Energy* 33 (19): 5256–5265.

Rahman, S.N.A., Masdar, M.S., Rosli, M.I. et al. (2016). Overview biohydrogen technologies and application in fuel cell technology. *Renew. Sust. Energ. Rev.* 66: 137–162. https://doi.org/10.1016/j.rser.2016.07.047.

Rai, P.K., Singh, S.P., Asthana, R.K., and Singh, S. (2014). Biohydrogen production from sugarcane bagasse by integrating dark- and photo-fermentation. *Bioresour. Technol.* 152: 140–146. https://doi.org/10.1016/j.biortech.2013.10.117.

Ren, N., Guo, W., Liu, B. et al. (2011). Biological hydrogen production by dark fermentation: challenges and prospects towards scaled-up production. *Curr. Opin. Biotechnol.* 22 (3): 365–370. https://doi.org/10.1016/j.copbio.2011.04.022.

Rudin, A. and Choi, P. (2013). Biopolymers. *Elements Polymer Sci. Eng.* 43: 521–535. https://doi.org/10.1016/b978-0-12-382178-2.00013-4.

Saratale, G., Saratale, R., Kim, S., and Kumar, G. (2018). Screening and optimization of pretreatments in the preparation of sugarcane bagasse feedstock for biohydrogen production and process optimization. *Int. J. Hydrog. Energy* 43 (25): 11470–11483. https://doi.org/10.1016/j.ijhydene.2018.01.187.

Saxena, I.M. and Brown, R.M. (2001). Biosynthesis of cellulose. In: *Molecular Breeding of Woody Plants*, Proceedings of the International Wood Biotechnology Symposium (IWBS) (eds. N. Morohoshi and A. Komamine), 69–76. https://doi.org/10.1016/s0921-0423(01)80057-5.

Sereewatthanawut, I., Prapintip, S., Watchiraruji, K. et al. (2008). Extraction of protein and amino acids from deoiled rice bran by subcritical water hydrolysis. *Bioresour. Technol.* 99 (3): 555–561. https://doi.org/10.1016/j.biortech.2006.12.030.

Serov, A., Zenyuk, I.V., Arges, C.G., and Chatenet, M. (2018). Hot topics in alkaline exchange membrane fuel cells. *J. Power Sources* 375: 149–157. https://doi.org/10.1016/j.jpowsour.2017.09.068.

Sivaramakrishna, D., Sreekanth, D., Himabindu, V., and Narasu, M.L. (2010). Thermo-acidophilic biohydrogen production from rice bran de-oiled wastewater by selectively enriched mixed culture. *Int. J. Energy Environ.* 1 (4): 657–666.

Sopian, K. and Wan Daud, W.R. (2006). Challenges and future developments in proton exchange membrane fuel cells. *Renew. Energy* 31 (5): 719–727. https://doi.org/10.1016/j.renene.2005.09.003.

Tandon, M., Thakur, V., Tiwari, K., and Jadhav, S. (2018). Enterobacter ludwigii strain IF2SW-B4 isolated for bio-hydrogen production from rice bran and de-oiled rice bran. *Environ. Technol. Innov.* 10: 345–354. https://doi.org/10.1016/j.eti.2018.03.008.

Vendruscolo, F. (2014). Starch: a potential substrate for biohydrogen production. *Int. J. Energy Res.* 39 (3): 293–302. https://doi.org/10.1002/er.3224.

Wang, S., Ma, Z., and Su, H. (2018). Two-step continuous hydrogen production by immobilized mixed culture on corn stalk. *Renew. Energy* 121: 230–235. https://doi.org/10.1016/j.renene.2018.01.015.

Wei, J., Liu, Z.-T., and Zhang, X. (2010). Biohydrogen production from starch wastewater and application in fuel cell. *Int. J. Hydrog. Energy* 35 (7): 2949–2952. https://doi.org/10.1016/j.ijhydene.2009.05.035.

Yokoi, H., Saitsu, A., Uchida, H. et al. (2001). Microbial hydrogen production from sweet potato starch residue. *J. Biosci. Bioeng.* 91 (1): 58–63. https://doi.org/10.1016/s1389-1723(01)80112-2.

Yusuf, N.Y., Masdar, M.S., Nordin, D., and Husaini, T. (2015). Challenges in biohydrogen technologies for fuel cell application. *Am. J. Chem.* 5 (3A): 40–47. https://doi.org/10.5923/c.chemistry.201501.06.

Zhang, M., Fan, Y., Xing, Y. et al. (2007). Enhanced biohydrogen production from cornstalk wastes with acidification pretreatment by mixed anaerobic cultures. *Biomass Bioenergy* 31 (4): 250–254. https://doi.org/10.1016/j.biombioe.2006.08.004.

Zhao, L., Cao, G., Wang, A. et al. (2012). Fungal pretreatment of cornstalk with Phanerochaete chrysosporium for enhancing enzymatic saccharification and hydrogen production. *Bioresour. Technol.* 114: 365–369. https://doi.org/10.1016/j.biortech.2012.03.076.

Zhao, L., Cao, G., Sheng, T. et al. (2017). Bio-immobilization of dark fermentative bacteria for enhancing continuous hydrogen production from cornstalk hydrolysate. *Bioresour. Technol.* 243: 548–555. https://doi.org/10.1016/j.biortech.2017.06.161.

4

Strategies for Obtaining Biofuels Through the Fermentation of C5-Raw Materials: Part 1

Alexandre S. Santos[1], *Lílian A. Pantoja*[2], *Mayara C. S. Barcelos*[3], *Kele A. C. Vespermann*[3] *and Gustavo Molina*[3]

[1] *Department of Basic Science, Federal University of Jequitinhonha and Mucuri Valleys UFVJM, Diamantina, MG, Brazil*
[2] *Institute of Science and Technology, Federal University of Jequitinhonha and Mucuri Valleys UFVJM, Diamantina, MG, Brazil*
[3] *Laboratory of Food Biotechnology, Institute of Science and Technology, UFVJM, Diamantina, MG, Brazil*

4.1 The Nature of Pentoses

Pentoses are five carbon monosaccharides (C5) found as components of polysaccharides, called hemicelluloses, which are present in the cell walls of all terrestrial plants. Hemicelluloses represent 28.5–37.2% of the components of the plant cell wall (Pauly and Keegstra 2008) and are formed by units of L-arabinose (C5), L-fucose (C6), D-galactose (C6), D-glucose (C6), D-mannose (C6), L-ramnose (C6), D- xylose (C5), uronic acids (C6), acetic acid, and ferulic acid (Figure 4.1) combined in different proportions and polymeric arrangements (Zhou et al. 2017).

According to the proportion of one or more monosaccharides in the structure of hemicellulose, this will be classified as arabinogalactans, arabinoxylan, galactoglucomannan, glucomannan, glucuronoarabinoxylan, glucuronoxylan, mannan, xylan, or xyloglucan. The types of hemicelluloses and their abundance vary widely between different plant species (Table 4.1).

The biotic or abiotic decomposition of hemicelluloses releases their monomeric units, monosaccharides, and other components (Figure 4.1). These monosaccharides can be used for natural purposes, through the biotic cycling of their elements, when they will become part of the cell structures or energy metabolism of other living organisms. Or they can still be used for industrial purposes, when they will be transformed into materials and chemicals.

Xylose is the pentose with the greatest potential for industrial use. The main reason for this is its availability. After glucose, xylose is the most abundant monosaccharide found in vegetables or agro-industrial residues (Table 4.2). This would justify its commercial exploitation on a large scale, especially if the destination of this raw material, as intended here, is the biofuel industry.

Bioenergy Research: Evaluating Strategies for Commercialization and Sustainability, First Edition.
Edited by Neha Srivastava and Manish Srivastava.

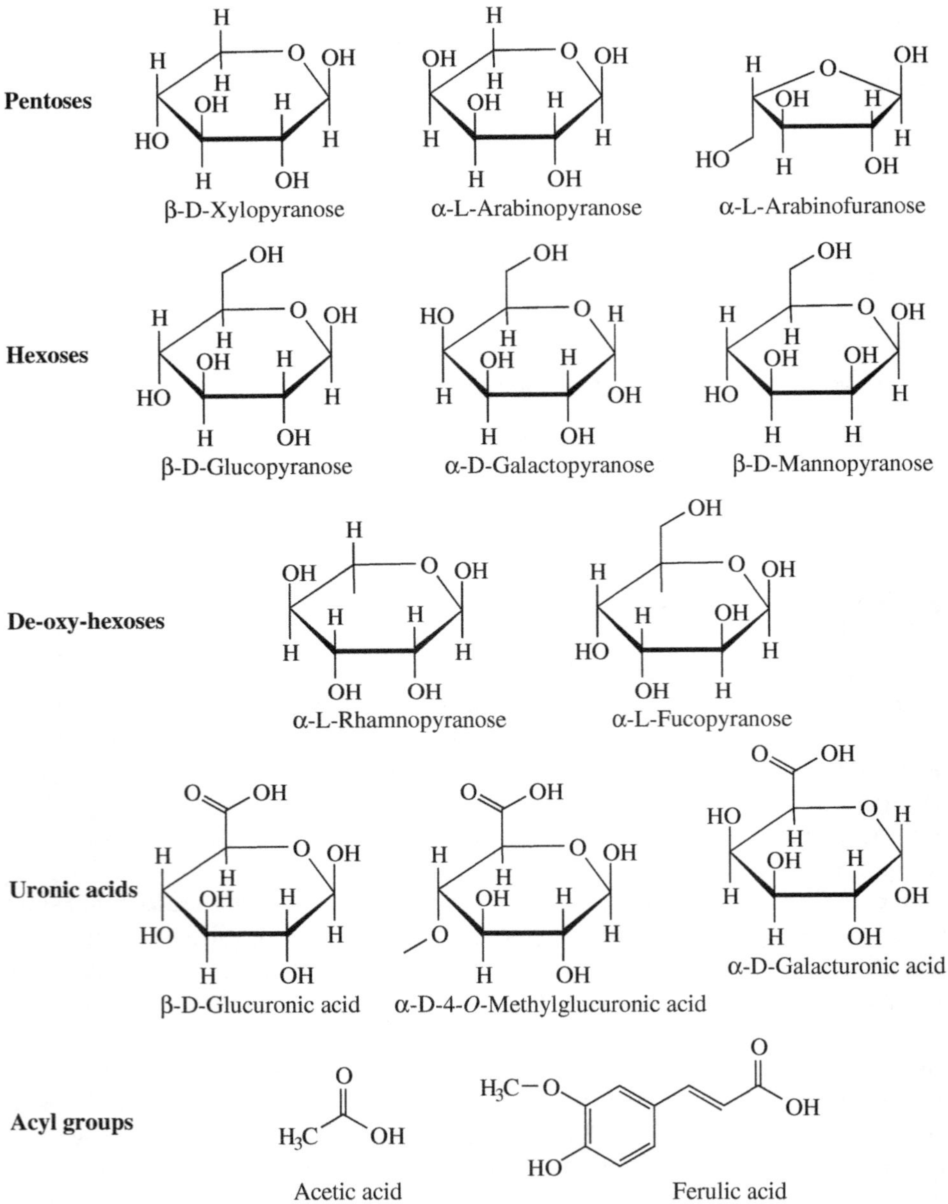

Figure 4.1 Monosaccharides and other chemical species found as components of the polymeric structure of hemicelluloses.

Table 4.1 Occurrence of types of hemicellulose in cell walls of different groups of plants.

Types of hemicellulose	Dicotyledon/ Hardwood	Conifers/ Softwood	Grasses
	% p/p dry mass		
Arabinoglucuroxylana	–	5–10	–
Galactoglucomanana	0–3	10–30	–
Glucomanana	2–5	–	0–5
Glucuronoarabinoxylan	5	2–15	20–50
Glucuronoxylan	15–30	–	–
Xyloglucan	20–25	10	2–5

Sources: Holtzapple (2003); Scheller and Ulvskov (2010); Gírio et al. (2010).

Table 4.2 Fundamental constituents of hemicelluloses from different biomasses of plant origin.

Component	Softwoods	Hardwood	Agricultural and agro-industrial residues
	% m/m – biomass dry weigth		
Xylose	5.3–10.6	11.7–27.3	12.3–35.3
Arabinose	1.0–4.2	0.3–4.0	1.1–11.4
Mannose	5.6–15.0	0.4–3.5	0–3.7
Galactose	1.9–4.3	0.3–2.3	0.1–4.4
Rhamnose	0–0.3	0.3–1.0	0–1.0
Uronic acids	1.8–6.0	2.3–6.3	1.2–3.0
Acetyl group	1.2–2.4	0.5–4.3	0.4–3.8

Sources: Gírio et al. (2010); Perez, D.D., Guillemain, A., Berthelot, A., N'guyen-The, N., De Perez, D.D., Guillemain et al. (2010).

4.2 Alcoholic Fermentation of C5

Alcoholic fermentation is the given name of the metabolic process in which energy and ethanol are produce through the partial oxidation of organic compounds, usually sugars, using other organic compounds as electron donors and acceptors (Wolfe 2015). Alcoholic fermentation, *stricto* sensu, occurs without oxygen consumption, as it is an evolutionary adaptation of some microorganisms for survival in an oxygen-free environment. However, as we will see ahead, the alcoholic fermentation of pentoses by wild microorganisms can be benefit from the use of some oxygen to balance the cell's redox potential.

The main metabolic pathways for pentose fermentation are similar in bacteria, yeasts and filamentous fungi, with significant exceptions and differences regarding transport, regulation, cofactor preference, and proportion of final products (McMillan 1993; Agbogbo and Coward-Kelly 2008; Hahn-Hägerdal et al. 1994; Gírio et al. 2010).

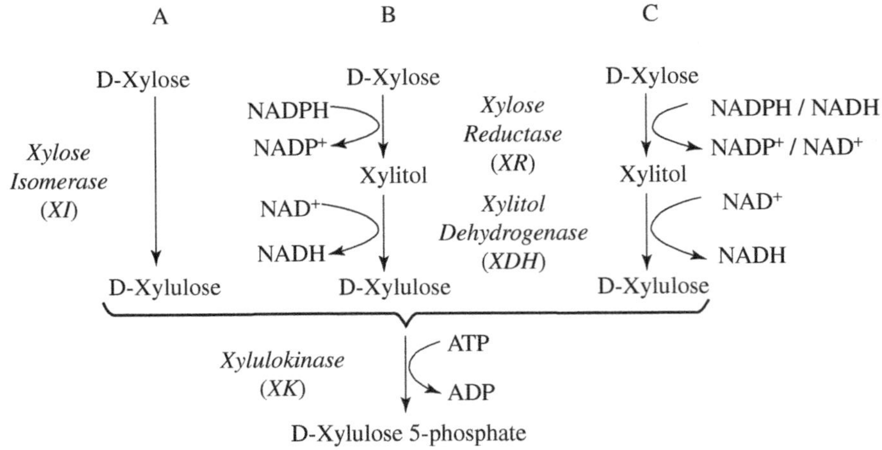

Figure 4.2 Pathways for the conversion of D-xylose to D-xylulose-5-phosphate. A- typical pathway found in bacteria; B- pathway found in fungi/yeasts, in which xylose reductase has specificity for NADPH; C- pathway found in fungi/yeasts that possess xylose reductase with dual affinity (NADPH and NADH) and variable preference, or presents two or more XRs with different specificities (NADPH or NADH). Source: Adapted from Agbogbo and Coward-Kelly (2008).

The entry of pentoses in the main metabolic pathways that lead to the formation of ethanol occurs after its transformation into D-xylulose. In bacteria and few fungi, D-xylose is transformed into D-xylulose in a single reaction step, catalyzed by the enzyme xylose isomerase (XI). For most filamentous fungi and yeasts, this transformation takes place in a two-reaction stage. D-xylose is converted to xylitol, in a reduction reaction, catalyzed by NAD(P)H-dependent xylose reductase (XR), and subsequently, xylitol is oxidizes to D-xylulose by the action of the enzyme NAD^+-dependent xylitol dehydrogenase (XDH) (Figure 4.2). Finally, in both cases, D-xylulose accesses the pentose phosphate pathway as D-xylulose-5-phosphate through the action of xylulose kinase (XK) using ATP as a source of phosphate and energy.

The L-arabinose is also taken to the pentose phosphate pathway after conversion to D-xylulose-5-phosphate (Figure 4.3). In fungi (Figure 4.3a), L-arabinose is reduced to L-arabitol by the action of NADPH-dependent L-arabinose (aldose) reductase (AR). In sequence, L-arabitol is oxidized to L-xylulose by the action of NAD^+-dependent L-arabitinol 4-dehydrogenase (LAD). In succession, by catalytic action of NAD(P)H-dependent L-xylulose reductase (LXR), L-xylulose is reduced to xylitol, which is then oxidized to D-xylulose by the action of NAD^+-dependent xylose dehydrogenase (XDH) (Chiang and Knight 1960). At the bacteria pathway (Figure 4.3b), the enzyme L-arabinose isomerase (AI) catalyzes the conversion of L-arabinose to L-ribulose. In the sequence, a ribulokinase (RK) phosphorylates L-ribulose into L-ribulose-5-phosphate and, finally, a ribulose-5-phosphate-4-epimerase (RPE) converts L-ribulose-5-phosphate into D-xylulose-5-phosphate (Richard et al. 2002).

The difference in selectivity/preference for cofactors between reductases and dehydrogenases that act in pentose assimilation pathways, described above, brings in an imbalance leads to the accumulation of $NADP^+$ and NADH. As the microorganisms need to solve this redox imbalance, this results in lower ethanol yields than those achieved with the fermentation of hexoses, something that has been observed in the fermentative

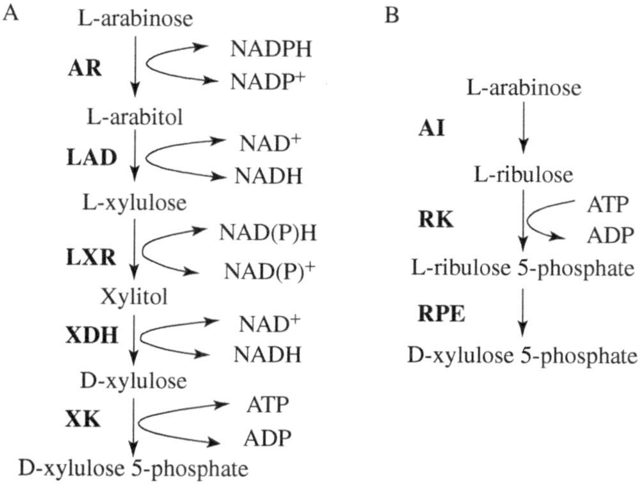

Figure 4.3 Conversion reactions of L-arabinose to D-xylulose-5-phosphate. (a) Pathway found in fungi; (b) Pathway found in bacteria. AR - L-arabinose (aldose) NADPH-dependent reductase; LAD - L-arabitinol 4-dehydrogenase NAD+-dependent; LXR - L-xylulose reductase NAD(P)H-dependent; XDH - Xylose dehydrogenase; NAD+-dependent; XK - xylulose kinase; AI - L-arabinose iomerase; RK - ribulokinase; RPE - ribulose-5-phosphate-4-epimerase.

processes carried out with many wild strains until now evaluated as fermenting pentoses (Hahn-Hägerdal et al. 1991; Jeffries and Jin 2004). In the absence of oxygen, redox balance is achieved by distributing the conversion of sugars into oxidized and reduced products (Hahn-Hägerdal et al. 1994), mainly organic acids and alcohols, respectively. In this case, part of xylose, when the only carbon source, is usually diverted to the production of NADPH, via the pentose phosphate pathway (Shin et al. 2019). These intracellular adjustments decrease the ethanol yield and usually lead to the accumulation of xylitol and synthesis of by-products. When oxygen is available, the reducing power is balanced with the participation of oxidative metabolism, which regenerates NAD^+, but alcoholic yield is compromised (Liang et al. 2014). Finding the oxygen supply condition that favors ethanol production has been a successful way of increasing the ethanol production (McMillan 1993; Silva et al. 2011). However, such a strategy would increase the cost of ethanol production to the extent that it requires appropriate instrumentation for the precise control of the amount of dissolved oxygen.

Pentoses, transformed into intermediates in the pentose phosphate pathway (Figure 4.4), either in the form of xylulose 5-phosphate or, later, as glyceraldehyde 3-phosphate and fructose-6-phosphate, will follow pathways that will lead them to the formation of ethanol in microbial species, which meet the appropriate metabolic conditions. Some fungi and most bacteria will also synthesize, to a greater or lesser extent scale, acetic acid, lactic acid, butyric acid, formic acid, glycerol, 1,2-propanediol, 2,3-butanediol, acetone, isopropanol, n-butanol, and/or succinate from pentoses (Rosenberg 1980; McMillan 1993; Isern et al. 2013; Gomes et al. 2019).

There are, among microorganisms, three main metabolic pathways responsible for the dissimilation of sugars that will eventually lead to the production of ethanol: (i) the Embden-Meyerhof-Parnas (EMP) pathway, used by some prokaryotes and most eukaryotes, including the yeasts *Saccharomyces cerevisiae* and *Scheffersomyces stipitis*; (ii) the

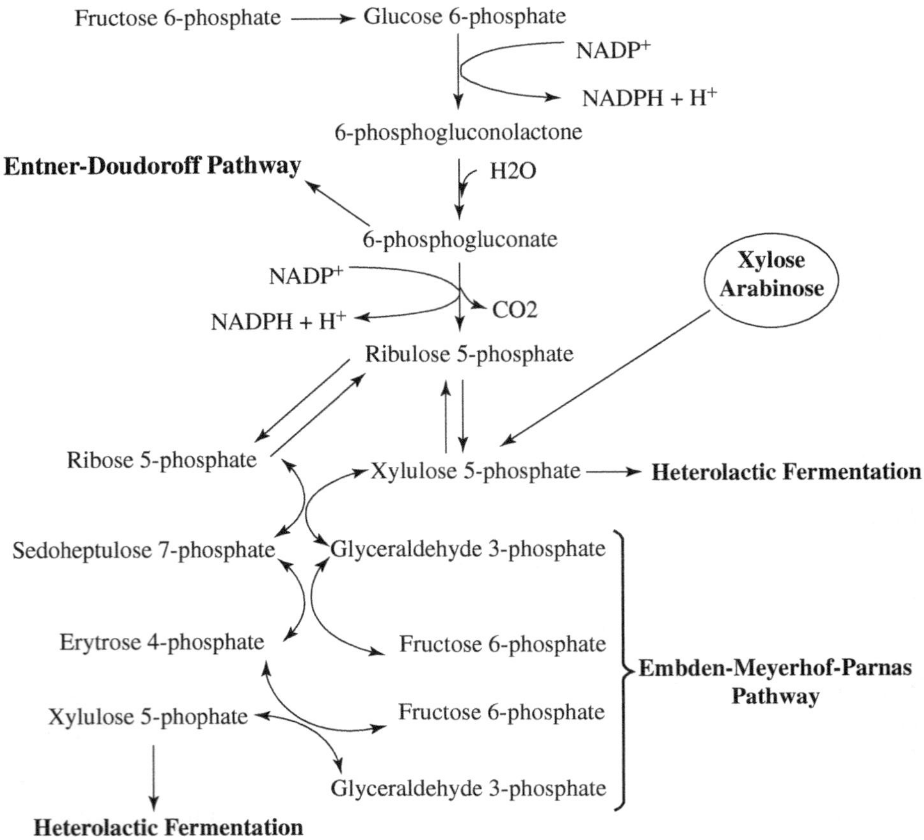

Figure 4.4 Pentose-phosphate pathways and its connection with the fermentative metabolism of sugars.

heterolactic or phosphoketolase (PPK) pathway, commonly found in lactic acid bacteria of the genus *Leuconostoc, Lactobacillus,* and *Oenococcus*; and (iii) the Entner-Doudoroff (ED) pathway, associated with the oxidative metabolism of some gram-negative bacteria of the genus *Rhizobium, Pseudomonas,* and *Agrobacterium*, for example, and which finds in the species *Zymomonas mobilis* the only representative that uses it for anaerobic production of ethanol. These metabolic pathways occur exclusively, or in combination, in different microbial species given rise to several other products in addition to ethanol. For this reason, the alcoholic yield varies greatly between the metabolic pathways that each species follows, also with the type of sugar used and other bioprocess' conditions. EMP is the recurrent pathway in pentose fermenting microorganisms, with no report of ethanol production by fermentation of pentoses in wild bacteria using exclusively the ED pathway to dissimilate sugars. However, here it is considered as a possible target of connection with the fermentation of pentoses through genetic modification of bacteria with industrial interest.

The EMP pathway (Figure 4.5) is generally called the glycolytic pathway, mainly by those working with eukaryotic cell metabolism. This metabolic pathway has glucose as its main substrate, but it can incorporate other hexoses and also pentoses. Intermediates in the EMP

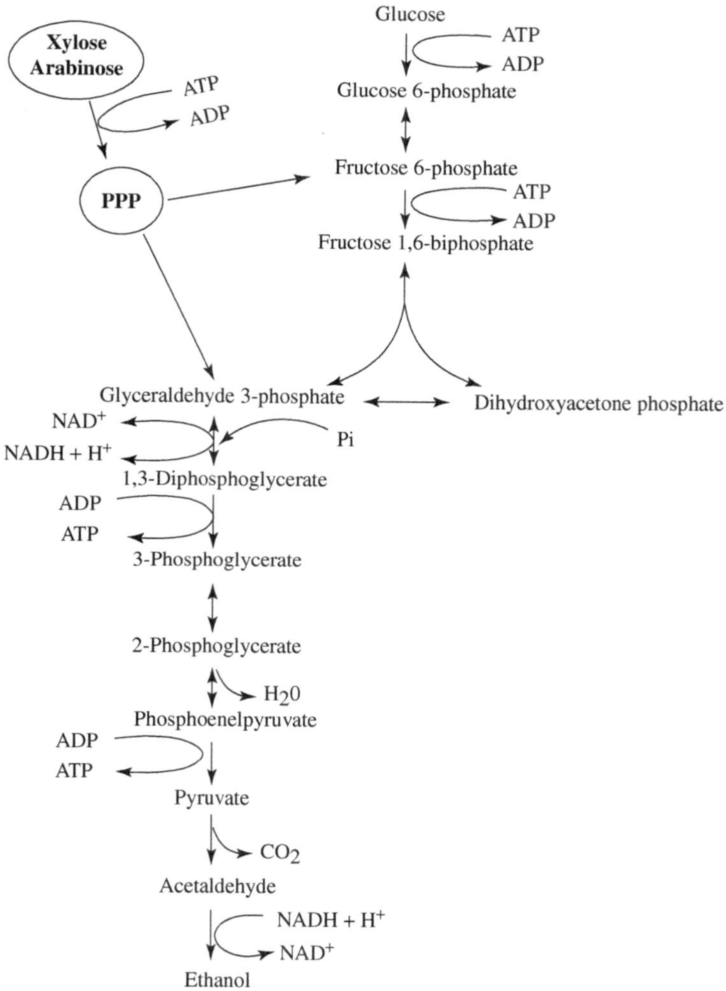

Figure 4.5 Embden-Meyerhof-Parnas pathway and alcoholic fermentation, integrated with assimilation of pentoses.

pathway undergo transformations until pyruvic acid is reached. From pyruvic acid various products can be originated, depending on the biological species, the availability or not of oxygen and the presence of inductors and repressors in the environment. The global reactions of the EMP pathway, associated with the alcoholic fermentation of hexoses and pentoses, can be represented by Eqs. (4.1) and (4.2).

$$1\ \text{hexose} \rightarrow 2\ \text{ethanol} + 2\ CO_2 + 2\ \text{ATP}. \tag{4.1}$$

$$1\ \text{pentose} \rightarrow 1.6\ \text{ethanol} + 1.6\ CO_2 + 1.6\ \text{ATP}. \tag{4.2}$$

The stoichiometry presented in Eq. (4.1) is the result of the theoretical conversion of 1 mol of hexose (glucose, fructose, galactose, and others) in 2 mol of glyceraldehyde-3-phosphate. In the EMP route, this results in the net production of 2 mol of ATP and 2 mol of ethanol.

Equation (4.2) shows the possible conversion of xylose or arabinose monosaccharides to ethanol. The fundamental difference between the stoichiometry presented in the two equations happens because the pentoses integrate the EMP pathway after passing through the pentose phosphate pathway, where they are converted into 2 mol of fructose 6-phosphate and 1 mol of glyceraldehyde 3-phosphate for every 3 mol of processed pentose. Incorporated into the EMP pathway, these 2 mol of fructose 6-phosphate will result in a balance of 4 mol of ethanol, 4 mol of CO_2, and 3 mol of ATP, and the 1 mol of glyceraldehyde 3-phosphate will be transformed into 1 mol of ethanol, 1 mol of CO_2, and 2 mol of ATP.

The heterolactic fermentation pathway (Figure 4.6) differs from the EMP pathway by the existence of a deviation in the route after the conversion of glucose into glucose 6-phosphate, forcing it to enter the pathway of pentose phosphate. The glucose 6-phosphate

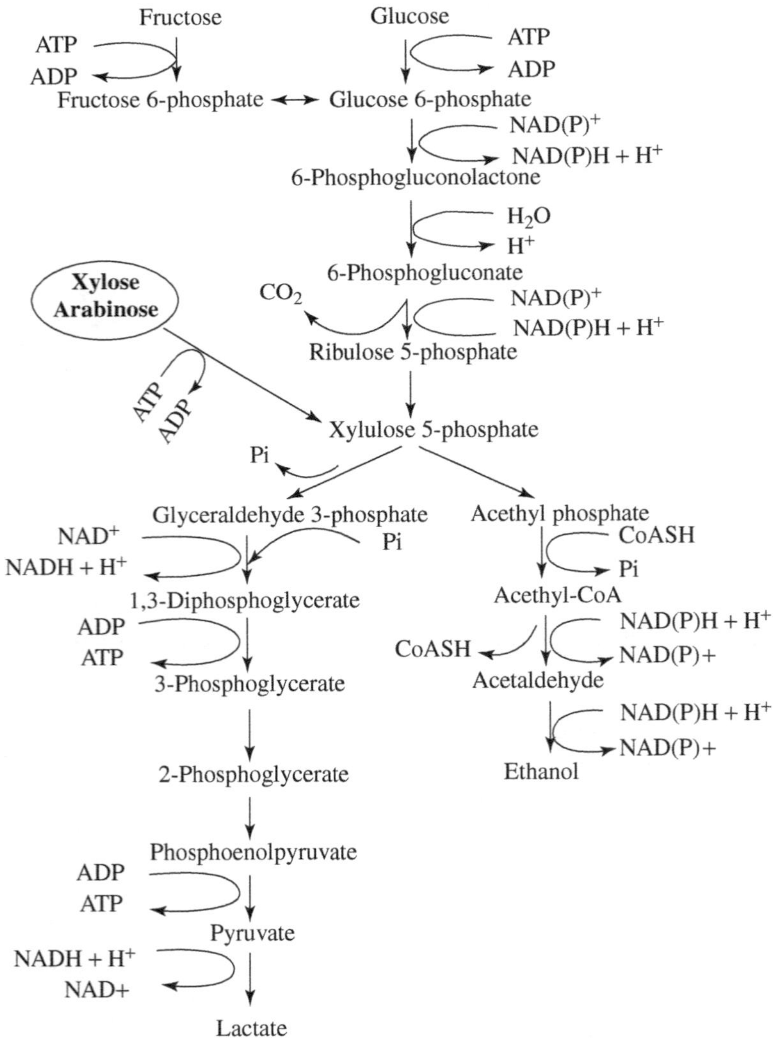

Figure 4.6 Heterolactic fermentative pathway associated with pentose fermentation.

is oxidized to 6-phosphogluconic acid, which, by oxidation and decarboxylation reactions, is transformed into ribulose 5-phosphate, and immediately afterward, into xylulose 5-phosphate. The xylulose 5-phosphate is then cleaved into glyceraldehyde 3-phosphate and acetyl phosphate by the action of the enzyme PPK, a key enzyme in the heterolactic pathway. This cleavage forks the path in routes that lead to the formation of lactic acid and ethanol. Glyceraldehyde 3-phosphate is converted to lactic acid and acetyl phosphate is converted to ethanol and other by-products. The global reactions of the PPK pathway, associated only with the alcoholic fermentation of hexoses and pentoses, can be represented by Eqs. (4.3) and (4.4).

$$1\ \text{hexose} \rightarrow 1\ \text{lactic acid} + 1\ \text{ethanol} + 1\ \text{CO}_2 + 1\ \text{ATP}. \tag{4.3}$$

$$1\ \text{pentose} \rightarrow 1\ \text{lactic acid} + 1\ \text{ethanol} + 1\ \text{ATP}. \tag{4.4}$$

The stoichiometry of ethanol and energy production through strict heterolactic fermentation of hexoses and pentoses is similar (Eqs. (4.3) and (4.4)). However, in the strict heterolactic fermentation of pentoses to ethanol, there is no production of CO_2, since the pentoses are incorporated into the route in the form of xylulose 5-phosphate, with no previous or subsequent decarboxylation (Figure 4.6).

The use of heterolactic fermentative bacteria for bioethanol production results in low alcohol yield, especially with high initial xylose concentrations (Tanaka et al. 2002), and would only be used in biorefinery units that target multiple products.

In the ED pathway (Figure 4.7), glucose 6-phosphate is converted to 6-phosphoglyconate, as in the pentose phosphate pathway, and then dehydrated to 2-keto-3-deoxy-6-phosphoglyconate by action of 6-phosphoglycolate dehydratase. In the sequence, 2-keto-3-deoxy-6-phosphoglyconate is cleaved in pyruvate and glyceraldehyde 3-phosphate by 2-keto-3-deoxy-6-phosphoglyconate aldolase and follows identical destinations to those that they would find in the EMP. There are reports of few bacteria that exclusively use the ED pathway. These bacteria lost some essential EMP pathway enzymes, such as phosphofructokinase 1. In this fermentative pathway, only 1 mol of ATP is produced per mole of hexose. The global reaction of the ED pathway, associated with alcoholic fermentation, is given by Eq. (4.5):

$$1\ \text{hexose} \rightarrow 2\ \text{ethanol} + 2\ \text{CO}_2 + 1\ \text{ATP}. \tag{4.5}$$

Several species of bacteria, especially of the genus *Pseudomonas*, are able to selectively use the ED and EMP pathways. They can use the ED pathway exclusively to metabolize glucose, while using EMP to metabolize fructose. Some species are able to combine the EMP and ED pathways to metabolize fructose, transforming it into glucose 6-phosphate through the action of phosphohexose isomerase (Wilkes et al. 2019).

The only bacterial species capable of fermenting sugars to ethanol using exclusively the ED route in anaerobiosis, *Zymomonas mobilis*, uses glucose, fructose, and sucrose with an alcoholic yield of up to 97%, but is unable to assimilate pentoses (Sprenger 1996). The existence of an incomplete pentose-phosphate pathway and the absence of enzymes responsible for the assimilation of pentoses would be responsible for not finding positive ED bacteria capable of producing ethanol from these sugars. However, there is no impediment to the construction of recombinants that incorporate this ability through metabolic engineering (Kalnenieks et al. 2014). The introduction of genes responsible for the xylose assimilation

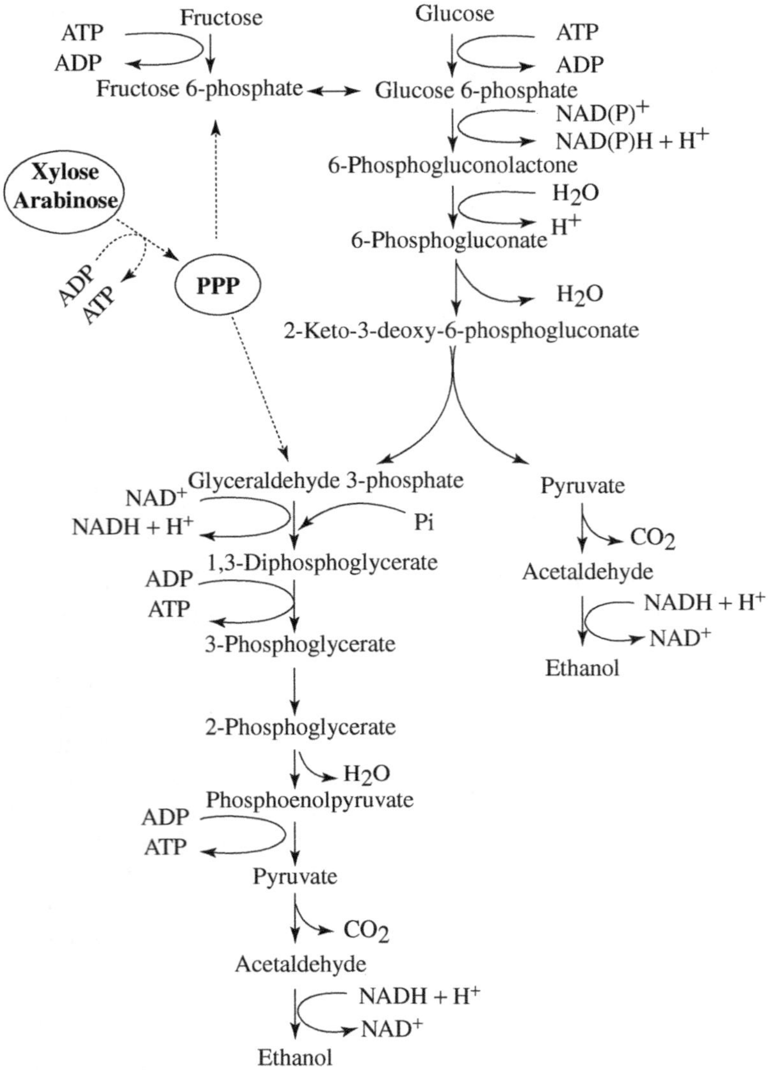

Figure 4.7 Entner-Doudoroff fermentative pathway and the artificial possibility of connection with the metabolism of pentoses.

pathway and others responsible for nonoxidative stages of the pentose phosphate pathway in *Z. mobilis* made alcoholic xylose fermentation possible in this bacterium (Zhang et al. 1995). With this strategy, it was possible to produce 5 mol of ethanol for each 3 mol of xylose (Table 4.3).

Apparently, the recurrence of the EMP pathway in hexoses and pentoses fermenting microorganisms is related to its high energy yield, explained in the observed ratio between the amount of ATP and fermented sugars (Table 4.3). This would guarantee these microorganisms a better reproductive performance in conditions of oxygen restriction. From another point of view, among the hexoses and pentoses fermenting microorganisms that

Table 4.3 End-products and theoretical yields in microbial fermentative pathways.

Pathway	Substrate	Ethanol	Lactic acid	CO_2	ATP
Embden-Meyerhof-Parnas	Glucose, fructose, galactose	2	0	2	2
	xylose, arabinose	1.6	0	1.6	1.6
Heterolactic Strict	glucose, fructose, galactose	1	1	1	1
	xylose, arabinose	1	1	0	1
Entner-Doudoroff	glucose, fructose	2	0	2	1
	xylose[a), arabinose[a)	1.6	0	1.6	1

a) values presented in alcoholic fermentation of xylose using genetically modified *Zymomonas mobilis.*

use EMP, the preference for the use of glucose over pentoses prevails. It is likely that the selective pressure that elected such organisms to constitute a majority have also privileged the energy efficiency, since more ATP is produced per mol of sugar when it is a hexose.

4.3 Lipid Biosynthesis from C5

Yeast, filamentous fungi, and microalgae seem to be the most suitable microorganisms for bio-oil production, because store lipids mostly in the form of triacylglycerol (80–90% of the neutral lipid fraction) (Beopoulos and Nicaud 2012; Xiong et al. 2012; Sajjadi et al. 2018). Most of bacteria, on the other hand, store lipids mainly as polyhydroxyalkanoate, polyhydroxybutirate, or wax ester (Beopoulos and Nicaud 2012; Xiong et al. 2012), which are not interesting for the production of biodiesel. For this reason, emphasis will be given to describe the biosynthesis of lipids in eukaryotes. The lipogenesis in this group includes fatty acid and triacylglycerol synthesis. De novo fatty acids synthesis that occurs in many eukaryotes microorganisms can use the same pentose incorporation routes accessed by pentose alcoholic fermenting microorganisms. But, unlike ethanolic fermentation, lipid synthesis is a fundamental part of the metabolism of all living beings, which need this component for cell maintenance and reproduction. However, not every microorganism is capable of producing and accumulating lipids from pentoses, or to do it with the same efficiency. Pentose assimilating oleaginous microorganisms are those who lipid content excessed 20% of its dry cell weight. Normally, five carbon skeleton from C5 sugars were destined to pentose phosphate pathway, where they are transformed into glyceraldehyde 3-phosphate or fructose 6-phosphate and connected to EMP pathway (Figure 4.8). On the aerobic EMP route, pyruvate is produced, which, in turn, flows into the tricarboxylic acid cycle (TCA) after being transformed into acetyl-CoA by mitochondrial pyruvate dehydrogenase. Some species, as *Rhodotorula toruloides* yeast, alternatively, can cleaves xylulose 5-phosphate to acetyl phosphate and glyceraldehyde 3-phosphate using the PPK enzyme (Tiukova et al. 2019). In all cases, in oil-rich microorganisms, acetyl-CoA from pyruvate is transformed into citrate on TCA pathway. In oleaginous microorganisms, the lipid overproduction is induced by the exhaustion or limitation of a primary nutrient, mainly nitrogen, that cause a deceleration in TCA cycle, resulting in accumulation of citrate, which is then exported to the cytosol,

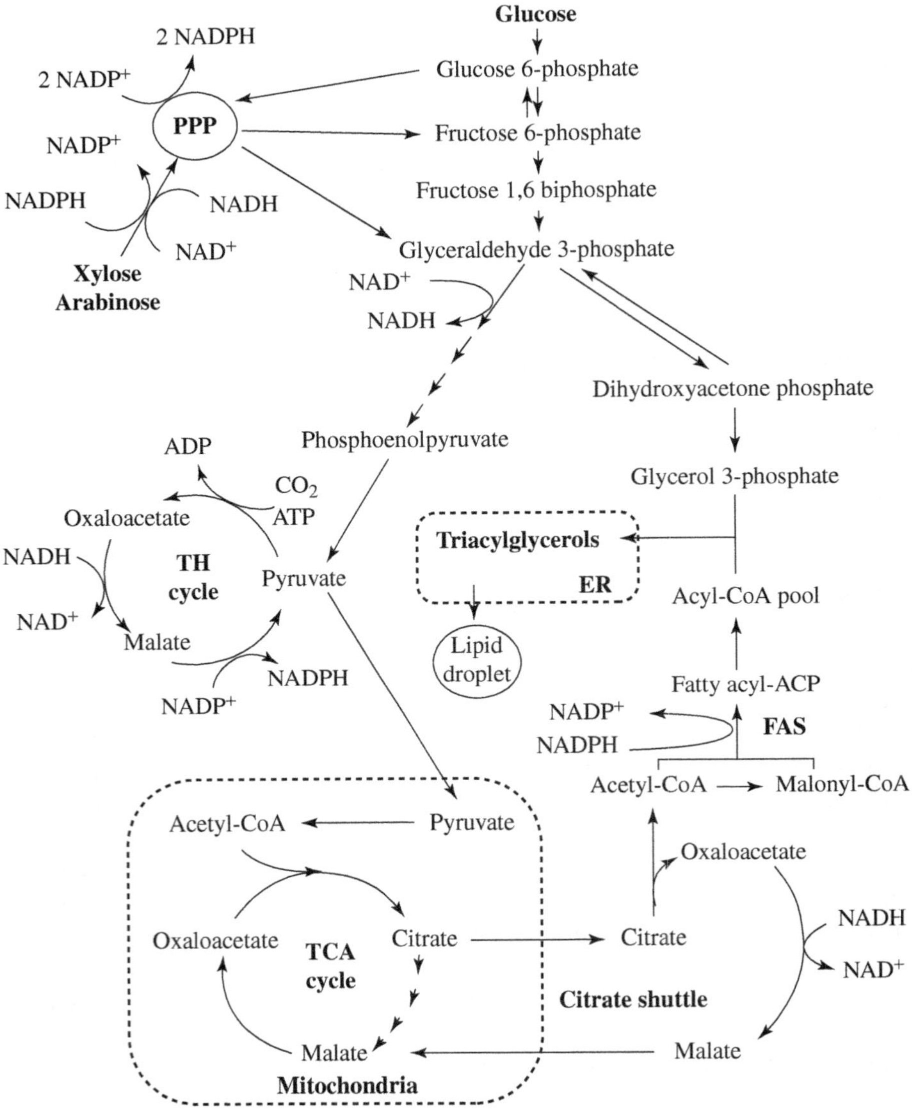

Figure 4.8 Pentose assimilation and lipid biosynthesis in yeast and filamentous fungi: an approximation. Abbreviations: PPP, pentose phosphate pathway; TH, transhydrogenase; TCA, tricarboxylic acid; ER, endoplasmic reticulum, ACP, acyl carrier protein; FAS, fatty acid synthase complex. Sources: Ratledge (2014); Liang and Jiang (2015); Yamada et al. (2017).

generally followed by malate antiport (Figure 4.8) (Beopoulos and Nicaud 2012; Koivuranta et al. 2018; Santek et al. 2018).

On cytosol, the citrate is cleaved in acetyl-CoA and oxaloacetate by an ATP:citrate lyase. To prepare the building blocks for starting the de novo lipid synthesis, the enzyme acetyl-CoA carboxylase converts acetyl-CoA and HCO_3^- to malonyl-CoA. At this point, the enzyme fatty acid synthase complex (FAS) plays a role of condensation and chain elongation of the emerging fatty acid by transferring carbon pairs from malonyl-CoA, first to acetyl-CoA, and before to others short acyl-CoA groups attached to acyl carrier protein (ACP), a constituent part of FAS. With each condensation and elongation cycle promoted by FAS, the fatty chain gains two carbons. This is repeated until fatty acids with up to 16 or 18 carbons are reached. The products of FAS action, after hydrolysis to acyl-CoA by a hydrolase or thioesterase, can be further modified by various elongases and desaturases enzymes to acyl-chains of different lengths, with or without saturation. The free fatty acids produced are then storage esterified with glycerol 3-phosphate by catalytic action of acyltransferases from endoplasmic reticulum (ER) to produce triacylglycerol (Figure 4.8) (Ratledge 2004). Fatty acid synthesis, as in any biosynthetic metabolism, needs a lot of energy. To produce 1 mol of stearic acid (C18 : 0) from 8 mol acetyl-CoA, for example, 8 mol ATP and 16 mol NADPH are required (Ratledge 2014). Therefore, microbial anabolism carried out in aerobiosis has a big advantage as of the high energy yield of cellular respiration compared to anaerobic fermentation.

In addition to the ATPs required for lipids biosynthesis, the necessary reducing power for fatty acid synthesis is provides by NADPH. When the carbon source used by yeast or filamentous fungi for growth is D-xylose or L-arabinose, part of the cellular NADPH is also demanded by xylose reductase (XR) and arabinose reductase (AR), respectively, for its assimilation. The more cited source of NADPH for fatty acid synthesis in oleaginous microbes is the cytosolic malic enzyme that, in conjunction with pyruvate carboxylase and malate dehydrogenase constitutes the transhydrogenase cycle (Ratledge 2014; Liang and Jiang 2015; Chen et al. 2015). In this cycle, the pyruvate is carboxylated to oxaloacetate which, soon after, is reduced to malate by a dependent NADH cytosolic malate dehydrogenase, and finally the malate is returned to pyruvate by the $NADP^+$ dependent malic enzyme in an oxidative-reductive decarboxylation. Nevertheless, it is not

Table 4.4 Fatty acid profile found in bio oil recovered from processes carried out with different yeast species using xylose or glucose-xylose combination as carbon source.

Lipids/ Microorganism	(C14:0)	(16 : 0)	(C18:0)	(C18:1)	(C18:2)	(C18:3)	(C22:0)	(C24:0)	Reference
Cryptococcus curvatus	6.8	21.6	14.7	56.9	–	–	–	–	Yamada et al. (2017)
Rhodosporidium toruloides	1.2	24.1	15	38.1	22.1	–	–	–	Yamada et al. (2017)
Pseudozyma hubeiensis	1.3	22.8	16.4	26.7	18.9	–	3.4	6.9	Tanimura et al. (2016)
Rhodotorula Toruloides	1.3	26.7	7.7	45.1	13.8	3	–	0.7	Tiukova et al. (2019)
Trichosporon pullulans		15	2	57	24	1			Beopoulos and Nicaud, (2012)
Cystobasidium iriomotense IPM46-17	0.8	30.2	8.1	36.3	22.7	0.2	0.3	1.2	Tanimura et al. (2018)

the only one process of NADPH regeneration. The glucose 6-phosphate dehydrogenase, 6-phosphogluconate dehydrogenase, both enzymes from pentose phosphate pathway, and a cytosolic $NADP^+$-dependent isocitrate dehydrogenase, found in many eukaryotic organisms, contribute also to supply this reduced coenzyme (Ratledge 2014; Chen et al. 2015; Tiukova et al. 2019).

The fatty acid profile produced by different oleaginous microbes may vary in type and quantity (Table 4.4), including influenced by the growth phase (Tiukova et al. 2019).

4.4 Conclusion

The existence of microorganisms capable of converting pentose into ethanol and bio-oil are responsible for the great biotechnological potential contained in the eventual destination of part of the plant biomass produced now in the earth to meet the large human demand for energy and fuels. Obviously, there are still issues of metabolic engineering, chemical engineering, mechanical engineering, and other devices that must be resolved in order to move toward the maximum possible conversion of pentoses into biofuels at the lowest cost. For now, this is the paradigm that prevails in production chains based on bioprocesses.

References

Agbogbo, F.K. and Coward-Kelly, G. (2008). Cellulosic ethanol production using the naturally occurring xylose-fermenting yeast, *Pichia stipitis*. *Biotechnol. Lett.* 30 (9): 1515–1524. doi: 10.1007/s10529-008-9728-z.

Beopoulos, A. and Nicaud, J.M. (2012). Yeast: a new oil producer? *OCL* 19 (1): 22–28. https://doi.org/10.1684/ocl.2012.0426.

Chen, H., Hao, G., Wang, L. et al. (2015). Identification of a critical determinant that enables efficient fatty acid synthesis in oleaginous fungi. *Sci. Rep.* 5: 11247. https://doi.org/10.1038/srep11247.

Chiang, C. and Knight, S.G. (1960). A new pathway of pentose metabolism. *Biochem. Biophys. Res. Commun.* 3 (5): 554–559. doi: 10.1016/0006-291X(60)90174-1.

Gírio, F.M., Fonseca, C., Carvalheiro, F. et al. (2010). Hemicelluloses for fuel ethanol: a review. *Bioresour. Technol.* 101 (13): 4775–4800. doi: 10.1016/j.biortech.2010.01.088.

Gomes, A.C., Rodrigues, M.I., de França Passos, D. et al. (2019). Acetone–butanol–ethanol fermentation from sugarcane bagasse hydrolysates: utilization of C5 and C6 sugars. *Electron. J. Biotechnol.* 42: 16–22. https://doi.org/10.1016/j.ejbt.2019.10.004.

Hahn-HäGerdal, B., Lindén, T., Senac, T., and Skoog, K. (1991). Ethanolic fermentation of pentoses in lignocellulose hydrolysates. *Appl. Biochem. Biotechnol.* 28 (1): 131–144.

Hahn-Hägerdal, B., Jeppsson, H., Olsson, L., and Mohagheghi, A. (1994). An interlaboratory comparison of the performance of ethanol-producing micro-organisms in a xylose-rich acid hydroysate. *Appl. Microbiol. Biotechnol.* 41 (1): 62–72. https://doi.org/10.1007/BF00166083.

Holtzapple, M.T. (2003). Hemicelluloses. In: *Encyclopedia of Food Sciences and Nutrition*, 2e (ed. B. Caballero), 3060–3071. Academic Press https://doi.org/10.1016/B0-12-227055-X/00589-7.

Isern, N.G., Xue, J., Rao, J.V. et al. (2013). Novel monosaccharide fermentation products in *Caldicellulosiruptor saccharolyticus* identified using NMR spectroscopy. *Biotechnol. Biofuels* 6 (1): 47. https://doi.org/10.1186/1754-6834-6-47.

Jeffries, T.W. and Jin, Y.S. (2004). Metabolic engineering for improved fermentation of pentoses by yeasts. *Appl. Microbiol. Biotechnol.* 63 (5): 495–509. https://doi.org/10.1007/s00253-003-1450-0.

Kalnenieks, U., Pentjuss, A., Rutkis, R. et al. (2014). Modeling of *Zymomonas mobilis* central metabolism for novel metabolic engineering strategies. *Front. Microbiol.* 5: 42. https://doi.org/10.3389/fmicb.2014.00042.

Koivuranta, K., Castillo, S., Jouhten, P. et al. (2018). Enhanced triacylglycerol production with genetically modified *Trichosporon oleaginosus*. *Front. Microbiol.* 9: 1337. https://doi.org/10.3389/fmicb.2018.01337.

Liang, Y.-J. and Jiang, J.-G. (2015). Characterization of malic enzyme and the regulation of its activity and metabolic engineering on lipid production. *RSC Adv.* 5: 45558–455770. https://doi.org/10.1039/C5RA04635A.

Liang, M., Damiani, A., He, Q.P., and Wang, J. (2014). Elucidating xylose metabolism of *Scheffersomyces stipitis* for lignocellulosic ethanol production. *ACS Sustainable Chem. Eng.* 2 (1): 38–48. https://doi.org/10.1021/sc400265g.

McMillan, J.D. (1993). *Xylose fermentation to ethanol. A review (No. NREL/TP-421-4944).* Golden, CO (United States): National Renewable Energy Lab.

Pauly, M. and Keegstra, K. (2008). Cell-wall carbohydrates and their modification as a resource for biofuels. *Plant J.* 54 (4): 559–568. https://doi.org/10.1111/j.1365-313X.2008.03463.x.

Perez, D.D., Guillemain, A., Berthelot, A. et al. (2010). Evaluation of forestry biomass quality for the production of second-generation biofuels. *Cell. Chem. Technol.* 44: 1–14.

Ratledge, C. (2004). Fatty acid biosynthesis in microorganisms being used for single cell oil production. *Biochimie* 86 (11): 807–815. https://doi.org/10.1016/j.biochi.2004.09.017.

Ratledge, C. (2014). The role of malic enzyme as the provider of NADPH in oleaginous microorganisms: a reappraisal and unsolved problems. *Biotechnol. Lett.* 36: 1557–1568. https://doi.org/10.1007/s10529-014-1532-3.

Richard, P., Putkonen, M., Väänänen, R. et al. (2002). The missing link in the fungal-arabinose catabolic pathway, identification of thel-xylulose reductase gene. *Biochemistry* 41 (20): 6432–6437. https://doi.org/10.1021/bi025529i.

Rosenberg, S.L. (1980). Fermentation of pentose sugars to ethanol and other neutral products by microorganisms. *Enzyme Microb. Technol.* 2 (3): 185–193.

Sajjadi, B., Chen, W.-Y., Raman, A.A.A., and Ibrahim, S. (2018). Microalgae lipid and biomass for biofuel production: a comprehensive review on lipid enhancement strategies and their effects on fatty acid composition. *Renewable Sustainable Energy Rev.* 97: 200–232. https://doi.org/10.1016/j.rser.2018.07.050.

Santek, M.I., Beluhan, B., and Santek, B. (2018). Production of microbial lipids from lignocellulosic biomass. In: *Advances in Biofuels and Bioenergy* (eds. M. Nageswara-Rao and J.R. Soneji). IntechOpen https://doi.org/10.5772/intechopen.74013e.

Scheller, H.V. and Ulvskov, P. (2010). Hemicelluloses. *Annu. Rev. Plant Biol.* 61 https://doi.org/10.1146/annurev-arplant-042809-112315.

Shin, M., Kim, J.W., Ye, S. et al. (2019). Comparative global metabolite profiling of xylose-fermenting *Saccharomyces cerevisiae* SR8 and *Scheffersomyces stipitis*. *Appl. Microbiol. Biotechnol.* 103 (13): 5435–5446. https://doi.org/10.1007/s00253-019-09829-5.

Silva, J.P.A., Mussatto, S.I., Roberto, I.C., and Teixeira, J.A. (2011). Ethanol production from xylose by *Pichia stipitis* NRRL Y-7124 in a stirred tank bioreactor. *Braz. J. Chem. Eng.* 28 (1): 151–156. https://doi.org/10.1590/S0104-66322011000100016.

Sprenger, G.A. (1996). Carbohydrate metabolism in *Zymomonas mobilis*: a catabolic highway with some scenic routes. *FEMS Microbiol. Lett.* 145 (3): 301–307. https://doi.org/10.1016/S0378-1097(96)00396-5.

Tanaka, K., Komiyama, A., Sonomoto, K. et al. (2002). Two different pathways for D-xylose metabolism and the effect of xylose concentration on the yield coefficient of L-lactate in mixed-acid fermentation by the lactic acid bacterium *Lactococcus lactis* IO-1. *Appl. Microbiol. Biotechnol.* 60 (1–2): 160–167. https://doi.org/10.1007/s00253-002-1078-5.

Tanimura, A., Takashima, M., Sugita, T. et al. (2016). Lipid production through simultaneous utilization of glucose, xylose, and L-arabinose by *Pseudozyma hubeiensis*: a comparative screening study. *AMB Expr.* 6: 58. https://doi.org/10.1186/s13568-016-0236-6.

Tanimura, A., Sugita, T., Endoh, R. et al. (2018). Lipid production via simultaneous conversion of glucose and xylose by a novel yeast, *Cystobasidium iriomotense*. *PLoS One.* 13 (9): e0202164. https://doi.org/10.1371/journal.pone.0202164.

Tiukova, I.A., Brandenburg, J., Blomqvist, J. et al. (2019). Proteome analysis of xylose metabolism in *Rhodotorula toruloides* during lipid production. *Biotechnol. Biofuels.* 12: 137. https://doi.org/10.1186/s13068-019-1478-8.

Wilkes, R.A., Mendonca, C.M., and Aristilde, L. (2019). A cyclic metabolic network in *Pseudomonas protegens* Pf-5 prioritizes the Entner-Doudoroff pathway and exhibits substrate hierarchy during carbohydrate co-utilization. *Appl. Environ. Microbiol.* 85 (1): e02084–e02018. https://doi.org/10.1128/AEM.02084-18.

Wolfe, A.J. (2015). Glycolysis for the microbiome generation. *Metab. Bact. Path*: 1–16.

Xiong, H., Chen, J., Wang, H., and Shi, H. (2012). Influences of volatile solid concentration, temperature and solid retention time for the hydrolysis of waste activated sludge to recover volatile fatty acids. *Biores. Technol.* 119: 285–292. https://doi.org/10.1016/j.biortech.2012.05.126.

Yamada, R., Yamauchi, A., Kashihara, T., and Ogino, H. (2017). Evaluation of lipid production from xylose and glucose/xylose mixed sugar in various oleaginous yeasts and improvement of lipid production by UV mutagenesis. *Biochem. Eng. J.* 128: 76–82. https://doi.org/10.1016/j.bej.2017.09.010.

Zhang, M., Eddy, C., Deanda, K. et al. (1995). Metabolic engineering of a pentose metabolism pathway in ethanologenic *Zymomonas mobilis*. *Science* 267 (5195): 240–243. https://doi.org/10.1126/science.267.5195.240.

Zhou, X., Li, W., Mabon, R., Broadbelt, L.J. A critical review on hemicellulose pyrolysis. *Energy Technol.* 5 (2017): 52–79. https://doi.org/ 10.1002/ente.201600327

5

Strategies for Obtaining Biofuels Through the Fermentation of C5-Raw Materials: Part 2

Alexandre Soares dos Santos[1], Lílian Pantoja[2], Kele A. C. Vespermann[3], Mayara C. S. Barcelos[3] and Gustavo Molina[3]

[1] *Department of Basic Science, Federal University of Jequitinhonha and Mucuri Valleys - UFVJM – Diamantina, MG, Brazil*
[2] *Institute of Science and Technology, Federal University of Jequitinhonha and Mucuri Valleys – UFVJM Diamantina, MG, Brazil*
[3] *Laboratory of Food Biotechnology – Institute of Science and Technology, UFVJM, Diamantina, MG, Brazil*

5.1 Introduction

The direct use of plants or their derivatives, such as sucrose, for the production of ethanol, or the use of microbial lipids and edible or nonedible oils for the production of biodiesel, comes up against environmental issues and the "food versus fuel" dilemma (Levering et al. 2015; Poontawee et al. 2017). On the other hand, the conversion of agricultural and agro-industrial residues into substrates for the production of biofuels is a path to be taken to reduce production costs, provide industrial competitiveness, and address synergy relationships between food and biofuel production chains (Liu et al. 2016; Li and Yang 2016). The lignocellulosic biomass, such as corn cobs, sugarcane bagasse, wheat straw, waste paper, sawdust, and oilseed cakes contains polymerized pentoses and hexoses, which can be used as the main input by industries of biofuels (Brito et al., 2018; Rodrigues et al. 2018; Matos et al. 2018; Santos et al. 2019). Despite the recognized interest in the content of hexoses from vegetal biomass, the lignocellulosic hydrolysates present high content of pentoses (reaching 25% in some cases) that can be harnessed by the biofuels industries (Zabed et al. 2016).

The deconstruction and consequent solubilization of hemicellulose monomer units, promoted by a process generically called pretreatment of lignocellulosic material, also results in the formation of by-products with variable toxicity for the microorganisms used to alcoholic fermentation (Parawira and Tekere 2011; Jönsson et al. 2013; Pan et al. 2019; Bhatia et al. 2020) or to oleaginous microorganisms growth (Kurosawa et al. 2015; Dias et al. 2012; Ayadi et al. 2019; Osorio-González et al. 2019). Such components with toxic or inhibitory properties can be classified into three main groups (Jonsson et al., 2013; Palmqvist and Hahn-Hägerdal 2000): (i) weak acids, such as formic, acetic, and levulinic acid; (ii) furan aldehydes, the main ones being 5-hydroxymethyl-furfural (5-HMF) and furfural; and (iii) phenolic compounds, such as vanillin, syringaldehyde, and 4-hydroxybenzoic acid.

Bioenergy Research: Evaluating Strategies for Commercialization and Sustainability, First Edition.
Edited by Neha Srivastava and Manish Srivastava.

The concentration of these chemical species in the hemicellulosic hydrolysate will depend on the composition of the lignocellulosic raw material and the severity of the pretreatment used (Rorke et al. 2017). Elevated temperatures, high concentrations of hydrolytic agents (acid and alkali catalysts), and prolonged reaction times are responsible for the severity of the process (Capolupo and Faraco 2016; Fengel and Wegener 1983).

Despite the pentoses' industrial potential, the main bottleneck for second-generation biofuels production is the technically difficult pretreatment due to the recalcitrance of the lignocellulosic biomass and the inability of some microorganisms to use these substrates in an advantageous way (Branco et al. 2020), it will be important to invest in research and technical advances for this area to attain an economically viable commercial biofuels production (Zabed et al. 2016).

5.2 Ethanol Production Using C5-Fermenter Strain

The world production of lignocellulosic ethanol reached 10 billion liters per year in 2019, and there are plans for it to continue increasing. It is estimated that the conversion to ethanol of the monosaccharides present in the hemicellulose fraction of lignocellulosic biomass (C5 fraction) could increase the production of second-generation ethanol by 50% (Nogué and Karhumaa 2015). Other important factors to consider for the use of pentoses as an input in the production of bioethanol are the know-how and infrastructure established to produce first-generation (1G) and second-generation (2G) fuel ethanol.

The 1G ethanol is obtained from sucrose and glucose originated from the processing of sugary and starch biomass, respectively. The 2G ethanol is obtained from glucose released from cellulose fraction contained in lignocellulosic biomass. Both fuel generations are based on the conversion of hexoses to ethanol. Then, we shall call 2G+ ethanol the one obtained from C6 + C5 (hexoses and pentoses) lignocellulosic fractions, to highlight the incorporation of the pentoses into the process. However, although there may be similarities between 1G, 2G, and 2G+ ethanol production platforms that allow convergence of some unit operations; there is a peculiarity in the pentose fermentation stage that still hinders its integration with industrial plants that ferment sucrose or glucose (Dias et al. 2012; Vasconcelos et al. 2020). This is because the microorganisms conventionally used in the alcohol industry for the fermentation of glucose or sucrose are not capable of converting pentoses into ethanol.

5.2.1 Pentose-Fermenting Microorganisms

The microorganisms most widely used for the production of fuel ethanol are *Saccharomyces cerevisiae*, a robust yeast that can convert hexoses (glucose, fructose, mannose, and galactose) into ethanol (alcoholic fermentation) in anaerobic or microaerobic environment with an initial high concentration sugar in wort (>18% w/v) (Stambuk et al. 2008). In addition, industrial strains of this yeast species have a high sugar assimilation rate (2–4 g sugar/g yeast/h), high fermentative efficiency (>90%), and considerable tolerance to ethanol accumulated in fermented wort (~12% v/v). However, this microbial species, like others for industrial use, does not ferment pentoses. Therefore, to integrate the pentose syrup

(C5 fraction), derived from the hydrolysis of the hemicelluloses present in lignocellulosic biomass, to the 2G ethanol (cellulosic) production line, we would have to adapt the plant with an independent pentose fermentation unit, using specific microorganisms for this purpose; or use the same fermentation infrastructure as glucose syrup (C6 fraction) existing on 1G or 2G ethanol plants using a microorganism capable of cofermenting hexoses and pentoses.

Several species of different genera of filamentous fungi, yeasts, and bacteria have been identified as capable of carrying out alcoholic fermentation of pentoses (Table 5.1). However, most of them find it difficult to ferment pentoses in the hemicellulosic hydrolysate environment.

The election of the ideal microorganism to serve the 2G+ ethanol platform with high performance is still delayed. As, in addition to the ability to convert pentoses and hexoses into ethanol, this organism will have to be able to do it in unfavorable conditions (Dien et al. 2003; Laluce et al. 2012), namely: (i) wort with a high concentration of sugars, which implies processes of osmotic stress and catabolic repression; (ii) presence of toxic by-products from the pretreatment used for the deconstruction of hemicellulose,

Table 5.1 Examples of microorganisms capable of converting xylose and or arabinose to ethanol in synthetic medium.

Yeasts	Filamentous fungi
Ambrosiozyma monospora	*Fusarium oxysporum*
Brettanomyces naardenensis	*Mucor*
Candida akabanensis	*Neurospora crassa*
Candida amazonensis	*Paecilomyces sp*
Candida aurigiensis	*Monilia sp*
Candida tropicalis	*Polyporus*
Clavispora opuntiae	*Rhizopus*
Dekkera bruxellensis	**Bacteria**
Galactomyces geotrichum	*Aerobacter hydrophila*
Kluyveromyces marxianus	*Bacillus macerans*
Meyerozyma guilliermondii	*Clostridium acetobytylicum*
Pachysolen tannophilus	*Clostridium thermosaccharolyticum*
Pichia kudriavzevii	*Erwinia chrysanthemi*
Scheffersomyces (Candida) shehatae	*Klebsiela oxytoca*
Scheffersomyces (Pichia) stipitis	*Klebsiela planticola*
Scheffersomyces stambukii	*Sarcina ventriculi*
Schizosaccharomyces	*Thermoanaerobacter ethanolicus*
Spathaspora passalidarum	*Caldicellulosiruptor saccharolyticus*

Sources: McMillan (1993); Chandel et al. (2011); Valinhas et al. (2018); Matos et al. (2018); Cadete et al. (2012); Finn et al. (1984); Dien et al. (1996); Matos et al. (2014); Hahn-Hägerdal et al. (1994); Isern et al. (2013); Galafassi et al. (2012); Sharma et al. (2018); Millati et al. (2005); Antunes et al. (2014); Singh et al. (2018); McMillan and Boynton (1994).

breakdown of cellulose crystallinity, and removal of lignin from lignocellulosic biomass; (iii) high concentrations of ethanol in the fermented wort, a precondition for the distillation process, but which is toxic to microorganisms; and (iv) ability to ferment under industrial pH, salinity, and temperature conditions, generally different from optimal conditions for microbial life. These, and the guarantee of safety for human manipulation, must be the characteristics sought in the ideal microorganism.

So far, the strategies to achieve, with the pentoses, the success obtained with the alcoholic fermentation of hexoses were the selection and adaptation of microorganisms naturally occurring and capable of fermenting pentoses and hexoses; and the genetic modification of industrially successful microorganisms in the fermentation of hexoses so that they also become pentose fermenters. Regarding the first path, there were many attempts to isolate, select, and improve microbial strains capable of converting pentoses into ethanol (Chart 1), but none of them reproduced the success that has been achieved with the fermentation of hexoses by *S. cerevisiae*. This did not mean wasting time or money. The knowledge about the metabolism and genetics of these microorganisms started the genetic interventions that have tried to create/modify robust and efficient microorganisms in the simultaneous conversion of hexoses and pentoses for the production of 2G+ ethanol.

These two strategies continue to coexist and feed on new discoveries and knowledge. Some wild and genetically modified microorganisms are being used in demonstration plants and commercial plants for the production of 2G+ ethanol, even with suboptimal yields, waiting for new technological advances that will increase efficiency and reduce costs (Chandel et al. 2018; Padella et al. 2019). On the other hand, nature may be hiding microorganisms that are more suitable for industrial fermentation of pentoses, or that have genes that favor the genetic improvement of known species. It is estimated that there are between 2 200 000 and 3 800 000 species of fungi on planet Earth (Hawksworth and Lücking 2017) and, only about 144 000 species have so far been named and classified (Cannon et al. 2018). It is possible and likely that new pentose fermenting microorganisms will be discovered and that they will make the 2G+ ethanol platform even more efficient and competitive.

While the ideal microorganism is not found in nature, several studies to improve the processes for obtaining 2G+ ethanol have been carried out (Hassan et al. 2019; Rojas-Chamorro et al. 2020; Dey et al. 2020; Malik et al. 2020). Among the main approaches to increase the fermentative performance of microorganisms in converting the pentoses present in hemicellulosic hydrolysates are:

a) use of wild microorganisms in detoxified hemicellulosic hydrolysate;
b) evolutionary adaptation of natural strains to the adverse conditions of the hemicellulosic hydrolysate; and
c) genetic manipulation of naturally robust microorganisms, but unable to ferment pentoses.

Chart 2 shows values of fermentative parameters for the production of 2G+ bioethanol using different strategies to solve the bottlenecks of this platform.

The detoxification of hemicellulosic hydrolysate promotes a reduction in the concentration of the toxic species by removal or transformation into harmless or less toxic chemical species (Bhatia et al. 2020). Several methods have been proposed for the detoxification of lignocellulosic hydrolysates. Physical methods include low pressure evaporation, membrane filtration, and adsorption; chemical methods involve the use of ion exchange resins, the use of alkalis and organic solvents; for biological detoxification microorganisms and enzymes are used (Kim et al 2018; Candido et al. 2020). Detoxification methods, combined or not, have been reported to be effective and promising in the removal of inhibitory compounds in processes for the production of ethanol from hemicellulosic hydrolysate (Brito et al. 2017; Deng and Aita 2018; Sarawan et al. 2019; Candido et al. 2020; Bhatia et al. 2020).

Evolutionary adaptation of wild species to the conditions of the industrial fermentative process to obtain hemicellulosic ethanol can lead to increased tolerance to the inhibitory components present in the hydrolysate, resulting in increased ethanol yields (McMillan 1993; Slininger et al. 2015; Kim 2018). Non-permanent phenotypic adaptations do not qualify as an instrument for improving a microbial strain and are not suitable for biotechnological applications. The evolutionary microbial adaptation, also called acclimatization, must encompass permanent genetic and phenotypic changes, promoted when successive generations of a microbial strain are subjected to a certain selective pressure. This approach can also be called evolutionary engineering, whose principles are similar, and which can also be combined with mutagenesis methods (Winkler and Kao 2014). Microorganisms undergoing the adaptation process show greater viability, biomass production and fermentative capacity when compared to nonadapted cells (Slininger et al. 2015; Van Dijk et al. 2019).

Genetic manipulation of microorganisms in order to make them able to ferment pentose usually starts from industrial strains conventionally adopted by the 1G and 2G ethanol production platforms. These industrial strains, in addition to efficient hexoses fermenters, are more tolerant to ethanol and inhibitory compounds present in the lignocellulosic hydrolysate (Zhang et al. 1995; Talukder et al. 2019; Stambuk et al. 2008; Cunha et al. 2019). Bacteria, such as *Zymomonas mobilis* and *Escherichia coli*, have also been evaluated for this purpose. However, *Saccharomyces cerevisiae* is undoubtedly the main target for genetic modification to design strains capable of fermenting pentoses quickly and efficiently. In addition, *S. cerevisiae* has the advantage of being well characterized and, therefore, genetically accessible, with multiple tools available for its genetic manipulation. Early interventions attempted to complement the genome of wild or industrial strains with genes encoding the key enzymes for incorporating xylose into the metabolism of the pentose pathway, which in turn would lead to the existing Embden-Meyrhof-Parnas pathway (Kwak et al. 2019). This strategy included the cloning and expression of the NAD(P)H-dependent xylose reductase and NAD+-dependent xylitol dehydrogenase genes from *Scheffersomyces (Pichia) stipitis*. However, the preference of oxidoreductases for different cofactors caused redox imbalance with consequent accumulation of xylitol and low ethanol yield. Naturally, it was thought to resolve the issue with the use of xylose reductase and xylitol dehydrogenase cloning that used the same cofactor, or a bacterial xylose isomerase, which was known to be able to directly convert xylose to xylulose without using cofactors (Sonderegger et al. 2004; Cunha et al. 2019). Even so, conversion rates from xylose to ethanol remained low. Probably, due to the lack of fine adjustments in the relationship between NADPH/NADP+ involved in the reduction of xylose and in the oxidative stage of the pentose phosphate pathway. Another problem faced for the production of ethanol from lignocellulosic hydrolysates was and is the selective conversion of glucose against other pentoses, mainly in positive Crabtree yeasts (Ha et al. 2011; Eiteman et al. 2008; Subtil and Boles 2012).

Most yeast designed to ferment xylose are not able to consume xylose until the glucose is completely depleted. This phenomenon was found to be partly due to the selectivity of membrane transporters (Sharma et al. 2018), but mainly by processes of catabolic repression or gene regulation (Ko and Lee 2018). Due to these and other setbacks, several other proposals for genetic intervention were and continue to emerge in an attempt to increase the specific consumption of xylose and the rate and yield of ethanol production. Examples are: (i) overexpression of the enzymes necessary for the conversion of D-xylulose to glycolysis intermediates; (ii) deletion of endogenous aldose reductase that converts xylose to xylitol; (iii) overexpression of heterologous transporters with preference for xylose or xylodextrins (Weber et al. 2010; Zhang et al. 2017; Sharma et al. 2018); (iv) introduction of key genes from the phosphoketolase pathway, common in acid-lactic bacteria (Sonderegger et al. 2004); or (v) disturbance of gene targets to reconfigure yeast metabolism seeking synergy for the cofermentation of pentoses and hexoses (Kim et al. 2013; Liu et al. 2020a,b). Evolutionary engineering, combined with metabolic engineering, has also been used to improve cell performance in ethanol production and increase the stability of recombinant strains (Kwak et al. 2019; Cai et al. 2012), as it selects recombinant strains against predefined selective pressures.

Despite the great work effort and the wide variety of engineering approaches and metabolic adaptation tested so far, the alcoholic fermentation of xylose present in hemicellulosic hydrolysates by genetically modified strains of *S. cerevisiae* shows lower yield and productivity values than glucose fermentation. Even so, research possibilities and efforts to consolidate 2G+ ethanol production processes have not been exhausted. We look forward to a new chapter on progress in this area of study in the coming years.

5.3 Microbial Lipid Production by C5-Fermenter Strains for Biofuel Advances

Biodiesel can be defined chemically as a mixture of methyl esters with long-chain fatty acids (Chatterjee and Mohan 2018), which is conventionally produced from the transesterification of animal fats, recycled cooking greases or oils, mainly vegetable oils, such as soy and rapeseed oil (Liu et al. 2020a,b). However, the great incentive in the production of biodiesel has put the food and oleochemical industries in conflict, in this way the production of lipids by fermentation can be an alternative to the traditional production (Tsigie et al. 2011). In addition, the use of these lipids as starting material for biodiesel brings with it a series of advantages, such as the short production period, ease of scale up, less labor, and land requirements for production and less dependence on climatic and seasonal factors (Mondala et al. 2015).

These microbial oils are produced by species called oleaginous species, found among yeasts, microalgae, and fungi, which accumulate at least 20% of their dry biomass as lipids (Pessôa et al. 2019). Microbial oils are considered potential feedstock for biodiesel since their fatty acid profile is similar to vegetable oils, rich in mono-unsaturated fatty acids, with triglycerides as main constituent (Juanssilfero et al. 2018b; Ma et al. 2018).

Among the oleaginous species, the yeasts of the genus *Yarrowia, Candida, Rhodotorula, Rhodosporidium, Cryptococcus, Trichosporon,* and *Lipomyces* are the most studied for the production of microbial oils, due to their higher lipid content, ease to scaling up (Poontawee et al. 2017), short growth cycle (Ma et al. 2018), and ability to grow at room temperature using sugars derived from lignocellulosic hydrolysates (Bandhu et al. 2019). The yeast *Rhodosporidium toruloides* has presented the ability of accumulate lipids up to 70% of its biomass dry weight, been extensively studied (Ma et al. 2018).

Most of the studies for the production of microbial oils are conducted with glucose as carbon source, however, the high costs of biodiesel from oleaginous microorganisms are directly linked to the cost of glucose (Tsigie et al. 2011). In this context, different approaches for the transformation of low-cost substrates into high-value products are being implemented, especially in the lignocellulosic biorefinery industry, where the use of alternative carbon sources, such as glycerol and lignocellulosic biomass, are important to reduce the inherent costs of the process and increase product competitiveness (Díaz-Fernández et al., 2019).

Lignocellulosic biomasses are widely available, inexpensive, and sustainable. Some studies have already conduct successful microbial lipid production employing waste paper (Nair et al. 2020), corn stover (Yu et al. 2020), wheat straw hydrolysate (Zhijia et al. 2020), dried sweet sorghum stalks (Antonopoulou et al. 2020), and oil palm biomass (Ahmad et al. 2019).

The hydrolysate of lignocellulosic materials are mainly constituted of xylose, glucose, and arabinose. In addition to some by-products such as lignin and acetic acid (HMF), furfural, and vanillin are also formed, which can be toxic and can inhibit the cell growth and sugar utilization by microorganisms during fermentation (Tsigie et al. 2011; Poontawee et al. 2017). Acetic acid has been already used as an alternative carbon source for the production of free fatty acids achieving satisfactory results (Xiao et al. 2013), however, it can inhibit the flux toward glycolysis and pentose phosphate pathway (Xu et al. 2017).

Thus, the low resistance of microorganisms to inhibitors and, in many cases, the inability to assimilate pentoses, are important barriers to the use of lignocellulosic biomass in a commercial scale (Liu et al. 2020a,b). Therefore, it is common to conduct previous studies with synthetic media, with sole or combined pentoses and hexoses, to evaluate the ability of oleaginous microorganisms to produce biomass and lipid before evaluating the use and lignocellulosic biomass (Guerfali et al. 2018), as describe above and summarized in Table 5.2.

Several oleaginous microorganisms present the ability to use different carbon sources for lipid biosynthesis (Carvalho et al. 2018). Some studies investigate the use of mixed carbon sources derived from lignocellulosic biomass (glucose and xylose) for the production of microbial oil. In some cases the microorganisms metabolize sugars sequentially, which can result in a prolonged lag time and incomplete consuming of xylose (Fang et al. 2016; Poontawee et al. 2017; Juanssilfero et al. 2018b). On the other hand, simultaneous utilization of both carbon sources also occurs and represents an important economical aspect in the use of lignocellulosic biomass (Tsigie et al. 2011; Anschau et al. 2014; Guerfali et al. 2018; Juanssilfero et al. 2018a,b).

In this sense, *Mucor circinelloides* was able to produce lipid in range from 6.2 to 43% using single sugars, disaccharide, alcohol, and polysaccharide, and empathized that similar lipid production were observed between growth on xylose (35%) and C6 sugars (42% and 43%) (Carvalho et al. 2018). *Rhodotorula graminis* was also able to growth and produce lipids from different carbon sources in the range of 6.58 g l^{-1} for glucose, 2.24 g l^{-1} for

Table 5.2 Lipid production using different oleaginous species with xylose or lignocellulosic hydrolysate as carbon source.

Strain	Strategy	Process condition	Lipid yield	References
Mortierella isabellina	On-site enzymatic lignocellulosic hydrolysis of steam exploded corn stover	28 °C, 160 rpm, 8 d, 43.32 g l^{-1} glucose and 7.06 g l^{-1} xylose	25.77 g l^{-1}	Fang et al. (2016)
Yarrowia lipolytica Po1g	Detoxified sugarcane bagasse hydrolysate and peptone as nitrogen source	26 °C, 160 rpm, pH 6.5, 20 g l^{-1} of total sugar, 4 d	6.68 g l^{-1} (58.5% w/w)	Tsigie et al. (2011)
Lipomyces starkeyi	Diferentes tamanhos de inoculo, effect of feeding strategies and ratio of glucose:xylose (30 : 70)	3-L bioreactor, 1 vvm, 400 rpm, and 28 °C, 3 g l^{-1}, repeated fed-batch, 60 g l^{-1} of sugars	41.8 g l^{-1}	Anschau et al. (2014)
Cryptococcus curvatus	High inoculum size in the presence of different toxic compounds (0.7 g l^{-1} furfural, furfural and the phenolic compounds 0.1 g l^{-1} p-hydroxybenzaldehyde, 0.1 g l^{-1} syringaldehyde and 0.1 g l^{-1}vanillin)	30 °C, 150 rpm, pH 5.5, xylose 20 g l^{-1}, 10% v/v of inoculum	0.045 g l^{-1}/h	Yu et al. (2014)
Rhodotorula glutinis	High cell density, optimized C : N ratio and two-stage nitrogen fed-batch culture with undetoxified corncob hydrolysate	5-L bioreactor, 30 °C, pH 6, 20 g l^{-1} of total sugars	33.5 g l^{-1}	Liu et al. (2015)
Activated sludge	Sugarcane bagasse hydrolysate with pre-adaptation phase with xylose (60 g l^{-1}) during 7 d	5-L bioreactor, 25 °C, 1vvm, 500 rpm, C : N ratio of 70 : 1	7.62 g l^{-1}	Mondala et al. (2015)
Lipomyces starkey	Different C : N ratio	30 °C, 250 rpm, 120 hr, 50 g l^{-1} of xylose, C : N ratio of 50	3.61 g l^{-1}	Liu et al. (2020a, 2020b)
Trichosporon cutaneum	Corncob acid hydrolysate, Different C : N ratio, different inoculum size and addition of micronutrients	28 °C, 160 rpm, 8 d, pH 6, 5% of inoculum, 45.7 g l^{-1} of sugars, C : N ratio of 100	10.4 g l^{-1}	Chen et al. (2013)
Mucor circinelloides URM 4182	Different carbon sources and direct transesterification reaction	26 °C, 250 rpm and 96 hr, 40 g l^{-1} of xylose	1.9 g l^{-1}	Carvalho et al. (2018)

Rhodotorula graminis	Different carbon sources	30 °C, 200 rpm, 6 d, xylose 50 $g\,l^{-1}$	2.24 $g\,l^{-1}$	Galafassi et al. (2012)
L. starkey	Different carbon sources or a combination of them	30 °C, 200 rpm, pH 5.8, 70 $g\,l^{-1}$ of xylose, 168 hr	12.71 $g\,l^{-1}$ (53%)	Gong et al. (2012)
Lipomyces doorenjongii, L. orientalis and *L. starkey*	Mixed carbon source	30 °C, 190 rpm, 50 $g\,l^{-1}$ glucose +50 $g\,l^{-1}$ xylose	17.45, 17.59, and 24.94 $g\,l^{-1}$, respectively	Juanssilfero et al. (2018b)
Lipomyces starkey	Different inoculum size (0.6–0.8, 4–5, and 14–18 OD_{600nm})	30 °C, 190 rpm, 50 $g\,l^{-1}$ glucose +50 $g\,l^{-1}$ xylose	14.3, 21.6, and 20.3 $g\,l^{-1}$, respectively	Juanssilfero et al. (2018a)
Mortierella isabellina	Different inoculum size and nitrogen sources	28 °C, 160 rpm, pH 6, yeast extract and ammonium sulfate (1 : 1, 50.4 mM) as nitrogen source, 100 $g\,l^{-1}$ xylose 10^8 ml^{-1} spore concentration	18.5 $g\,l^{-1}$	Gao et al. (2013)
Rhodosporidium fluviale DMKU-SP314	Mixed carbon source ratio of 2 : 1 of glucose and xylose	28 °C, 150 rpm, 216 hr, pH 5.5, 70 $g\,l^{-1}$ of total sugars	7.9 $g\,lL^{-1}$	Poontawee et al. (2017)
Trichosporon cutaneum	Mixed carbon source ratio of 1 : 1 of glucose and xylose	30 °C, pH 6, 200 hr, 40 $g\,l^{-1}$ of carbon source, C : N ratio of 100	4.58 $g\,l^{-1}$	Guerfali et al. (2018)
Yarrowia lipolytica	Expression of genes encoding xylose reductase, xylitol dehydrogenase and genes encoding DGA	28 °C, xylose and glycerol	7.64 $g\,l^{-1}$	Tai and Stephanopoulos (2013)
Rhodotorula mucilaginosa	Overexpression of cytosolic malic enzyme in a fed-batch process	15-L bioreactor, 32 °C, 150 rpm, pH 4.5, 1 vvm, 20 $g\,l^{-1}$ of xylose derived from sugarcane bagasse, 56 hr	4.29 $g\,l^{-1}$ (31% w/w)	Bandhu et al. (2019)

Source: Modified from Guerfali et al. (2018).

xylose, 2.94 $g\,l^{-1}$ for galactose, 3.36 $g\,l^{-1}$ for mannose, 1.96 for cellobiose, and 2.8 $g\,l^{-1}$ for glycerol, indicating the potential of this strain for fermentation of different types of residues (Galafassi et al. 2012).

The glucose: xylose ratio can also impact directly in the lipid concentration. *Lipomyces starkey*, for instance, presented higher lipid content (30.1 $g\,l^{-1}$) with glucose: xylose ratio of 30 : 70 against only 14 $g\,l^{-1}$ when glucose: xylose 70 : 30 ratio was used under the same fed-batch strategy (Anschau et al. 2014).

In some cases, higher lipid production is possible with xylose as only carbon source. *L. starkey* produced 3.03 $g\,l^{-1}$ of lipids with glucose compared to 3.61 $g\,l^{-1}$ obtained with xylose using a C : N (carbon nitrogen ratio) of 50 (Liu et al. 2020a,b). This possibility favors the use of lignocellulosic biomass with a higher percentage of xylose. Gong et al. (2012) reported the production of 0.57 g of lipids per gram of cell dry weight when *L. starkey* grew using xylose (70 $g\,l^{-1}$) as sole carbon source. In the same work, the authors revealed that lipid production was 0.5 and 0.55 $g\,g^{-1}$ when cellobiose-xylose (47 and 23 $g\,l^{-1}$) and cellobiose-xylose (60 and 30 $g\,l^{-1}$) combinations were used as carbon source.

The accumulation of cell lipids on hydrophilic substrates cultivation is improved by nutrient limitations, usually nitrogen limitation. In this case an adequate C : N ratio is necessary for cell growth and accumulation of lipids by oleaginous species (Chen et al. 2013). Several studies have shown that high C : N ratio actually favor the production of lipids (Chen et al. 2013; Guerfali et al. 2018; Liu et al. 2020a,b).

In addition, it is important to assess the ability of oleaginous species to resist toxic compounds present in lignocellulosic hydrolysates, so as not to require a detoxification step for these hydrolysates. As an example, Tsigie et al. (2011) demonstrated a reduction in the growth of *Y. lipolytica* when using nondetoxified sugarcane bagasse hydrolysate (5.88 $g\,l^{-1}$) in comparison with detoxified sugarcane bagasse hydrolysate (11.42 $g\,l^{-1}$), due to the presence of furfural and HMF. In contrast, the strain *Rhodosporidium fluviale* DMKU-SP314 demonstrate high tolerance to acetic acid, HMF, and vanillin (Poontawee et al. 2017).

The use of a high initial inoculum concentration can be an alternative to improve productivity, since it can promote greater cell growth and anticipate the production of lipids, reducing process costs. This fact was demonstrated by Juanssilfero et al. (2018a) in study with *L. starkey* with inoculum varying between 0.6–0.8 and 16–18 OD_{600nm}, with reduction of fermentation time from five to three days. In another study, larger inoculums (10% v/v) could reduce the toxic effect of furfural and the phenolic compounds on the cell biomass and lipid content of *C. curvatus* (Yu et al. 2014). Also, for filamentous fungi *M. isabellina*, the inoculum size was effective in increasing the production of lipids (~8.2–18.5 $g\,l^{-1}$) in combination with the increase in nitrogen concentration (from 7.58 to 50.4 mM). In addition, the use of ammonium sulfate and yeast extract (1 : 1) as nitrogen sources also influenced this increase, which is probably due to the supply of all required metal ions and micronutrients by the yeast extract, whereas ammonium sulfate is assimilated more quickly, providing such an improvement (Gao et al. 2013). In all cases, it can be considered that the increase in the size of the inoculum possibly provides better transfer of oxygen and nutrients, in addition to increasing the tolerance of microorganisms, whether toxic compounds or the concentration of xylose itself.

In another example, the increase in the size of the inoculum of *Rhodotorula glutinis* cultivated in undetoxified corncob hydrolysate reaching a lipid yield 13.2% greater than in cultivation with detoxified hydrolysate, 23.5 and 18.2 $g\,l^{-1}$, respectively, demonstrating not only high resistance of high density cell cultivation to toxic compounds present in the hydrolysate (acetic acid, furfural, and 5-HMF) but also the capacity to metabolize them (Liu et al. 2015). Sequentially, the same strain was cultivated in a 5-L bioreactor using high cell density (10%), C : N ratio of 75, two-stage nitrogen feeding strategy achieving a production of 33.5 $g\,l^{-1}$ of lipids (Liu et al. 2015).

Anschau et al. (2014) evaluated different inoculum sizes, effect of feeding strategies and ratio of glucose: xylose on the production of lipids by *L. starkey*, as previously described, the use of higher cell density as an initial inoculum favored the production of lipids, as well as the use of greater amount of xylose (sugar ratio of 70 : 30, glicose:xylose) and application of repeated fed-batch. The combination of these strategies resulted in a lipid production of 41.8 $g\,l^{-1}$ after 237 hours (Anschau et al. 2014).

In another study, the combination of optimal fermentation parameters, as inoculum concentration 5%, initial pH 6.0, $MgSO_47H_2O$ 0.3 $g\,l^{-1}$, $CuSO_45H_2O$ 0.003 $g\,l^{-1}$, $MnSO_4H2O$ 0.003 $g\,l^{-1}$, and KCl 0.4 $g\,l^{-1}$ at temperature 28 °C incubate for eight days provided the production of 10.4 $g\,l^{-1}$ of lipids by *Trichosporon cutaneum* (Chen et al. 2013), demonstrating that several factors impact together to favor the formation of lipids and that studies of process optimization are effective in improving production processes.

As for the broad market potential of these products, genetic engineering techniques are substantially used to increase lipid production, including the expression of heterologous genes or routes and overexpression of genes that encoded specify enzymes. Some reviews have already been elaborated on the genetic improvements for the production of microbial lipids and are recommended for reading (Kosa and Ragauskas 2011; Liang and Jiang 2013; Donot et al. 2014; Gong et al. 2014; Levering et al. 2015; Cheon et al. 2016).

Thus, the overexpression of a cytosolic malic enzyme from *R. glutinis* was carried out in *R. mucilaginosa* through chromosomal integration, in order to increase the amount of reducing power, especially NADPH, which is highly required in the synthesis of lipids in oleaginous yeasts. Cultivating in a 15-l bioreactor on sugarcane bagasse derived xylose, the mutant strain obtained an increase in lipid yield from 3.75 to 4.29 $g\,l^{-1}$ in relation to the wild strain, in addition to presenting an increase in the proportion of monounsaturated fatty acids (from 48.92 to 65.95% w/w) that contribute to the production of better quality biodiesel (Bandhu et al. 2019).

To use lignocellulosic materials such as xylose as a carbon source, Tai and Stephanopoulos (2013) inserted genes encoding xylose reductase and xylitol dehydrogenase and increased the expression of genes encoding diacylglycerol acyltransferase (DGA) in *Y. lipolytica* using cofermentation with xylose and glycerol in a 2-L bioreactor (operated under 2.5 vvm aeration, pH 6.8, 28 °C and 250 rpm). The *Y. lipolytica* MTYL081 mutant strain produced 7.64 g lipids/L (42% of 18 $g\,l^{-1}$ and productivity of 0.033 g lipid/L/h) from xylose and glycerol. Authors observed that ~80% of lipids were produced after glycerol depletion and during the xylose phase, indicating successful conversion of xylose into lipids from *Y. lipolytica* (Stephanopoulos and Tai, 2016).

A fatty acid profile rich in mono-unsaturated fatty acids contribute to the production of biodiesel with optimum kinematic fuel properties, such as such as low viscosity, cloud, and freezing point, especially due to the presence of oleic acid (Soccol et al. 2017), however, it is prompted to auto-oxidation (Ananthi et al. 2019). Most of the studies described here have reported the production of microbial oils composed mainly of oleic acid (C18:1) and palmitic acid (C16:0) and to a lesser extent by stearic acid (C18:0) and linoleic acid (C18:2), similar to palm oil composition, demonstrating the technical applicability of the microbial oils produced (Galafassi et al. 2012; Chen et al. 2013; Gao et al. 2013; Anschau et al. 2014; Guerfali et al. 2018; Bandhu et al. 2019; Liu et al. 2020a,b).

An ideal biodiesel can be produced from oils composed of methyl esters of both saturated and mono-unsaturated fatty acids, and with low percentage of poly-unsaturated fatty acid, obtaining the combination of excellent fluidity and oxidative stability (Guerfali et al. 2018).

5.4 Concluding Remarks

Hydrolyzed lignocellulosic biomass has great potential for application as a substrate for the production of biofuels, focused on bioethanol or microbial lipids to be used as starting material in the synthesis of biodiesel, since several studies have shown strains able to use xylose as sole carbon source, and in many cases, in simultaneous fermentation with glucose, enhancing the use of all potential sugar present in vegetal biomass. Another important point to be highlighted is the identification of strains also resistant to toxic compounds formed during the hydrolysis of lignocellulosic biomass, making them ideal biocatalysts for such value-adding fermentative processes.

As analyzed, even though the great potential in the use of pentoses as raw materials for bioprocesses, several investments in scientific and technological research must still be carried out, aiming at circumventing the existing problems and raising the biofuels obtained in these processes to competitive and economically viable levels.

References

Ahmad, F.B. et al. (2019). The prospect of microbial oil production and applications from oil palm biomass. *Biochem. Eng. J.* 143: 9–23.

Ananthi, V. et al. (2019). Enhanced microbial biodiesel production from lignocellulosic hydrolysates using yeast isolates. *Fuel.* 256: 115932.

Anschau, A., Xavier, M.C., Hernalsteens, S., and Franco, T.T. (2014). Effect of feeding strategies on lipid production by Lipomyces starkeyi. *Bioresour. Technol.* 157: 214–222. https://doi.org/10.1016/j.biortech.2014.01.104.

Antonopoulou, I., Spanopoulos, A., and Matsakas, L. (2020). Single cell oil and ethanol production by the oleaginous yeast *Trichosporon fermentans* utilizing dried sweet sorghum stalks. *Renew. Energy* 146: 1609–1617. https://doi.org/10.1016/j.renene.2019.07.107.

Antunes, F.A.F., Chandel, A.K., Milessi, T. et al. (2014). Bioethanol production from sugarcane bagasse by a novel Brazilian pentose fermenting yeast *Scheffersomyces shehatae* UFMG-HM 52.2: evaluation of fermentation medium. *Int. J. Chem. Eng.* https://doi.org/10.1155/2014/180681.

Ayadi, I., Belghith, H., Gargouri, A., and Guerfali, M. (2019). Utilization of wheat bran acid hydrolysate by *Rhodotorula mucilaginosa* Y-MG1 for microbial lipid production as feedstock for biodiesel synthesis. Hindawi. *BioMed. Res. Int.*: 3213521. https://doi.org/10.1155/2019/3213521.

Bandhu, S., Bansal, N., Dasgupta, D. et al. (2019). Overproduction of single cell oil from xylose rich sugarcane bagasse hydrolysate by an engineered oleaginous yeast *Rhodotorula mucilaginosa* IIPL32. *Fuel* 254: 115653. https://doi.org/10.1016/j.fuel.2019.115653.

Bhatia, S.K., Jagtap, S.S., Bedekar, A.A. et al. (2020). Recent developments in pretreatment technologies on lignocellulosic biomass: effect of key parameters, technological improvements, and challenges. *Bioresour. Technol.*: 122724. https://doi.org/10.1016/j.biortech.2019.122724.

Branco, R.H., Amândio, M.S., Serafim, L.S., and Xavier, A.M. (2020). Ethanol production from hydrolyzed Kraft pulp by mono-and co-cultures of yeasts: the challenge of C6 and C5 sugars consumption. *Energies* 13 (3): 744.

Brito, P.L., de Azevedo Ferreira, C.M., Silva, A.F.F. et al. (2018). Hydrolysis, detoxification and alcoholic fermentation of hemicellulose fraction from palm press fiber. *Waste Biomass Valori.* 9 (6): 957–968. https://doi.org/10.1007/s12649-017-9882-4.

Cadete, R.M., Melo, M.A., Dussán, K.J. et al. (2012). Diversity and physiological characterization of D-xylose-fermenting yeasts isolated from the Brazilian Amazonian Forest. *PLoS One* 7 (8) https://doi.org/10.1371/journal.pone.0043135.

Cai, Z, Zhang, B, Li, Y. (2012). Engineering Saccharomyces cerevisiae for efficient anaerobic xylose fermentation: reflections and perspectives. *Biotechnol. J.* 7 (1): 34–46. https://doi.org/10.1002/biot.201100053.

Candido, J.P., Claro, E.M.T., de Paula, C.B.C. et al. (2020). Detoxification of sugarcane bagasse hydrolysate with different adsorbents to improve the fermentative process. *World J. Microbiol. Biotechnol.* 36 (3): 1–12. doi: 10.1007/s11274-020-02820-7.

Cannon, P., Aguirre-Hudson, B., Aime, M. C., Ainsworth, A. M., Bidartondo, M. I., & Gaya, E. (2018). Definition and diversity. State of the Word's Fungi. Royal Botanic Gardens, Kew, 4–11.

Capolupo, L. and Faraco, V. (2016). Green methods of lignocellulose pretreatment for biorefinery development. *Appl. Microbiol. Biotechnol.* 100 (22): 9451–9467. https://doi.org/10.1007/s00253-016-7884-y.

Carvalho, A.K.F., Bento, H.B., Rivaldi, J.D., and de Castro, H.F. (2018). Direct transesterification of *Mucor circinelloides* biomass for biodiesel production: effect of carbon sources on the accumulation of fungal lipids and biofuel properties. *Fuel* 234: 789–796. https://doi.org/10.1016/j.fuel.2018.07.029.

Chandel, A.K., Chandrasekhar, G., Radhika, K. et al. (2011). Bioconversion of pentose sugars into ethanol: a review and future directions. *Biotechnol. Mol. Biol. Rev.* 6 (1): 008–020.

Chandel, A.K., Antunes, F.A., Terán-Hilares, R. et al. (2018). Bioconversion of hemicellulose into ethanol and value-added products: commercialization, trends, and future opportunities. In: *Advances in Sugarcane Biorefinery*, 97–134. Elsevier https://doi.org/10.1016/B978-0-12-804534-3.00005-7.

Chatterjee, S. and Mohan, S.V. (2018). Microbial lipid production by *Cryptococcus curvatus* from vegetable waste hydrolysate. *Bioresour. Technol.* 254: 284–289. https://doi.org/10.1016/j.biortech.2018.01.079.

Chen, X.F., Huang, C., Yang, X.Y. et al. (2013). Evaluating the effect of medium composition and fermentation condition on the microbial oil production by *Trichosporon cutaneum* on corncob acid hydrolysate. *Bioresour. Technol.* 143: 18–24. https://doi.org/10.1016/j.biortech.2013.05.102.

Cheng, K.K., Cai, B.Y., Zhang, J.A. et al. (2008). Sugarcane bagasse hemicellulose hydrolysate for ethanol production by acid recovery process. *Biochem. Eng. J.* 38 (1): 105–109. https://doi.org/10.1016/j.bej.2007.07.012.

Cheon, S., Kim, H.M., Gustavsson, M., and Lee, S.Y. (2016). Recent trends in metabolic engineering of microorganisms for the production of advanced biofuels. *Curr. Opin. Chem. Biol.* 35: 10–21. https://doi.org/10.1016/j.cbpa.2016.08.003.

Cunha, J.T., Soares, P.O., Romaní, A. et al. (2019). Xylose fermentation efficiency of industrial *Saccharomyces cerevisiae* yeast with separate or combined xylose reductase/xylitol dehydrogenase and xylose isomerase pathways. *Biotechnol. Biofuels* 12 (1): 20. https://doi.org/10.1186/s13068-019-1360-8.

Deng, F. and Aita, G.M. (2018). Fumaric acid production by *Rhizopus oryzae* ATCC® 20344™ from lignocellulosic syrup. *Bioenergy Res.* 11 (2): 330–340. https://doi.org/10.1007/s12155-018-9899-y.

Dey, P., Pal, P., Kevin, J.D., and Das, D.B. (2020). Lignocellulosic bioethanol production: prospects of emerging membrane technologies to improve the process–a critical review. *Rev. Chem. Eng.* 36 (3): 333–367. https://doi.org/10.1515/revce-2018-0014.

Dias, M.O.S., Junqueira, T.L., Cavalett, O. et al. (2012). Integrated versus stand-alone second generation ethanol production from sugarcane bagasse and trash. *Bioresour. Technol.* 103 (1): 152–161. https://doi.org/10.1016/j.biortech.2011.09.120.

Díaz-Fernández, D. et al. (2019). Microbial lipids from industrial wastes using xylose-utilizing *Ashbya gossypii* strains. *Bioresour. Technol.* 293: 122054. https://doi.org/10.1016/j.biortech.2019.122054.

Dien, B.S., Kurtzman, C.P., Saha, B.C., and Bothast, R.J. (1996). Screening for L-arabinose fermenting yeasts. *Appl. Biochem. Biotechnol.* 57 (1): 233. https://doi.org/10.1007/BF02941704.

Dien, B.S., Cotta, M.A., and Jeffries, T.W. (2003). Bacteria engineered for fuel ethanol production: current status. *Appl. Microbiol. Biotechnol.* 63 (3): 258–266. https://doi.org/10.1007/s00253-003-1444-y.

Donot, F., Fontana, A., Baccou, J.C. et al. (2014). Single cell oils (SCOs) from oleaginous yeasts and moulds: production and genetics. *Biomass Bioenergy* 68: 135–150. https://doi.org/10.1016/j.biombioe.2014.06.016.

Eiteman, M.A., Lee, S.A., and Altman, E. (2008). A co-fermentation strategy to consume sugar mixtures effectively. *J. Biol. Eng.* 2: 3. https://doi.org/10.1186/1754-1611-2-3.

Fang, H., Zhao, C., and Chen, S. (2016). Single cell oil production by *Mortierella isabellina* from steam exploded corn stover degraded by three-stage enzymatic hydrolysis in the context of on-site enzyme production. *Bioresour. Technol.* 216: 988–995. https://doi.org/10.1016/j.biortech.2016.06.051.

Fengel, D. and Wegener, G. (1983). *Wood: Chemistry, Ultrastructure, Reactions*. Germany: Verlag Kessel.

Finn, R.K., Bringer, S., and Sahm, H. (1984). Fermentation of arabinose to ethanol by *Sarcina ventriculi*. *Appl. Microbiol. Biotechnol.* 19 (3): 161–166. https://doi.org/10.1007/BF00256448.

Galafassi, S., Cucchetti, D., Pizza, F. et al. (2012). Lipid production for second generation biodiesel by the oleaginous yeast *Rhodotorula graminis*. *Bioresour. Technol.* 111: 398–403. https://doi.org/10.1016/j.biortech.2012.02.004.

Gao, D., Zeng, J., Zheng, Y. et al. (2013). Microbial lipid production from xylose by *Mortierella isabellina*. *Bioresour. Technol.* 133: 315–321. https://doi.org/10.1016/j.biortech.2013.01.132.

Gong, Z., Wang, Q., Shen, H. et al. (2012). Co-fermentation of cellobiose and xylose by *Lipomyces starkeyi* for lipid production. *Bioresour. Technol.* 117: 20–24. https://doi.org/10.1016/j.biortech.2012.04.063.

Gong, Y., Wan, X., Jiang, M. et al. (2014). Metabolic engineering of microorganisms to produce omega-3 very long-chain polyunsaturated fatty acids. *Prog. Lipid Res.* 56: 19–35. https://doi.org/10.1016/j.plipres.2014.07.001.

Guerfali, M., Ayadi, I., Belhassen, A. et al. (2018). Single cell oil production by *Trichosporon cutaneum* and lignocellulosic residues bioconversion for biodiesel synthesis. *Process Saf. Environ. Prot.* 113: 292–304. https://doi.org/10.1016/j.psep.2017.11.002.

Ha, S.-J., Galazka, J.M., Rin Kim, S. et al. (2011). Engineered *Saccharomyces cerevisiae* capable of simultaneous cellobiose and xylose fermentation. *Proc. Natl. Acad. Sci.* 108 (2): 504–509. https://doi.org/10.1073/pnas.1010456108.

Hahn-Hägerdal, B., Jeppsson, H., Olsson, L., and Mohagheghi, A. (1994). An interlaboratory comparison of the performance of ethanol-producing micro-organisms in a xylose-rich acid hydroysate. *Appl. Microbiol. Biotechnol.* 41 (1): 62–72. https://doi.org/10.1007/BF00166083.

10.1016/j.rser.2018.11.041Hassan, S.S., Williams, G.A., and Jaiswal, A.K. (2019). Moving towards the second generation of lignocellulosic biorefineries in the EU: drivers, challenges, and opportunities. *Renewable Sustainable Energy Rev.* 101: 590–599.

Hawksworth, D.L. and Lücking, R. (2017). Fungal diversity revisited: 2.2 to 3.8 million species. The fungal kingdom, 79–95. https://doi.org/10.1128/microbiolspec.FUNK-0052-2016

Isern, N.G., Xue, J., Rao, J.V. et al. (2013). Novel monosaccharide fermentation products in Caldicellulosiruptor saccharolyticus identified using NMR spectroscopy. *Biotechnol. Biofuels* 6 (1): 47. https://doi.org/10.1186/1754-6834-6-47.

Jönsson, L.J., Alriksson, B., and Nilvebrant, N.O. (2013). Bioconversion of lignocellulose: inhibitors and detoxification. *Biotechnol. Biofuels* 6 (1): 16. https://doi.org/10.1186/1754-6834-6-16.

Juanssilfero, A.B., Kahar, P., Amza, R.L. et al. (2018a). Effect of inoculum size on single-cell oil production from glucose and xylose using oleaginous yeast *Lipomyces starkeyi*. *J. Biosci. Bioeng.* 125 (6): 695–702. https://doi.org/10.1016/j.jbiosc.2017.12.020.

Juanssilfero, A.B., Kahar, P., Amza, R.L. et al. (2018b). Selection of oleaginous yeasts capable of high lipid accumulation during challenges from inhibitory chemical compounds. *Biochem. Eng. J.* 137: 182–191. https://doi.org/10.1016/j.bej.2018.05.024.

Kim, D. (2018). Physico-chemical conversion of lignocellulose: inhibitor effects and detoxification strategies: a mini review. *Molecules* 23 (2): 309. https://doi.org/10.3390/molecules23020309.

Kim, S.R., Park, Y.C., Jin, Y.S., and Seo, J.H. (2013). Strain engineering of *Saccharomyces cerevisiae* for enhanced xylose metabolism. *Biotechnol. Adv.* 31 (6): 851–861. https://doi.org/10.1016/j.biotechadv.2013.03.004.

Ko, J.K. and Lee, S.M. (2018). Advances in cellulosic conversion to fuels: engineering yeasts for cellulosic bioethanol and biodiesel production. *Curr. Opin. Biotechnol.* 50: 72–80. https://doi.org/10.1016/j.copbio.2017.11.007.

Kosa, M. and Ragauskas, A.J. (2011). Lipids from heterotrophic microbes: advances in metabolism research. *Trends Biotechnol.* 29 (2): 53–61. https://doi.org/10.1016/j.tibtech.2010.11.002.

Kurosawa, K., Laser, J., and Sinskey, A.J. (2015). Tolerance and adaptive evolution of triacylglycerol-producing Rhodococcus opacus to lignocellulose-derived inhibitors. *Biotechnol. Biofuels* 8: 76. https://doi.org/10.1186/s13068-015-0258-3.

Kwak, S., Jo, J.H., Yun, E.J. et al. (2019). Production of biofuels and chemicals from xylose using native and engineered yeast strains. *Biotechnol. Adv.* 37 (2): 271–283. https://doi.org/10.1016/j.biotechadv.2018.12.003.

Laluce, C., Schenberg, A.C.G., Gallardo, J.C.M. et al. (2012). Advances and developments in strategies to improve strains of *Saccharomyces cerevisiae* and processes to obtain the Lignocellulosic ethanol–a review. *Appl. Biochem. Biotechnol.* 166 (8): 1908–1926. https://doi.org/10.1007/s12010-012-9619-6.

Levering, J., Broddrick, J., and Zengler, K. (2015). Engineering of oleaginous organisms for lipid production. *Curr. Opin. Biotechnol.* 36: 32–39. https://doi.org/10.1016/j.copbio.2015.08.001.

Li, S. and Yang, X. (2016). Biofuel production from food wastes. *Handbook Biofuel. Prod.*: 617–653. https://doi.org/10.1016/b978-0-08-100455-5.00020-5.

Liang, M.H. and Jiang, J.G. (2013). Advancing oleaginous microorganisms to produce lipid via metabolic engineering technology. *Prog. Lipid Res.* 52 (4): 395–408. https://doi.org/10.1016/j.plipres.2013.05.002.

Liu, Y., Wang, Y., Liu, H., and Zhang, J.A. (2015). Enhanced lipid production with undetoxified corncob hydrolysate by *Rhodotorula glutinis* using a high cell density culture strategy. *Bioresour. Technol.* 180: 32–39. https://doi.org/10.1016/j.biortech.2014.12.093.

Liu, J., Yuan, M., Liu, J.N. et al. (2016). Microbial conversion of mixed volatile fatty acids into microbial lipids by sequencing batch culture strategy. *Bioresour. Technol.* 222: 75–81. https://doi.org/10.1016/j.biortech.2016.09.100.

Liu, T., Peng, B., Huang, S., and Geng, A. (2020a). Recombinant xylose-fermenting yeast construction for the co-production of ethanol and cis, cis-muconic acid from lignocellulosic biomass. *Bioresou. Technol. Rep.* 9: 100395. https://doi.org/10.1016/j.biteb.2020.100395.

Liu, Z., Feist, A.M., Dragone, G., and Mussatto, S.I. (2020b). Lipid and carotenoid production from wheat straw hydrolysates by different oleaginous yeasts. *J. Clean. Prod.* 249: 119308. https://doi.org/10.1016/j.jclepro.2019.119308.

Ma, X. et al. (2018). Microbial lipid production from food waste saccharified liquid and the effects of compositions. *Energy Convers. Manag.* 172: 306–315.

Malik, K., Salama, E.S., Kim, T.H., and Li, X. (2020). Enhanced ethanol production by *Saccharomyces cerevisiae* fermentation post acidic and alkali chemical pretreatments of cotton stalk lignocellulose. *Int. Biodeterior. Biodegrad.* 147: 104869. https://doi.org/10.1016/j.ibiod.2019.104869.

Matos, I.T.S.R., Cassa-Barbosa, L.A., Galvão, R.D.S.M., Nunes-Silva, C.G., and Astolfi Filho, S. (2014). Isolation, taxonomic identification and investigation of the biotechnological potential of wild-type Meyerozyma guilliermondii associated with Amazonian termites able to ferment D-xylose.

Matos, J.P., Souza, K., Santos, A.S., and Pantoja, L.A. (2018). Fermentação alcoólica de hidrolisado hemicelulósico de torta de girassol por *Galactomyces geotrichum* UFVJM-R10 e

Candida akabanensis UFVJM-R131. *Quím. Nova* 41: 23–29. https://doi.org/10.21577/0100-4042.20170146.

McMillan, J.D. (1993). *Xylose fermentation to ethanol. A review* (No. NREL/TP-421-4944). National Renewable Energy Lab., Golden, CO (United States).

McMillan, J.D. and Boynton, B.L. (1994). Arabinose utilization by xylose-fermenting yeasts and fungi. *Appl. Biochem. Biotechnol.* 45 (1): 569–584. https://doi.org/10.1007/BF02941831.

Millati, R., Edebo, L., and Taherzadeh, M.J. (2005). Performance of *Rhizopus*, *Rhizomucor*, and *Mucor* in ethanol production from glucose, xylose, and wood hydrolyzatehydrolysates. *Enzym. Microb. Technol.* 36 (2–3): 294–300. https://doi.org/10.1016/j.enzmictec.2004.09.007.

Mondala, A., Hernandez, R., French, T. et al. (2015). Enhanced microbial oil production by activated sludge microorganisms from sugarcane bagasse hydrolyzatehydrolysate. *Renew. Energy* 78: 114–118. https://doi.org/10.1016/j.renene.2014.12.073.

Nair, A.S. et al. (2020). Production of microbial lipids from optimized waste office paper hydrolysate, lipid profiling and prediction of biodiesel properties. *Renew. Energy.* 148: 124–134.

Nogué, V.S. and Karhumaa, K. (2015). Xylose fermentation as a challenge for commercialization of lignocellulosic fuels and chemicals. *Biotechnol. Lett.* 37 (4): 761–772. https://doi.org/10.1007/s10529-014-1756-2.

Osorio-González, C.S., Hegde, K., Brar, S.K. et al. (2019). Challenges in lipid production from lignocellulosic biomass using *Rhodosporidium* sp.; a look at the role of lignocellulosic inhibitors. *Biofuels. Bioprod. Bioref.* 13: 740–759. https://doi.org/10.1002/bbb.1954.

Padella, M., O'Connell, A., and Prussi, M. (2019). What is still limiting the deployment of cellulosic ethanol? Analysis of the current status of the sector. *Appl. Sci.* 9: 4523. https://doi.org/10.3390/app9214523.

Palmqvist, E. and Hahn-Hägerdal, B. (2000). Fermentation of lignocellulosic hydrolysates. I: inhibition and detoxification. *Bioresour. Technol.* 74 (1): 17–24. https://doi.org/10.1016/S0960-8524(99)00160-1.

Pan, C., Liu, Z., Yao, L. et al. (2019). Effects of ethanol pretreatment on dissolution and structural changes of lignin from steam-exploded wheat straw. *Bioresources* 14 (3): 6958–6969.

Parawira, W. and Tekere, M. (2011). Biotechnological strategies to overcome inhibitors in lignocellulose hydrolysates for ethanol production. *Crit. Rev. Biotechnol.* 31 (1): 20–31. https://doi.org/10.3109/07388551003757816.

Pessôa, M.G., Vespermann, K.A., Paulino, B.N. et al. (2019). Newly isolated microorganisms with potential application in biotechnology. *Biotechnol. Adv.* 37 (2): 319–339. https://doi.org/10.1016/j.biotechadv.2019.01.007.

Poontawee, R., Yongmanitchai, W., and Limtong, S. (2017). Efficient oleaginous yeasts for lipid production from lignocellulosic sugars and effects of lignocellulose degradation compounds on growth and lipid production. *Process Biochem.* 53: 44–60. https://doi.org/10.1016/j.procbio.2016.11.013.

Rodrigues, P.R., Araújo, M.F.L., Rocha, T.L. et al. (2018). Evaluation of buriti endocarp as lignocellulosic substrate for second generation ethanol production. *Peer J.* 6: e5275. https://doi.org/10.7717/peerj.5275.

Rojas-Chamorro, J.A., Romero-García, J.M., Cara, C. et al. (2020). Improved ethanol production from the slurry of pretreated brewers' spent grain through different co-fermentation strategies. *Bioresour. Technol.* 296: 22367. https://doi.org/10.1016/j.biortech.2019.122367.

Rorke, D.C., Suinyuy, T.N., and Kana, E.G. (2017). Microwave-assisted chemical pre-treatment of waste sorghum leaves: process optimization and development of an intelligent model for determination of volatile compound fractions. *Bioresour. Technol.* 224: 590–600. https://doi.org/10.1016/j.biortech.2016.10.048.

Santos, C.M., Brito, P.L., Santos, A.T.O. et al. (2019). Production of lignocellulosic ethanol from waste paper: review on production technology. *Inter. J. Develop. Res.* 9: 28383–28390.

Sarawan, C., Suinyuy, T.N., Sewsynker-Sukai, Y., and Gueguim, E.K. (2019). Optimized activated charcoal detoxification of acid-pretreated lignocellulosic substrate and assessment for bioethanol production. *Bioresour. Technol.* 286: 121403–121403. https://doi.org/10.1016/j.biortech.2019.121403.

Sharma, A., Nain, V., Tiwari, R. et al. (2018). Optimization of fermentation condition for co-production of ethanol and 2, 3-butanediol (2, 3-BD) from hemicellolosic hydrolysates by *Klebsiella oxytoca* XF7. *Chem. Eng. Commun.* 205 (3): 402–410. https://doi.org/10.1080/00986445.2017.1398743.

Singh, N., Puri, M., Tuli, D.K. et al. (2018). Bioethanol production potential of a novel thermophilic isolate *Thermoanaerobacter* sp. DBT-IOC-X2 isolated from Chumathang hot spring. *Biomass Bioenergy* 116: 122–130. https://doi.org/10.1016/j.biombioe.2018.05.009.

Slininger, P.J., Shea-Andersh, M.A., Thompson, S.R. et al. (2015). Evolved strains of *Scheffersomyces stipitis* achieving high ethanol productivity on acid- and base-pretreated biomass hydrolysate at high solids loading. *Biotechnol. Biofuels.* 8: 60. https://doi.org/10.1186/s13068-015-0239-6.

Soccol, C.R., Neto, C.J.D., Soccol, V.T. et al. (2017). Pilot scale biodiesel production from microbial oil of *Rhodosporidium toruloides* DEBB 5533 using sugarcane juice: performance in diesel engine and preliminary economic study. *Bioresour. Technol.* 223: 259–268. https://doi.org/10.1016/j.biortech.2016.10.055.

Sonderegger, M., Schümperli, M., and Sauer, U. (2004). Metabolic engineering of a phosphoketolase pathway for pentose catabolism in *Saccharomyces cerevisiae*. *Appl. Environ. Microbiol.* 70 (5): 2892–2897. https://doi.org/10.1128/AEM.70.5.2892-2897.2004.

Stambuk, B.U., Eleutherio, E.C.A., Florez-Pardo, L.M., Souto-Maior, A.M., and Bon, E.P. (2008). Brazilian potential for biomass ethanol: challenge of using hexose and pentose cofermenting yeast strains.

Subtil, T. and Boles, E. (2012). Competition between pentoses and glucose during uptake and catabolism in recombinant *Saccharomyces cerevisiae*. *Biotechnol. Biofuels.* 5: 14. https://doi.org/10.1186/1754-6834-5-14.

Tai, M. and Stephanopoulos, G. (2013). Engineering the push and pull of lipid biosynthesis in oleaginous yeast Yarrowia lipolytica for biofuel production. *Metab. Eng.* 15: 1–9. https://doi.org/10.1016/j.ymben.2012.08.007.

Talukder, A.A., Adnan, N., Siddiqa, A. et al. (2019). Fuel ethanol production using xylose assimilating and high ethanol producing thermosensitive *Saccharomyces cerevisiae* isolated from date palm juice in Bangladesh. *Biocatal. Agric. Biotechnol.* 18: 101029. https://doi.org/10.1016/j.bcab.2019.101029.

Tsigie, Y.A., Wang, C.Y., Truong, C.T., and Ju, Y.H. (2011). Lipid production from *Yarrowia lipolytica* Po1g grown in sugarcane bagasse hydrolysate. *Bioresour. Technol.* 102 (19): 9216–9222. https://doi.org/10.1016/j.biortech.2011.06.047.

Valinhas, R.V., Pantoja, L.A., Maia, A.C.F. et al. (2018). Xylose fermentation to ethanol by new *Galactomyces geotrichum* and *Candida akabanensis* strains. *Peer J.* 6: e4673. https://doi.org/10.7717/peerj.4673.

Van Dijk, M., Erdei, B., Galbe, M. et al. (2019). Strain-dependent variance in short-term adaptation effects of two xylose-fermenting strains of *Saccharomyces cerevisiae*. *Bioresour. Technol.* 292: 121922. https://doi.org/10.1016/j.biortech.2019.121922.

Vasconcelos, M.H., Mendes, F.M., Ramos, L. et al. (2020). Techno-economic assessment of bioenergy and biofuel production in integrated sugarcane biorefinery: identification of technological bottlenecks and economic feasibility of dilute acid pretreatment. *Energy* 199: 117422. https://doi.org/10.1016/j.energy.2020.117422.

Weber, C., Farwick, A., Benisch, F. et al. (2010). Trends and challenges in the microbial production of lignocellulosic bioalcohol fuels. *Appl. Microbiol. Biotechnol.* 87 (4): 1303–1315. https://doi.org/10.1007/s00253-010-2707-z.

Winkler, J.D. and Kao, K.C. (2014). Recent advances in the evolutionary engineering of industrial biocatalysts. *Genomics* 104 (6): 406–411. https://doi.org/10.1016/j.ygeno.2014.09.006.

Xiao, Y., Ruan, Z., Liu, Z. et al. (2013). Engineering *Escherichia coli* to convert acetic acid to free fatty acids. *Biochem. Eng. J.* 76 (2013): 60–69.

Xu, X., Xian, M., and Liu, H. (2017). Efficient conversion of acetate into hloroglucinol by recombinant *Escherichia coli*. *RSC Adv.* 7: 50942–50948.

Yu, X., Zeng, J., Zheng, Y., and Chen, S. (2014). Effect of lignocellulose degradation products on microbial biomass and lipid production by the oleaginous yeast *Cryptococcus curvatus*. *Process Biochem.* 49 (3): 457–465. https://doi.org/10.1016/j.procbio.2013.10.016.

Yu, Y., Xu, Z., Chen, S., and Jin, M. (2020). Microbial lipid production from dilute acid and dilute alkali pretreated corn Stover via *Trichosporon dermatis*. *Bioresour. Technol.* 290 https://doi.org/10.1016/j.biortech.2019.122253122253.

Zabed, H., Sahu, J.N., Boyce, A.N., and Faruq, G. (2016). Fuel ethanol production from lignocellulosic biomass: an overview on feedstocks and technological approaches. *Renewable Sustainable Energy Rev.* 66: 751–774.

Zhang, M., Eddy, C., Deanda, K. et al. (1995). Metabolic engineering of a pentose metabolism pathway in ethanologenic *Zymomonas mobilis*. *Science* 267 (5195): 240–243. https://doi.org/10.1126/science.267.5195.240.

Zhang, C., Acosta-Sampson, L., Yu, V.Y., and Cate, J.H. (2017). Screening of transporters to improve xylodextrin utilization in the yeast *Saccharomyces cerevisiae*. *PLoS One* 12 (9) https://doi.org/10.1371/journal.pone.0184730.

Zhijia, L., Adam, M.F., Giuliano, D., and Solange, I.M. (2020). Lipid and carotenoid production from wheat straw hydrolysates by different oleaginous yeasts. *J. Clean. Prod.* https://doi.org/10.1016/j.jclepro.2019.119308119308.

6

An Overview of Microalgal Carotenoids: Advances in the Production and Its Impact on Sustainable Development

Rahul Kumar Goswami, Komal Agrawal and Pradeep Verma

Bioprocess and Bioenergy Laboratory, Department of Microbiology, Central University of Rajasthan, Ajmer, Rajasthan, India

6.1 Introduction

Carotenoid, a photosynthetic pigment, is the broadest category of secondary metabolites pigment present inside microorganisms, photosynthetic plants, or microalgae. It is an important pigment for photosynthesis and is the largest family of metabolites and subfamily of isoprenoids. It is derived from 8 isoprenes (5-carbon) units that are polymerized enzymatically (Agrawal et al 2020; Bhardwaj et al. 2020; Cardozo et al. 2007; Chaturvedi et al. 2020; Chu 2012; Goswami et al. 2020a,b). It is generally categorized into primary carotenoid and secondary carotenoid. Approximately, 1100 carotenoids are been identified in 600 different organisms (e.g. fungi, plants bacteria, cyanobacteria, macro, and microalgae), fruits, and vegetables. It is a fat-soluble molecule, absorbs light spectrum at 450–550 nm and is present in different color ranges, such as red, orange, and yellow. Redness of carotenoid is dependent on the presence of conjugated double bond (Novoveská et al. 2019) and its structure was defined by Swiss scientist Paul Karrer. Previously it was used as a coloring agent in food industries, however, later research found that it has many medicinal properties and its consumption is good for human health.

Carotenoid is a precursor of vitamin A (Chu 2012; Saini and Keum 2017) and has many therapeutic values, e.g., antioxidant activities, anti-aging activities, and protects humans from oxidative stress. As commercially synthesized carotenoid had many side effects to humans, later researchers focused on the production of naturally occurring carotenoids (Dufossé et al. 2005; Gong and Bassi 2016; Saini and Keum 2017). Many organisms are capable of producing carotenoid intracellularly, but their productivity rate is very slow. Therefore, scientists are focusing research to find an appropriate feedstock and organism that has the ability to produce high amounts of different varieties of natural carotenoid. They found that carotenoids are present in the highest amount inside microalgal cells. Due to their medicinal properties, it is highly used in human consumption. The market value of carotenoid has increased by 1.5 billion $ (2014) to 1.8 billion $ (2019) (Mata-Gómez et al. 2014; Saini and Keum 2017).

Bioenergy Research: Evaluating Strategies for Commercialization and Sustainability, First Edition.
Edited by Neha Srivastava and Manish Srivastava.

6.1.1 Interaction and Understanding of Carotenoid

Carotenoids are diverse isoprenoid metabolites, which are lipid-soluble/lipophilic molecules, fat-soluble molecules, and present in various microorganisms or plants. It is divided into two main groups: xanthophyll such as astaxanthin, lutein (which contains oxygen molecules inside their structure), and hydrocarbon containing carotenoids such as carotene (not containing any oxygenated molecules) and present in cis or transform (Cohen 2017; Saini and Keum 2017). Carotenoids are mainly present in three or four colors: yellow, red, brown, and orange. The main function of carotenoid is to help in the photosynthesis process in photosynthetic organisms (Contreras et al. 2020). Generally, carotenoids show C-40 backbone structure of isoprene units and is sensitive toward light, temperature, and oxygen, and as a result, its handling and storage becomes very difficult. Although a diverse range of carotenoid is present inside organisms, only 30 different carotenoids help in the photosynthesis process (Varela et al. 2015). In photosynthetic organisms, carotenoids exist inside thylakoids membranes and are attached to the light-harvesting complex (LHCs) (Nisar et al. 2015). Based on their function in the photosynthesis process, it is categorized into primary (1°) or secondary (2°) types of carotenoids. Primary carotenoids (lutein) play a major role in transferring energy to chlorophyll (Ye et al. 2008), and help in mounting the light-absorbing spectrum of photosynthetic organisms, while secondary carotenoids (astaxanthin and canthaxanthin) are a protective element for plant or microalgae. In stress conditions, microalgae produces secondary carotenoids that form protective layers and protect the cells from oxidative damages or other stressful conditions (Begum et al. 2016; Sun et al. 2018). In-plant or algae, generally, carotenoids, are present as diester or ester forms. So, extraction of carotenoid requires the saponification process. Due to their antioxidant properties, it can guard the cells by oxidative stress, free radicals, and prevent the cell for lipid peroxidation or encourage the stability and functionality of the photosynthetic system of plants (Sun et al. 2018)

6.1.2 Differentiation between Natural or Chemically Synthesized Carotenoids

Due to their medicinal property, carotenoids are consumed by humans. The synthetic carotenoids are artificially produced by different chemical methods, such as dehydration, elimination homodimerization, and selective condensation of carbonyl compound. Generally, chemically carotenoids are producing by using Grignard composites or combining two phosphonium salt (C15 molecules) and one dialdehyde molecules (C10) (witting reaction), and then it is isomerized and produces carotene, astaxanthin, and lycopene (Supamattaya et al. 2005; Bogacz-Radomska and Harasym 2018; Novoveská et al. 2019). Synthetic carotenoids synthesis is a fast process compared to natural carotenoids and can be produced using a low-cost substrate, thereby reducing the manufacturing cost. However, synthetic carotenoids are not as effective as compared to naturally produced carotenoids. Natural carotenoids are produced from photosynthetic organisms, such as algae, fungi plants. The biorefinery process is costly, or productivity rate is very slow, but their antioxidant efficiency, or other nutraceutical properties, and health-promoting properties are higher than chemically synthesize carotenoids. The naturally occurring carotenoids are less toxic and as a result, are preferred by consumers and are used as health supplements. On the other hand, chemically synthesized carotenoids are toxic and are used as preservative agents, coloring agents, or animal feed (Novoveská et al. 2019). However, the production of natural-occurring carotenoids is a slow process and high cost is associated with its harvesting and biorefinery technology. But their increasing demands by the society are influencing the researchers to focus the research on isolating novel microorganism, innovating new low-cost biorefinery technologies and processes for the production of natural carotenoids. As a result, some microalgae strains were reported, showing positive responses for the production of a high amount of carotenoids as well as its productivity rate, which can be enhanced by modification of their metabolic pathways (Sun et al. 2018). In this chapter, the role of microalgae for the production of different types of carotenoid, their carotenogenesis pathways/carotenoid production will be discussed. In addition, how its production can be enhanced by applying different approaches, such as physiological and modification in their metabolic pathways, significance on human health, limitation on carotenoids production, and their present market scenario will also be discussed.

6.2 Diverse Category of Carotenoids

The composition of carotenoid varies from organisms to organisms. But mainly it is categorized into two groups, i.e., oxygen-containing xanthophylls (e.g. astaxanthin and zeaxanthin) and hydrocarbon containing carotene lacking oxygen (e.g. β carotene and lycopene) (Cohen 2017; Saini and Keum 2017). It has also been categorized based on color, such as red, yellow, and orange carotenoids. Different types of carotenoids are defined in Sections 6.2.1–6.2.4.

6.2.1 β-Carotene

β-carotene are important biological active biomolecules. It has an active form of provitamin A, (used as health supplements). The main benefits of naturally synthesized β-carotene, is that it can be easily absorbed by the body and has positive effects as compared to the synthetic form (Novoveská et al. 2019). The increasing demand of natural β-carotene has influenced the interest in the extraction of this valuable pigments from different natural sources (plants, microorganisms, and vegetables and fruits). The microalgae *Dunaliella salina* (*D. salina*) has been reported to be a potent organism for the commercial production of β-carotenes (Bogacz-Radomska and Harasym 2018). The structure of β-carotene is represented in Figure 6.1a. β-carotene has been used as coloring agents, antioxidants, and have many medicinal benefits where it shows anticancer, anti-aging, and immunomodulatory activities (Chen et al. 1993; Sathasivam et al. 2012, 2019; Sathasivam and Ki 2018).

6.2.2 Lutein

Lutein is an important primary carotenoid and its vital pigment is present inside macula lutea (or yellow spot) in the retina and lens of eyes. The structure of lutein is shown in Figure 6.1b. Lutein is synthesis from lycopene and is present in several organisms such

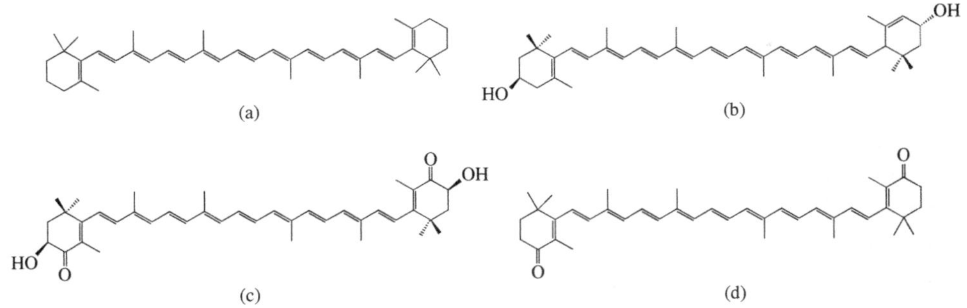

Figure 6.1 Different types of carotenoid; (a) β-carotene, (b) lutein, (c) astaxanthin, and (d) canthaxanthin. Source: Dufossé (2009). © 2009, Springer.

as fish, the eyes of many animals, or photosynthetic organism (cyanobacteria, microalgae, and plants) (Del Campo et al. 2000, 2007; Hu et al. 2018). The maximum amount of lutein is present in marigold flowers (Guedes et al. 2011) and the same has been used for its commercially production as well (Guedes et al. 2011). Lutein has several health benefits and has been used as a coloring agent in food industries. Some microalgae contain lutein in their chloroplast but in very low amount. Lutein production in microalgae is influenced by temperature, pH, oxidative stress, and light (Del Campo et al. 2007; Sathasivam and Ki 2018; Sathasivam et al. 2019).

6.2.3 Astaxanthin

It is a secondary carotenoid (xanthophyll) and oxygenated derivates of carotenoids. Astaxanthin (3,3′-dihydroxy-β, β-1-carotene-4,4′-dione) is a red color or pinkish-red carotenoid present in many organisms such as fish, vegetables, as well as in photosynthetic organisms, such as microalgae, cyanobacteria, and terrestrial plants (Hu et al. 2018). The structure of astaxanthin is shown in Figure 6.1c. Astaxanthin show antioxidant activities 500-fold higher as compared to other carotenoids like β-carotene, lutein, zeaxanthin, etc., it is also called "super vitamin E" (Cysewski and Lorenz 2004). It has more abundant secondary metabolites. It is composed of long, nonsaturated C-40 carbon atoms. Commercially, astaxanthin is produced by microalgae *Haematococcus pluvialis* (*H. pluvialis*) (Johnson and An 1991; Ip and Chen 2005). Astaxanthin plays a major role in microalga cells; it can protect from UV radiation, oxidative stress, and photooxidation. In humans, astaxanthin shows antioxidant activities, anti-aging activities, and protects from peroxidation and damage of low-density lipoprotein. Its shows potent activities against cancerous cells, metabolic syndrome, eye disease, and other neurodegenerative diseases (Hu et al. 2018; Sathasivam et al. 2019).

6.2.4 Canthaxanthin

It is a type of secondary carotenoid present in animals or egg yolk or many microorganisms. It is generally used as a dietary supplement for humans, poultry animals, etc. Several reports suggest that canthaxanthin increases the concentration of vitamin E in the liver

(Surai et al. 2003). The canthaxanthin structure is shown in Figure 6.1d. Canthaxanthin are synthesized from isopentenyl pyrophosphate (IPP) pathways and it helps in the synthesis of astaxanthin. Canthaxanthin has many medicinal properties, e.g., it can be used as antioxidative agents that reduces the free reactive oxygen and reduces early aging symptoms. Several reports suggest that canthaxanthin have anti-inflammatory compounds and neuroprotective effects (Chan et al. 2009). Several microalgae contain high amounts of canthaxanthin inside their cells. *Chlorella zofingiensis* (*C. zofingiensis*) are reported for the highest production of canthaxanthin. Different stress conditions might influence canthaxanthin production in microalgae cells (Pelah et al. 2004; Abe et al. 2007).

6.3 Microalgae Prospects for the Production of Carotenoids

Carotenoid are essential components present in many photosynthetic microorganisms. Due to their medicinal properties, humans can use it as food supplements, cosmetics, or as nutraceutical foods. Carotenoids show many potent activities such as antioxidant, anti-aging, and anticancerous activity (Cardozo et al. 2007; Chu 2012). As a result, carotenoids are commercially produced using different compounds, however, certain types of carotenoids are chemically synthesized as well. The process is fast compared to natural carotenoid production and it also reduces the cost by using low-cost substrates, although it exhibits some toxic effects on a human being; as a result, natural carotenoids are more preferred over the synthetic versions (Novoveská et al. 2019). The main benefits of using natural carotenoids are that is nontoxic and shows more healthy effects than chemically synthesized carotenoids. Due to the increasing demand of natural carotenoids, the researchers are focusing on an alternate option for large-scale production of carotenoid. After several research, scientists found that microalgae are a promising candidate for large-scale production of natural carotenoid. The microalgae gain attention due to their photosynthetic activities or presence of different varieties of carotenoid. It can capture carbon dioxide (CO_2) with the help of sunlight and water and increase their biomass (Sathasivam and Ki 2018). Microalgae contain different varieties of carotenoids inside their cells, and microalgal biomass contains 8–14% of different varieties of carotenoid inside their cells (Priyadarshani and Rath 2012; Mulders et al. 2015). These carotenoids are essential components of microalgae cells; it helps in the photosynthesis process as well as protecting the microalgae from different stress conditions. It is an innovative approach to the production of carotenoid from microalgae cells. It contains a diverse range of carotenoids, and they grow faster than terrestrial plants. Their cultivation process and biorefinery process is easy. It can utilize atmospheric CO_2 and wastewater for their growth. The main problem arising from carotenoid production is low carotenoid content of microalgal cells, but this problem can be resolved by applying proper conditions or modifying the metabolic pathways of microalgae. Currently, several microalgae strains are used for the production of natural carotenoid, e.g., *D. salina* use as for production of β-carotene and astaxanthin (Borowitzka et al. 1990; Hejazi et al. 2004; Pisal and Lele 2005), *Hematococcus* sp. is used for the production of astaxanthin (Ambati et al. 2014). β-carotene, astaxanthin, and lutein are important carotenoids for humans, because of its medicinal activities. Many countries or food safety departments approved the microalgal-based carotenoid, and it is commercially available in markets. The production process is costly, and it increases the price of natural carotenoids. This problem can be resolved by selection and screening of high carotenoids producing microalga strain. The proper cultivation conditions, growth-influencing conditions, and metabolic modifications can enhance the production of carotenoids. The production cost is also reduced by designing a low-cost cultivation system. The low-cost biorefinery may reduce the market price of carotenoids, and humans can use as daily food supplements. In this below section, we describe the biosynthesis of carotenoids in microalgae cells, carotenoid enhancing conditions, and metabolic engineering of microalgae strains.

6.3.1 Bio-Formation of Carotenoids inside Microalgae/Carotenogenesis inside Microalgae Cells

Carotenoid is constituent of terpenoids and the primary carotenoid is present inside the microalgae cells, mainly in a photosynthetic apparatus, like chloroplast or thylakoids, whereas secondary carotenoid is present inside lipid vesicles, the stroma of plastids, and in the cytosol of microalgae. Mainly, in algae, carotenoids help the photosynthesis process and protect it from different stress conditions. It also increases the absorption spectrum of photosynthetic pigments. The biosynthesis of a wide range of carotenoids is followed in the IPP pathways. Generally, xanthophyll is synthesized in the chloroplast with the help of many nuclear genes and is then exported to other cellular compartments, whereas astaxanthin is accumulated inside the cytoplasm of microalgae (León et al. 2007; Bonet et al. 2016). Primary carotenoid is an essential pigment present as in the photosynthetic apparatus, and helps in the process of photosynthesis. It is a structural part of microalgae cells, whereas secondary carotenoids are synthesized in stress conditions by the carotenogenesis process. The carotenogenesis process is a complex process, as several genes and enzymes are involved in the synthesis of carotenoid. In general, carotenogenesis follows IPP pathways, and two different pathways are found in carotenoid synthesis in organisms: mevalonate pathways that occur in the cytosol of organisms and the nonmevalonate 1-deoxy-D-xylulose-5-phosphate pathway in the chloroplast (DOXP pathway or MEP pathway) (Lichtenthaler et al. 1997). In microalgae, the biosynthesis of carotenoid may vary from species to species, but it follows the common carotenogenesis pathway (IPP). IPP is a fundamental intermediate in isoprenoid biosynthesis, or its isomer dimethylallyl diphosphate (DMAPP) acts as the precursor for the synthesis of carotenoid in microalgae. The carbon-containing precursor, DMAPP and IPP, are synthesized from glyceraldehyde-3-phosphate (G3P) and pyruvate following methylerythritol 4-phosphate (MEP) pathway or either by acetyl CoA following the mevalonic pathway (MVA). Many reports suggest that microalgae and cyanobacterial cells follow the MEP pathway for the synthesis of intermediate precursor IPP and DMAPP (Zhao et al. 2013). After the synthesis of intermediate precursor IPP, it isomerized into DMAPP, and later it is formed as a polyprenyl pyrophosphate chain by the condensation process with the help of the prenyltransferase enzyme. Condensation is initiated from head to tail that leads to the formation of polyprenyl pyrophosphate. The C5 IPP is converted into polyprenyl pyrophosphate (C10, C15, and C20) with the help of enzyme prenyltransferase. This pyrophosphate chain forms geranylgeranyl pyrophosphate (GGPP), which is an essential molecule for carotenoid production (Takaichi 2011). Two GGPP are condensed, and form phytoene (C40 molecules) with the help of phytoene synthase (PSY) enzyme. It causes rate-limiting steps of carotenoid synthesis (Jin et al. 2003; Takaichi 2011; Ambati et al. 2014). Phytoene is then converted into F-carotene, with the help of phytoene desaturase. The F-carotene is further desaturated with the help of F-carotene desaturase enzymes and then forms pro lycopene. The pro lycopene is isomerized and forms lycopene with the help of highly specific enzyme carotenoid isomerase. Then lycopene is split into two sub-pathways, in one sub-pathway α-carotene is produced, and in another sub-pathway β-carotene is produced. For the synthesis of α-carotene, lycopene is catalyzed by β-cyclase and e-cyclase enzyme. Then this α-carotene is hydroxylated by two enzymes: carotene β-hydroxylase and carotene e-hydroxylase and forms lutein. The fortunate of lycopenes are subjected to activities or β-cyclase and ε-cyclase, whereas another sub-pathway is the β-carotene synthesis pathway. The β-carotene is synthesized with the help of lycopene β-cyclase. This β-carotene is a form of zeaxanthin created by the hydroxylation process. This hydroxylation of β-carotene is processed by the help of enzyme carotene β-hydroxylase. This zeaxanthin is converted into violaxanthin by epoxidation reaction. The epoxidation reaction depends on the presence of oxygen and cofactors, such as NADPH, ferredoxin, and flavin adenosine dinucleotide (FAD) (Varela et al. 2015). This is basic pathways that are generally followed by all microalgae species. However, some microalgae strain, such as *C. zofingiensis*, *Scendesmus* sp., and *H. pluvialis*, can accumulate a rare carotenoid that contains both OH^- and oxygenated groups, e.g., astaxanthin (Varela et al. 2015; Chakdar and Pabbi 2017). Generally, astaxanthin is synthesized from violaxanthin and zeaxanthin by the help of β-carotene ketolase. But *H. pluvialis* can produce astaxanthin via two pathways: one by violaxanthin and zeaxanthin, and another pathway that is the conversion of β-carotene into echinenone with the help of β-carotene ketolase, and then echinenone is converted into canthaxanthin with the help of β-carotene ketolase, and adonirubin, and then canthaxanthin is converted into highly valuable compound astaxanthin with the help of β-carotene hydroxylase (Lemoine and Schoefs 2010; Han et al. 2013; Saini et al. 2020). The overall synthesis of carotenoid in microalgae is explained in Figure 6.2.

6.3.2 Potent Microalgae Strain for Carotenoid Production

The strain *Chlorella* sp. are used for the production of food supplements, such as PUFA (polyunsaturated fatty acid), protein, and carbohydrates, but carotenoids are commercially produced using many potent microalgae strains, such as *Haematococcus* sp., (astaxanthin), *D. salina* (β-carotene), *Chlorella protothecoides* (lutein), and *C. zofingiensis* (astaxanthin) (Del Campo et al. 2000; Ip and Chen 2005; Ahmed et al. 2015; Sathasivam et al. 2019). The microalgae strain has the potential for production of carotenoids, while some microalgae are used commercially (Goswami et al. 2010; Batista et al. 2012) and some microalgae strain are produced using the minimum amount of carotenoids, however, the productivity rate can be enhanced by optimizing their growth conditions or providing some stress, which enhances the productivity. The brief details of potent microalgae are discussed in Section 6.3.2.1–6.3.2.3. Figures 6.3 and 6.4.

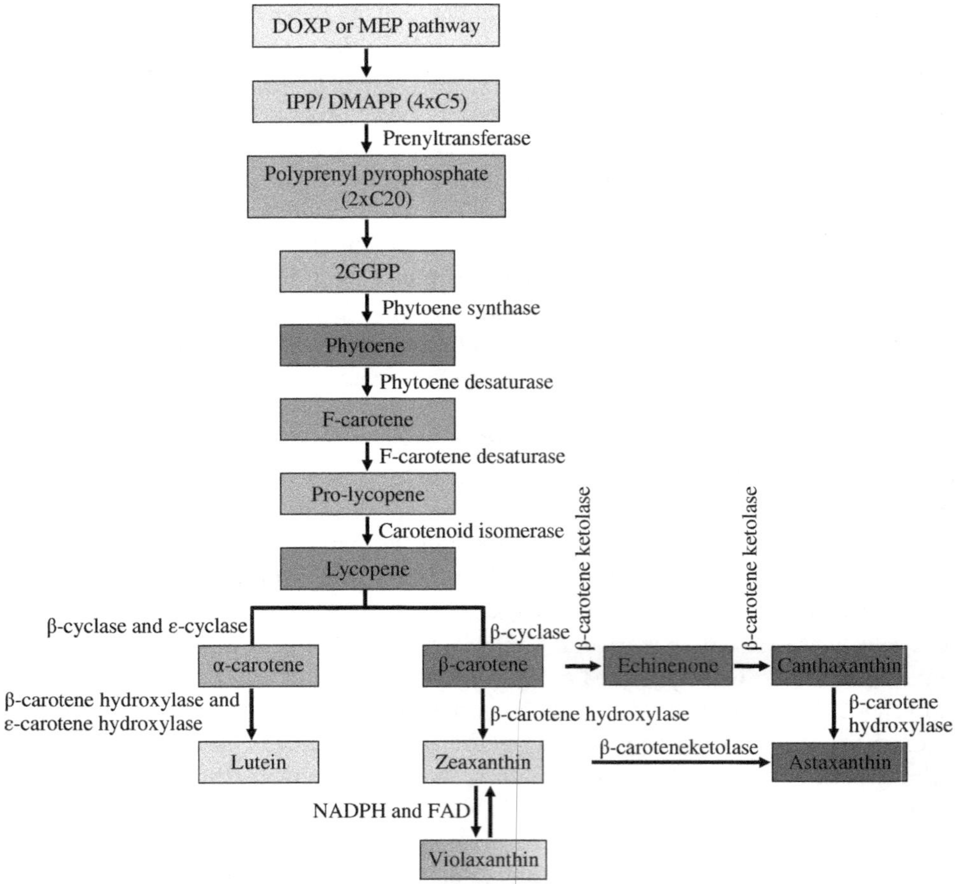

Figure 6.2 Systematic representation of carotenogenesis process of β-carotene, lutein, and astaxanthin. Sources: Varela et al. (2015); Chakdar and Pabbi (2017); Lemoine and Schoefs (2010); Han et al. (2013); Saini et al. (2020).

6.3.2.1 *Haematococcus pluvialis*

It is small unicellular, motile, thin-walled, flagellated microalgae used for commercial production of astaxanthin, lutein, and other carotenoids. *H. pluvialis* contains the highest amount of astaxanthin inside their cells (Liu et al. 2018). It follows both astaxanthin synthesis pathways: (violaxanthin and zeaxanthin) and (canthaxanthin) (Pelah et al. 2004). Astaxanthin accumulation in *Haematococcus* sp. occurs in two stages: first stage (green stage), where it is grown and increases the biomass, and then second stage (astaxanthin accumulating stage), due to environmental stress conditions, the green stage or flagellated stage is converted into a red, nonflagellated, resting stage where their motility is reduced and aplanospores are formed; this aplanospores contain astaxanthin. *Haematococcus* sp. contains 4–5% of astaxanthin in their total dried biomass (Boussiba et al. 1999). By condition optimization, genetic modification in *Haematococcus* sp. can enhance the production of astaxanthin (Lemoine and Schoefs 2010; Han et al. 2013; Saini et al. 2020).

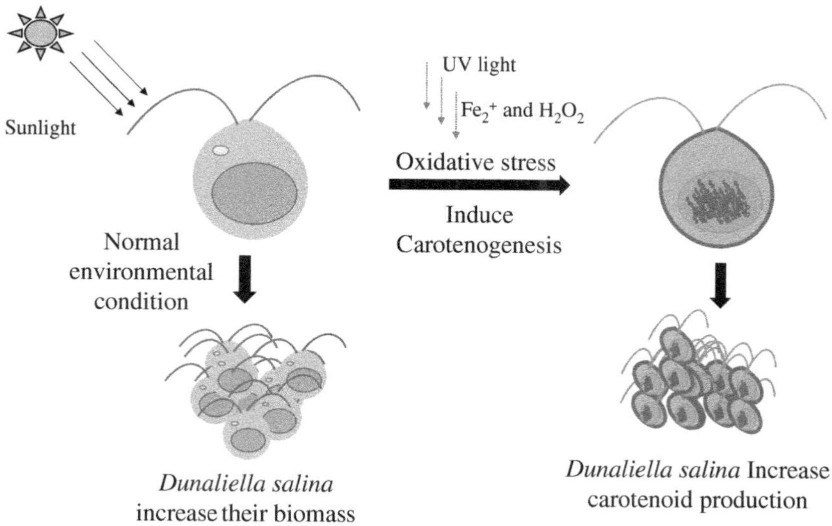

Figure 6.3 Shows how the oxidative stress induces the carotenogenesis in microalgae *Dunaliella salina* for β-carotene production. Source: Based on Mojaat et al. (2008).

6.3.2.2 *Dunaliella salina*

Dunaliella salina is a green, halophilic, phototrophic, biflagellate microalgae, used for the commercial production of β-carotene. It contains a high amount of β-carotene inside its cells. Their dry biomass contains 10–14% of β-carotene (Sathasivam et al. 2012). It produces both cis or trans type of β-carotene. In normal growth conditions, *D. salina* divides and increase their biomass, but during stress condition, their physiological balance disbalances and releases β-carotenes in order to neutralize stress condition. The β-carotene can be enhanced by genetic modification of *D. salina* sp. genes, optimizing their growth parameters (Anila et al. 2016). Several reports suggest that environmental stress conditions such as high salinity, extreme temperatures, and a high light intensity can help lead to an accumulation of β-carotene inside the *D. salina* cells (as shown in Figure 6.3) (Borowitzka et al. 1990; Ahmed et al. 2015; Zhu et al. 2020).

6.3.2.3 Other Microalgae Species Used for the Production of Carotenoids

Many microalgae have had an ability for production of carotenoid, and some are used for commercial products such as *D.salina* and *Haematococcus* sp. (Goswami et al. 2010; Batista et al. 2012; Sathasivam and Ki 2018; Sathasivam et al. 2019), but there are several microalgae species also produce a large amount of carotenoid inside their cells, but their commercial production has not yet started. *Chlamydomonas zofingiensis*, *Muriellopsis* sp., and *Scenedesmus* sp. are reported for the production of lutein and are still not commercialized. These organisms are already used for the production of lutein in a pilot scale (Del Campo et al. 2000; Blanco et al. 2007; Cordero et al. 2011), but their production process is not suitable for an industrial level. Table 6.1 shows the different kind carotenoids produced by microalgae and their functions.

Table 6.1 Different kinds of carotenoid produced by various type of microalgae and their function.

	Microalgae	Function of carotenoid	References
Astaxanthin	*Scenedesmus* *Chlorococcum* *Haematococcus pluvialis* *Chlorella zofingiensis* *Chromochloris zofingiensis*	1. High antioxidant activities than other carotenoid 2. Having anticancerous activity 3. Feed supplement 4. Generate immune response	Pirastru et al. (2012) Yuan et al. (2002) Pelah et al. (2004) Minyuk et al. (2020)
β-carotene	*Dunaliella salina* *Chlorococcum* sp. *Tetraselmis sp.* *Dunaliella bardawil* *Dunaliella salina* GY-H13 Dunaliella salina V-101	1. Food, pharmaceutical, and cosmetic industries 2. Precursor of provitamin A 3. Antioxidants and anticancerous properties 4. Generates immune response 5. Coloring agent.	Hejazi et al. (2004) Raja et al. (2007) Borowitzka et al. (1990) Yuan et al. (2002) Ahmed et al. (2015) Mogedas et al. (2009) Zhu et al. (2020) Anila et al. (2016)
Lutein	*Muriellopsis* sp. *Chlorella sorokiniana* *Chlorella zofingiensis* *Scenedesmus almeriensis* *Chlorococcum* Chlamydomonas sp. *JSC4* *Scenedesmus* sp. UFPS01	1. Food dyes 2. Feed addictive 3. Enhanced pigmentation in chicken and egg 4. Decrease the rate of cataract or eye-related problem	Del Campo et al. (2001) Cordero et al. (2011) Sánchez et al. (2008) Wei et al. (2008) Yuan et al. (2002) Ma et al. (2020) Contreras et al. (2020)
Canthaxanthin	*Scenedesmus sp.* *Dactylococcus. Dissociatus MT1* *Chlorococcum* *Chlorella zofingiensis* *Chromochloris zofingiensis* *Scenedesmus komareckii* *D. salina*	1. Food dye 2. For coloring chicken skin and egg yolks 3. Nutraceuticals foods with anticarcinogenic and antioxidant properties	Pirastru et al. (2012) Grama et al. (2014) Yuan et al. (2002) Pelah et al. (2004) Minyuk et al. (2020) Hanagata and Dubinsky (1999)

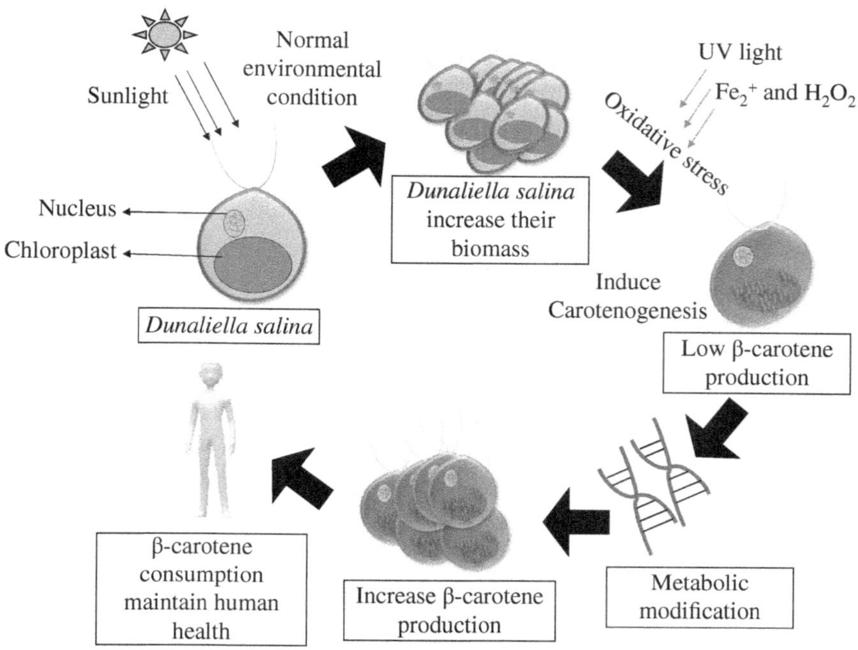

Figure 6.4 Systematic representation of overall process of β-carotene production. Sources: Sandesh Kamath et al. (2008); Ahmed et al. (2015); Anila et al. (2016).

6.3.3 Enhancement of Carotenoid Productivity by Optimizing Various Physiological Condition/Physiological Approaches for Enhancement of Carotenoid Production inside Microalga Cells

As per the study, several microalgal strains can produce carotenoids, but their production rate is low, and their synthesis of carotenoid inside cells is slow. However, commercial production requires a maximum amount of carotenoid synthesis by the microalgae strain. The commercial production of carotenoid requires selection and screening of proper microalgal strains (Del Campo et al. 2007; Sathasivam et al. 2019; Contreras et al. 2020). After selection, appropriate media is required for their growth and carotenoid production. Then different stress is provided to algae cells for the accumulation of carotenoids. The carotenoid production can be enhanced by optimizing the growth parameters. The effect of growth parameters and stress conditions for carotenoid production is shown in Figure 6.4 and discussed Section 6.3.3.1–6.3.3.5.

6.3.3.1 Role of Nutrient Deficient Stress for Carotenogenesis

The carbon (C) and nitrogen (N) are an important source for the development and growth of microalgae cells. The CO_2 is naturally utilized by microalgae from the atmosphere, or externally provided by different sugar components, such as glucose and sodium acetate (Bhatnagar et al. 2011), whereas nitrogen is provided as a nitrate form in media. Nitrogen is an essential nutrient for the synthesis of nucleic acids, proteins, and biologically important pigments and helps in the regulation of general metabolic pathways of microalgae cells. The C/N ratio in the media can enhance the production of pigment inside microalgae cells

(Contreras et al. 2020). Providing external C sources in media can enhance the astaxanthin production of astaxanthin, whereas in the absence of lights, citrate, malate, and pyruvate enhances the production of astaxanthin in *C. zofingiensis* (Hu et al. 2018). In the N depletion condition, *C. zofingiensis* and *Coelastrella* sp. synthesizes high amounts of canthaxanthin inside their cells (Pelah et al. 2004; Abe et al. 2007), whereas several reports suggest that low N content can reduce the lutein accumulation inside microalgae (Guedes et al. 2011). The absence of N in media creates stress conditions for microalgae, as many reports suggest that nitrogen stress conditions can increase the accumulation of carotenoid inside the microalgae cells, e.g., N depletion condition can enhance the accumulation of β-carotene is *D. salina* (Pisal and Lele 2005). The combination of C/N ratio are directly correlating the growth and development and accumulation of carotenoid inside the algae cells, so proper concentration of C and N should be optimized for production of carotenoids.

6.3.3.2 Lights and Temperature Stress for Induction of Carotenogenesis

Light and temperature are essential factors for the growth of microalgae as well as accumulation of carotenoids. Several reports suggest that light colors and intensity-based stress can enhance the production of different kinds of carotenoids inside microalgae cells. *C. zofingiensis*-enhanced astaxanthin production under light stress condition (Pelah et al. 2004). In continuous light supply to *H. pluvialis* culture, it can enhance the accumulation of astaxanthin (Domınguez-Bocanegra et al. 2004). But the high light intensity, also decreased the carotenoid accumulation in *Chlamydomonas reinhardtii*. The continuous light intensity may denature the carotenoid synthesis enzymes or cause photoinhibition in microalgae cells. However, the high light intensity is used in the generation of stress to microalgae cells for enhancement of lutein. High temperature can accumulate high content of lutein inside *D. salina*, but it also decreases the biomass productivity. Temperature shows the major effect on *D. salina* for an accumulation of β-carotene production. The optimum temperature for β-carotene synthesis in *D. salina* is 24–29 °C (Bhosale 2004; Lamers et al. 2010). Several reports suggest that high temperature can denature the enzyme involved in the synthesis of carotenoids. The temperature and light are related to production and accumulation of carotenoid in microalgae species; however, this parameter can be maintained by visualizing the optimum light intensity and temperature.

6.3.3.3 Role of Oxidative Stress in Carotenogenesis

Oxidative stress is an environmental stress condition provided to microalgae for an accumulation of a different kind of carotenoid inside microalgae. The oxidative stress is created through using different salts or chemicals, such as Fe_2^+ or ferrous salt, or by photooxidation (Mojaat et al. 2008). The ferrous salt responsible for the formation of hydroxyl radicals in microalgae (Hu et al. 2018). Ferrous salt-based oxidative stress is used for accumulation of canthaxanthin in *C. zofingiensis,* but this approach is not appropriate for an accumulation of astaxanthin (Pelah et al. 2004). According to the report of Ip and Chen (2005), during the stress conditions, *C. zofingiensis* generates hydroxyl ions, which enhanced the astaxanthin accumulation. Fe_2^+-based oxidative stress is useful for an accumulation of β-carotene in *Dunaliella salina* (Mojaat et al. 2008). The Fe_2^+-based oxidative stress also induces the synthesis of lutein in microalgae (Mogedas et al. 2009; Fernández-Sevilla et al. 2010; Guedes et al. 2011). Another method for the generation of oxidative stress is photooxidation. Photooxidation is generated by high light irradiance (UV), and this photooxidation-based stress can produce free active oxygen molecules; this generates oxidative stress, resulting in its trigger the carotenogenesis process, which enhances the synthesis of astaxanthin. Other than Fe_2^+, H_2O_2 is also used for the generation of oxidative stress. H_2O_2 generates hydroxyl radicals, which induced the carotenogenesis process in *C. zofingiensis,* and *Chlorococcum,* with the resultant increase of the synthesis of secondary carotenoids, such as astaxanthin. The combination of NaOCl, Fe_2+, and H_2O_2 generates free radical or reactive oxygen, which provides the oxidative stress to microalgae, and induces their carotenogenesis pathways, which results in its enhanced synthesis of lutein, whereas Fenton reaction, which is generated by Fe_2+ and H_2O_2 are induced or enhanced in the production of astaxanthin in the microalgae *H. pluvialis* (Hu et al. 2018).

6.3.3.4 Approaches which Enhance Carotenogenesis by Heterotrophic and Mixotrophic Cultivation of Microalgae

Heterotrophic cultivation and mixotrophic system are used for the production of microalgal biomass, which enhances carotenoid production. In heterotrophic cultivation (dark condition) different C sources are used, such as glucose, acetate, glycerol, and other C sources (Bhatnagar et al. 2011), whereas in a mixotrophic cultivation system, both light and carbon source is provided externally. In the absence of light (heterotrophic), microalgae utilize this organic C source and increase their biomass or enhance carotenoid production. However, in the presence of light, microalgae fix naturally carbon by photosynthesis, while in absence of light, it utilizes organic carbon present in media (Zhan et al. 2017), and in heterotrophic cultivation, can enhance lutein productivity (Shi et al. 2000). But several reports suggest that some microalgae are not compatible in a heterotrophic mode of cultivation. but that it can compatible for mixotrophic mode. Heterotrophic cultivation systems, especially, are not compatible for *D. salina* growth, for accumulation of β-carotene, however, it shows a maximum yield of β-carotene in mixotrophic cultivation (Morowvat and Ghasemi 2016), whereas *H. pluvialis* can grow in both heterotrophic conditions and mixotrophic conditions and synthesize the astaxanthin (Dufossé 2009). While *Chlorella pyrenoidosa* also grew in heterotrophic condition and enhances the lutein production (Shi et al. 2000), *C. zofingiensis* also produces and accumulates astaxanthin in heterotrophic conditions by utilizing glucose as a carbon source (Ip and Chen 2005). The heterotrophic and mixotrophic mode of cultivation is a good approach to the production of carotenoid, but it depends on the growth compatibility of microalgae in both conditions.

6.3.3.5 Cohesive Cultivation System in Microalgae for Enhancement of Carotenoid

Cohesive cultivation or coculture cultivation or mutual cultivation techniques are a good approach for the cultivation of microalgae and accumulation of carotenoids. In these, two or more types of organisms are grown simultaneously in the same culture or reactor and show mutual benefits to each other, e.g., bacteria (*Agrobacterium aurantiacum)* (Dufossé 2009) and microalgae can grow in the same bioreactors, whereas microalgae possess photosynthesis and provide oxygen to bacteria cells. This bacterium utilizes oxygen for their growth and releases CO_2 as a by-product, and this CO_2 can be captured by microalgae for their growth (Sun et al. 2018). The bacteria also induces the different stress conditions to microalgae, which may enhance the accumulation of carotenoids in microalga cells. These mutual benefits may reduce the cultivation process costs and this approach can be used for large-scale production of carotenoids. However, it also shows the negative effect on microalga, or may produce some toxic metabolites. Therefore, proper microorganisms selection is required for cohesive cultivation system.

6.3.4 Metabolic and Genetic Modification in Microalgae for Enhancement of Carotenoid Production

Carotenoid production from microalgae is a good approach, and several microalgae strains are also used for commercial-scale production. Though many microalgae species produces a low amount of carotenoid, as a result, alteration in the physiological approaches is used for enhancement of carotenoid in microalgae, whereas genetic or metabolic modification is a good approach for enhancement of carotenoid production. For metabolic modification, different approaches are used. The genetic engineer is the most prominent technology, which can customize the genes and enhance the function of particular microalgal genes (Cordero et al. 2011; Ahmed et al. 2015; Anila et al. 2016). Several microalga species such as *Chlorella sorokiniana* (*C. sorokiniana*), *D. salina*, and *H. pluvialis*, etc., genes are modified to enhance the production of highly valuable biomolecules (Sandesh Kamath et al. 2008; Cordero et al. 2011; Anila et al. 2016). Different mutagenic techniques are used for enhanced carotenoid in microalgae cells; random mutagenesis is a technique used for enhancement and selection of the highest pigment production microalgae strain. This approach is more convenient than other approaches. In this, UV rays or other radiation are used as a mutagen. Random mutagenesis in *C. sorokiniana* and *H. pluvialis* enhance the lutein and astaxanthin production (Cordero et al. 2011; Gómez et al. 2016). Another approach such as adaptive laboratory evolution is used in *Phaeodactylum tricornutum*, which enhances the carotenoid content. In adaptive evolution techniques, the high light intensity is generated by using a light-emitting diode (LED). This LED generates oxidative stress, which leads to enhancement of pigments (Yi et al. 2015). However, in microalgae, metabolic modification is induced by using genetic engineering and in this, mainly modifies the protein or enzyme, which is involved in the synthesis of carotenoid. Two genes are generally modified, which are associated with β-carotene to astaxanthin production. β-carotene hydroxylase enzyme converts β-carotene to zeaxanthin and β-carotene ketolase, which converts zeaxanthin into astaxanthin (Breitenbach et al. 1996).

Genetic engineering in metabolic pathways involves an up- and downregulation of transcription and translation factors (Vickers et al. 2014; Stephens et al. 2015; Saini et al. 2020). Genetic modification can enhance the pigment's production, but it also creates an overproduction of pigment, which causes feedback inhibition, for avoiding the feedback inhibition requires simultaneously avoiding the pigment transport mechanism. In this, synthesized carotenoid are immediately transported into storage sites and avoid the feedback inhibition. According to Lemoine and Schoefs (2010), phytoene desaturase is an enzyme that converts phytoene into F-carotene, (rate-limiting steps of carotenogenesis). Modification or upregulation of phytoene desaturase can enhance the synthesis of carotenoid. Genetic modification of phytoene desaturase in *C. zofingiensis* can enhance 32.1% of total carotenoids, whereas in *Hematococcus pluvialis*, it enhances 26% of astaxanthin (Steinbrenner and Sandmann 2006; Liu et al. 2014; Saini et al. 2020). However, modification in single enzymes cannot enhance the total carotenoid production. So, many modifications are required for the enhancement of carotenoid, and the coordination of different enzymes can enhance the overall production of carotenoids. Genetic modification in microalgae also depends on which types of carotenoids are required. For example, overproduction of β-carotene requires downregulation of the LCYE gene, which results in the inhibition of α-carotene production. So genetic engineering or metabolic modification is an important tool for enhancement of carotenoid production from microalgae, however, more work is required in this area to accomplish the highest carotenoid productivity from microalgae (Varela et al. 2015; Bajhaiya et al. 2016, 2017; Saini et al. 2020), and different carotenogenesis induction approaches in microalgae for enhancement of carotenoid is shown in Figure 6.4 and summarized in Table 6.2.

6.4 Significance of Carotenoid in Human Health

Carotenoid has many medicinal benefits on human health, and it can be used as food supplements, medicine, or as pharmaceutical drugs. Different kinds of carotenoid have different medicinal properties (Goswami et al. 2010; Batista et al. 2012; Sathasivam et al. 2019). Microalgal-based natural carotenoid shows more positive effects on human health. They are available in markets and sold as dietary supplements. Majorly, carotenoids show antioxidant activities, antiaging, and anticancerous activity. Thus, Section 6.4.1–6.4.4 defines the significance of carotenoid in human health.

6.4.1 Anti-Inflammatory and Antioxidant Properties

All kinds of carotenoid show antioxidant activity, the microalgae-based carotenoid is scavenging the free radical present inside the human body. Astaxanthin shows robust antioxidant activities against free radical or oxidative stress. This oxidative stress is responsible for causing cancers, inflammatory, neurodegenerative diseases, and other metabolic diseases. This oxidative stress is also responsible for aging on human. So, β-carotene, astaxanthin, phytoene, lutein, and other carotenoid have strong potential for reduction of oxidative stress (Del Campo et al. 2007; Guedes et al. 2011; Sathasivam et al. 2019). Astaxanthin shows 500 times more antioxidant compared to another carotenoid or tocopherol (Cysewski and Lorenz 2004), whereas β-carotene also shows a positive effect on human health, and it can protect humans by UV-induced erythema (Heinrich et al. 2003), as well as the generation of oxidative stress can cause major inflammatory disease in humans. Carotenoid has the potential for the treatment of inflammatory-based diseases. It shows anti-inflammatory activity and activates the immune system (Sathasivam and Ki 2018; Sathasivam et al. 2019).

6.4.2 Anticancerous Activity and their Potential of a Generation of an Immune Response

Cancer is a leading disease causing death in humans. Several approaches are used for treatment of this deadly disease. However, their treatment is costly as well, and its shown to be negative effect on the human. Carotenoid consumption has many anticancerous activities

Table 6.2 Different methods of carotenogenesis induction in microalgae for carotenoid production.

Microalgae	Type of carotenoids	Carotenogenesis induction method	References
Dunaliella salina	Neoxanthin violaxanthin	UV-C radiation and plant hormones	Ahmed et al. (2015)
Tetraselmis suecica	violaxanthin	UV-C radiation and plant hormones	Ahmed et al. (2015)
Dunaliella salina	β-carotene	Salinity stress	Borowitzka et al. (1990)
Chlorella sorokiniana	Lutein	Random mutagenesis and effect by physiological factor, such as irradiance, temperature, nitrogen salinity	Cordero et al. (2011)
Muriellopsis sp.	Lutein	Irradiance, temperature, and salinity	Del Campo et al. (2000)
Dactylococcus. Dissociates MT1	Canthaxanthin	Osmotic stress, oxidative stress, and nitrate starvation	Grama et al. (2014)
Chlorococcum	Astaxanthin, Lutein	Glucose supplementation	Yuan et al. (2002)
Chlorella zofingiensis	Canthaxanthin astaxanthin	Salt stress	Pelah et al. (2004)
Chromochloris zofingiensis	Astaxanthin and canthaxanthin	Nitrogen stress	Minyuk et al. (2020)
Chlamydomonas sp. JSC4	Lutein	Temperature	Ma et al. (2020)
Dunaliella salina GY-H13	B carotene	Nitrogen deficiency and transcription activation of β-carotene biosynthetic gene	Zhu et al. (2020)
Scenedesmus sp. *UFPS01*	Total carotenoid	C/N ratio	Contreras et al. (2020)
C. zofingiensis	Canthaxanthin	Nitrogen depletion	Pelah et al. (2004), Abe et al. (2007)
Chlorella sorokiniana	Lutein	Random mutagenesis	Cordero et al. (2011)
Dunaliella salina V-101	β-carotene	Genetic modification in *bkt* gene	Anila et al. (2016)
Hematococcus pluvialis	Astaxanthin	Mutation by 1-methyl 3-nitro 1-nitrosoguanidine and ethyl methane sulphonate	Sandesh Kamath et al. (2008)
Hematococcus pluvialis	Astaxanthin	Transformation of phytoene desaturase	Steinbrenner and Sandmann (2006)

and their daily consumption can reduce the chances of cancer in humans. As per reports of the National Cancer Institute, the consumption of β-carotene shows an anticarcinogenic effect in humans (Priyadarshani and Rath 2012). The consumption of carotenoids can also reduce the free radicals of oxygen that are responsible for causing cancer in the human body. Some reports states that the high UV rays can generate photooxidation, which shows a negative effect in human skin that' leads the causing of skin cancers. The carotenoid consumption reduces the photooxidation of skin, and reduces the chance of skin cancers. Carotenoid has many medicinal benefits; several researchers report that microalgae-based carotenoid, especially astaxanthin, has potential for generation of an immune response. Their strong antioxidant and anti-inflammatory properties have the potential of increasing the action of immunoglobin A, G, and M, and help in the production of t-cell antibodies (Cysewski and Lorenz 2004; Perusek and Maeda 2013; Sathasivam and Ki 2018; Sathasivam et al. 2019).

6.4.3 As Provitamin

β-carotene is a synthesis from *Dunaliella salina*, and it is a type of carotene that shows antioxidant activities. Besides the antioxidant activities, it also a provitamin (Novoveská et al. 2019). The main function of a provitamin is that it can be converted into vitamin A, and the main function of vitamin A is in reducing eye problems. Vitamin A is required for good vision and it is present inside the eye; this vitamin A is responsible for the synthesis of rhodopsin (an eye pigment). This rhodopsin is responsible for good vision, maintaining the mucous in eyes. So, β-carotene can work as provitamin and its consumption can reduce eye problems (Perusek and Maeda 2013).

6.4.4 Other Significance of Microalgae Carotenoids

Carotenoids have many advantages in human health, and their strong antioxidant activities can make it a valuable compound. Due to their medicinal potential, the human can consume them as food supplements, as nutraceutical food products, or for a cosmetic purpose. The carotenoid-based cosmetics have many benefits, such that it reduces the early aging problem, adds glow to the skin by reducing the oxidative stress of free radicals. It protects the skin from photooxidation (Hu et al. 2018; Sathasivam et al. 2019). Other than for humans, carotenoid is used for aquaculture, poultry, and in the food industry. In the food industry, different kinds of carotenoids are used as coloring agents. Carotenoid are natural coloring agents, and they have different color shades, such as reds, oranges, and yellows. The benefits of using carotenoid as coloring additives are that it can add to the coloring of food supplements and also provide health benefits (Cysewski and Lorenz 2004). The carotenoid can be used as high-grade animal food and feedstocks (Yaakob et al. 2014).

6.5 Opportunities and Challenges in Carotenoid Production

Naturally, carotenoid is a promising candidate for the improvement of health. It has several health benefits that make it a nutraceutical food product. Due to its health benefits, it has been consumed as food supplements by humans, thereby increasing the demand for natural-based carotenoid. Microalgae produces different kinds of carotenoids. Due to their photosynthetic efficiency, carotenoid synthesis efficiency, adaption efficiency, and fast biomass productivity rate make it a promising candidate for carotenoid production. But only a few microalgae species are capable of producing the high amount of carotenoid, such as *D. salina* (β-carotene) and *H. pluvialis* (astaxanthin) (Goswami et al. 2010; Batista et al. 2012; Sathasivam and Ki 2018; Sathasivam et al. 2019). Due to their high carotenoid producing capabilities they have been reported as a promising candidate for upcoming commercial production, and several countries have already used it for commercial production. However, microalgae-based lutein and other types of carotenoid are still not produced on a large scale. *Chlorella* sp. and *Dunaliella* sp. have been successfully used for the production of all types of carotenoid on the laboratory scale (Abdelaziz et al. 2013; Minyuk et al. 2020). The major limitation of using microalgae is their biorefinery process as well as low carotenoid production efficiency. The extra biorefinery steps are required (saponification) for extraction of carotenoid (Sun et al. 2018). This causes a limitation of carotenoid production, but this limitation can be overcome by the selection of appropriate microalgae, optimizing their growth media culture and condition, their use of low-cost cultivation system, and modification of their metabolic pathways by using genetic engineering (Steinbrenner and Sandmann 2006). These approaches provide an opportunity for the use of microalgae to large-scale production of microalgae. By using this approach, several (*C. zofingiensis* and *Muriellopsis* sp.) (Del Campo et al. 2001; Pelah et al. 2004) microalgae species are ready to produce pilot-scale production of lutein. Wastewater and atmospheric CO_2 can be used as a nutrient source for microalgae cultivation and can reduce the overall operational cost of carotenoid production.

6.6 Present Drifts and Future Prospects

Presently, carotenoid consumption is increasing, and several synthetic- and natural-based carotenoids are available on the market. The carotenoid markets increase day by day. US, Asian, and European countries have already set their carotenoid markets. According to reports, the United States is leading other countries in carotenoid production, but Asian markets are also increasing their carotenoid production rate (Guedes et al. 2011). Mainly three types of carotenoid are present in the market, which is commercially used for human consumption, that is, astaxanthin, β-carotene, and lutein. The lutein is produced from using marigold flowers (Guedes et al. 2011), whereas β-carotene and astaxanthin are produced by using microalgae *D. salina* and *H. pluvialis* (Priyadarshani and Rath 2012; Ambati et al. 2014; Leu and Boussiba 2014; Sathasivam et al. 2019). The astaxanthin is a high-value nutraceutical carotenoid and the market price is approximately ($15 000/kg pigment), which is mostly produced in China, Israel, and Chile (Leu and Boussiba 2014). The growing market demand for carotenoid may increase to USD 1.53 trillion in 2021. The use of β-carotene increases year by year. Presently, β-carotene production rate is 1088 621.69 kg tons per year. The market price increases day by day, and that effects the sale price, which ranges between USD 300/kg to USD 3000/kg. The sale price is dependent on quality, type, and concentration of products (Priyadarshani and Rath 2012; Contreras et al. 2020). However, the minimum production of natural carotenoid can increase the market price. So, microalgae-based carotenoid production is a good approach for the large-scale production of carotenoid. But their large-scale production technology requires proper modification, which can enhance the production rate of carotenoid in microalgae as well in low operational cost. This carotenoid production approach may increase the carotenoid production from microalgae and reduce the market price of natural carotenoid. Whereas astaxanthin and β-carotene are already available in the market, microalgal-based lutein production is not economically viable. However, various species are used for large-scale production, but they are not as efficient. *Scenedesmus almeriensis* and *C. zofingiensis* need genetic modifications of their metabolic pathways, which can enhance lutein production (Del Campo et al. 2001; Yuan et al. 2002; Sánchez et al. 2008; Wei et al. 2008; Cordero et al. 2011; Ma et al. 2020).

6.7 Conclusion

Microalgal-based natural carotenoid shows superior response over synthetic carotenoids; their strong antioxidant, anti-inflammatory, anticarcinogenic properties make a promising applicants for production on the large scale of various kind of carotenoids. Some types of carotenoid (β-carotene and astaxanthin) production are already established on a commercial scale and are available on the market. However, lutein and other carotenoid production using microalgae have not been established for large-scale production, as the low productivity of carotenoids make the process costly. So, for large-scale production of lutein and other carotenoids, several things are required, including proper selection of highly producing lutein strain, optimization of media or culture condition, and genetic modification in their metabolic pathways. The use of these approaches may enhance lutein and other carotenoid production in microalgae, and it may help establish the large-scale production of carotenoid. However, low-cost cultivation system by using wastewater and atmospheric CO_2 also reduces the operation cost of carotenoid production from microalgae and this low operation cost process may reduce the market price of astaxanthin and β-carotene.

References

Abdelaziz, A.E.M., Leite, G.B., and Hallenbeck, P.C. (2013). Addressing the challenges for sustainable production of algal biofuels: II. Harvesting and conversion to biofuels. *Environ. Technol.* 34: 1807–1836.

Abe, K., Hattori, H., and Hirano, M. (2007). Accumulation and antioxidant activity of secondary carotenoids in the aerial microalga *Coelastrella striolata* var. multistriata. *Food Chem.* 100: 656–661.

Agrawal, K., Bhatt, A., Bhardwaj, N., Kumar, B., and Verma, P. (2020). Algal biomass: potential renewable feedstock for biofuels production-part I. In: *Biofuel Production Technologies: Critical Analysis for Sustainability*, Springer, Singapore, 203–237.

Ahmed, F., Fanning, K., Netzel, M., and Schenk, P.M. (2015). Induced carotenoid accumulation in *Dunaliella salina* and *Tetraselmis suecica* by plant hormones and UV-C radiation. *Appl. Microbiol. Biotechnol.* 99: 9407–9416.

Ambati, R.R., Moi, P.S., Ravi, S., and Aswathanarayana, R.G. (2014). Astaxanthin: sources, extraction, stability, biological activities and its commercial applications - a review. *Mar. Drugs.* 12: 128–152.

Anila, N., Simon, D.P., Chandrashekar, A. et al. (2016). Metabolic engineering of *Dunaliella salina* for production of ketocarotenoids. *Photosynth. Res.* 127: 321–333.

Bajhaiya, A.K., Dean, A.P., Driver, T. et al. (2016). High-throughput metabolic screening of microalgae genetic variation in response to nutrient limitation. *Metabolomics.* 12: 9.

Bajhaiya, A.K., Moreira, J.Z., and Pittman, J.K. (2017). Transcriptional engineering of microalgae: prospects for high-value chemicals. *Trends Biotechnol.* 35: 95–99.

Batista, A.P., Nunes, M.C., Fradinho, P. et al. (2012). Novel foods with microalgal ingredients–effect of gel setting conditions on the linear viscoelasticity of *Spirulina* and *Haematococcus* gels. *J. Food Eng.* 110: 182–189.

Begum, H., Yusoff, F.M.D., Banerjee, S. et al. (2016). Availability and utilization of pigments from microalgae. *Crit. Rev. Food Sci. Nutr.* 56: 2209–2222.

Bhardwaj, N., Agrawal, K., and Verma, P. (2020). Algal Biofuels: An economic and effective alternative of fossil fuels. In: *Microbial Strategies for Techno-economic Biofuel Production*, Springer, Singapore, 207–227.

Bhatnagar, A., Chinnasamy, S., Singh, M., and Das, K.C. (2011). Renewable biomass production by mixotrophic algae in the presence of various carbon sources and wastewaters. *Appl. Energy* 88: 3425–3431.

Bhosale, P. (2004). Environmental and cultural stimulants in the production of carotenoids from microorganisms. *Appl. Microbiol. Biotechnol.* 63: 351–361.

Blanco, A.M., Moreno, J., Del Campo, J.A. et al. (2007). Outdoor cultivation of lutein-rich cells of *Muriellopsis* sp. in open ponds. *Appl. Microbiol. Biotechnol.* 73: 1259–1266.

Bogacz-Radomska, L. and Harasym, J. (2018). β-Carotene-properties and production methods. *Food Qual. Saf.* 2: 69–74.

Bonet, M.L., Canas, J.A., Ribot, J., and Palou, A. (2016). Carotenoids in nature. In: *Carotenoids in Nature: Biosynthesis, Regulation and Function*, vol. 79, 377–414.

Borowitzka, M.A., Borowitzka, L.J., and Kessly, D. (1990). Effects of salinity increase on carotenoid accumulation in the green alga *Dunaliella salina*. *J. Appl. Phycol.* 2: 111–119.

Boussiba, S., Bing, W., Yuan, J.-P. et al. (1999). Changes in pigments profile in the green alga *Haeamtococcus pluvialis* exposed to environmental stresses. *Biotechnol. Lett.* 21: 601–604.

Breitenbach, J., Misawa, N., Kajiwara, S., and Sandmann, G. (1996). Expression in *Escherichia coli* and properties of the carotene ketolase from *Haematococcus pluvialis*. *FEMS Microbiol. Lett.* 140: 241–246.

Cardozo, K.H.M., Guaratini, T., Barros, M.P. et al. (2007). Metabolites from algae with economical impact. *Comp. Biochem. Physiol. – C Toxicol. Pharmacol.* 146: 60–78.

Chakdar, H. and Pabbi, S. (2017). Algal pigments for human health and cosmeceuticals. In: *Algal Green Chemistry* (eds. R. Prasad Rastogi, D. Madamwar and A. Pandey), 171–188. Elsevier.

Chan, K., Mong, M., and Yin, M. (2009). Antioxidative and anti-inflammatory neuroprotective effects of astaxanthin and canthaxanthin in nerve growth factor differentiated PC12 cells. *J. Food Sci.* 74: H225–H231.

Chen, H., Pellett, L.J., Andersen, H.J., and Tappel, A.L. (1993). Protection by vitamin E, selinium, and β-carotene against oxidative damage in rat liver slices and homogenate. *Free Radic. Biol. Med.* 14: 473–482.

Chu, W.-L. (2012). Biotechnological applications of microalgae. *IeJSME.* 6: S24–S37.

Cohen, Z. (2017). Products from microalgae. In: *Handbook of Microalgal Mass Culture (1986)* (ed. A. Richmond), 421–454. CRC Press.

Contreras, J., Bibiana, J., Martinez, G., Barajas-Solano, A.F. (2020). Effect of de Carbon / Nitrogen ratio on the production of microalgae-based carotenoids Efecto de la relación carbono / Nitrógeno en la producción de carotenoids en microalgas. 22:1–12.

Cordero, B.F., Obraztsova, I., Couso, I. et al. (2011). Enhancement of lutein production in *Chlorella sorokiniana* (chorophyta) by improvement of culture conditions and random mutagenesis. *Mar. Drugs.* 9: 1607–1624.

Cysewski, G.R. and Lorenz, R.T. (2004). Industrial production of microalgal cell-mass and secondary products–species of high potential. In: *Handbook Microalgal Culture Biotechnology Applied Phycology* (ed. A. Richmond), 281.

Del Campo, J.A., Moreno, J., Rodrζguez, H. et al. (2000). Carotenoid content of chlorophycean microalgae: factors determining lutein accumulation in *Muriellopsis* sp.(Chlorophyta). *J. Biotechnol.* 76: 51–59.

Del Campo, J.A., Rodrıguez, H., Moreno, J. et al. (2001). Lutein production by *Muriellopsis* sp. in an outdoor tubular photobioreactor. *J. Biotechnol.* 85: 289–295.

Del Campo, J.A., GarcζGonzález, M., and Guerrero, M.G. (2007). Outdoor cultivation of microalgae for carotenoid production: current state and perspectives. *Appl. Microbiol. Biotechnol.* 74: 1163–1174.

Domζnguez-Bocanegra, A.R., Legarreta, I.G., Jeronimo, F.M., and Campocosio, A.T. (2004). Influence of environmental and nutritional factors in the production of astaxanthin from *Haematococcus pluvialis*. *Bioresour. Technol.* 92: 209–214.

Dufossé, L. (2009). Microbial and microalgal carotenoids as colourants and supplements. *Carotenoids* 5: 83–98.

Dufossé, L., Galaup, P., Yaron, A. et al. (2005). Microorganisms and microalgae as sources of pigments for food use: a scientific oddity or an industrial reality? *Trends Food Sci. Technol.* 16: 389–406.

Fernández-Sevilla, J.M., Acién Fernández, F.G., and Molina Grima, E. (2010). Biotechnological production of lutein and its applications. *Appl. Microbiol. Biotechnol.* 86: 27–40.

Gómez, P.I., Haro, P., Lagos, P. et al. (2016). Intraspecific variability among Chilean strains of the astaxanthin-producing microalga *Haematococcus pluvialis* (Chlorophyta): an opportunity for its genetic improvement by simple selection. *J. Appl. Phycol.* 28: 2115–2122.

Gong, M. and Bassi, A. (2016). Carotenoids from microalgae: a review of recent developments. *Biotechnol. Adv.* 34: 1396–1412.

Goswami, R.K., Mehariya, S., Verma, P., Lavecchia, R., and Zuorro, A. (2020a). Microalgae-based biorefineries for sustainable resource recovery from wastewater. *J. Water Process Eng.*, 101747.

Goswami, R.K., Mehariya, S., Karthikeyan, O.P., and Verma, P. (2020b). Advanced microalgae-based renewable biohydrogen production systems: a review. *Bioresour. Technol.*, 124301.

Goswami, G., Chaudhuri, S., and Dutta, D. (2010). The present perspective of astaxanthin with reference to biosynthesis and pharmacological importance. *World J. Microbiol. Biotechnol.* 26: 1925–1939.

Grama, B.S., Chader, S., Khelifi, D. et al. (2014). Induction of canthaxanthin production in a *Dactylococcus* microalga isolated from the algerian sahara. *Bioresour. Technol.* 151: 297–305.

Guedes, A.C., Amaro, H.M., and Malcata, F.X. (2011). Microalgae as sources of carotenoids. *Mar. Drugs* 9: 625–644.

Han, D., Li, Y., and Hu, Q. (2013). Astaxanthin in microalgae: pathways, functions and biotechnological implications. *Algae* 28: 131–147.

Hanagata, N. and Dubinsky, Z. (1999). Secondary carotenoid accumulation in *Scenedesmus komarekii* (Chlorophyceae, Chlorophyta). *J. Phycol.* 35: 960–966.

Heinrich, U., Gärtner, C., Wiebusch, M. et al. (2003). Supplementation with β-carotene or a similar amount of mixed carotenoids protects humans from UV-induced erythema. *J. Nutr.* 133: 98–101.

Hejazi, M.A., Holwerda, E., and Wijffels, R.H. (2004). Milking microalga *Dunaliella salina* for β-carotene production in two-phase bioreactors. *Biotechnol. Bioeng.* 85: 475–481.

Hu, J., Nagarajan, D., Zhang, Q. et al. (2018). Heterotrophic cultivation of microalgae for pigment production: a review. *Biotechnol. Adv.* 36: 54–67.

Ip, P.-F. and Chen, F. (2005). Production of astaxanthin by the green microalga *Chlorella zofingiensis* in the dark. *Process Biochem.* 40: 733–738.

Jin, E., Polle, J.E.W., Lee, H.-K. et al. (2003). Xanthophylls in microalgae: from biosynthesis to biotechnological mass production and application. *J. Microbiol. Biotechnol.* 13: 165–174.

Johnson, E.A. and An, G.-H. (1991). Astaxanthin from microbial sources. *Crit. Rev. Biotechnol.* 11: 297–326.

Lamers, P.P., van de Laak, C.C.W., Kaasenbrood, P.S. et al. (2010). Carotenoid and fatty acid metabolism in light-stressed *Dunaliella salina*. *Biotechnol. Bioeng.* 106: 638–648.

Lemoine, Y. and Schoefs, B. (2010). Secondary ketocarotenoid astaxanthin biosynthesis in algae: a multifunctional response to stress. *Photosynth. Res.* 106: 155–177.

León, R., Couso, I., and Fernández, E. (2007). Metabolic engineering of ketocarotenoids biosynthesis in the unicelullar microalga *Chlamydomonas reinhardtii*. *J. Biotechnol.* 130: 143–152.

Leu, S. and Boussiba, S. (2014). Advances in the production of high-value products by microalgae. *Ind. Biotechnol.* 10: 169–183.

Lichtenthaler, H.K., Schwender, J., Disch, A., and Rohmer, M. (1997). Biosynthesis of isoprenoids in higher plant chloroplasts proceeds via a mevalonate-independent pathway. *FEBS Lett.* 400: 271–274.

Liu, J., Sun, Z., Gerken, H. et al. (2014). Genetic engineering of the green alga *Chlorella zofingiensis*: a modified norflurazon-resistant phytoene desaturase gene as a dominant selectable marker. *Appl. Microbiol. Biotechnol.* 98: 5069–5079.

Liu, Z., Zeng, X., Cheng, J. et al. (2018). The efficiency and comparison of novel techniques for cell wall disruption in astaxanthin extraction from *Haematococcus pluvialis*. *Int. J. Food Sci. Technol.* 53: 2212–2219.

Ma, R., Zhao, X., Ho, S.-H. et al. (2020). Co-production of lutein and fatty acid in microalga *Chlamydomonas* sp. JSC4 in response to different temperatures with gene expression profiles. *Algal Res.* 47: 101821.

Mata-Gómez, L.C., Montañez, J.C., Méndez-Zavala, A., and Aguilar, C.N. (2014). Biotechnological production of carotenoids by yeasts: an overview. *Microb. Cell Factories* 13: 12.

Minyuk, G., Sidorov, R., and Solovchenko, A. (2020). Effect of nitrogen source on the growth, lipid, and valuable carotenoid production in the green microalga *Chromochloris zofingiensis*. *J. Appl. Phycol.* 32: 923–935.

Mogedas, B., Casal, C., Forján, E., and Vílchez, C. (2009). β-Carotene production enhancement by UV-A radiation in *Dunaliella bardawil* cultivated in laboratory reactors. *J. Biosci. Bioeng.* 108: 47–51.

Mojaat, M., Pruvost, J., Foucault, A., and Legrand, J. (2008). Effect of organic carbon sources and Fe2+ ions on growth and β-carotene accumulation by *Dunaliella salina*. *Biochem. Eng. J.* 39: 177–184.

Morowvat, M.H. and Ghasemi, Y. (2016). Culture medium optimization for enhanced β-carotene and biomass production by *Dunaliella salina* in mixotrophic culture. *Biocatal. Agric. Biotechnol.* 7: 217–223.

Mulders, K.J.M., Weesepoel, Y., Bodenes, P. et al. (2015). Nitrogen-depleted *Chlorella zofingiensis* produces astaxanthin, ketolutein and their fatty acid esters: a carotenoid metabolism study. *J. Appl. Phycol.* 27: 125–140.

Nisar, N., Li, L., Lu, S. et al. (2015). Carotenoid metabolism in plants. *Mol. Plant* 8: 68–82.

Novoveská, L., Ross, M.E., Stanley, M.S. et al. (2019). Microalgal carotenoids: a review of production, current markets, regulations, and future direction. *Mar. Drugs* 17: 1–21.

Pelah, D., Sintov, A., and Cohen., E. (2004). The effect of salt stress on the production of canthaxanthin and astaxanthin by *Chlorella zofingiensis* grown under limited light intensity. *World J. Microbiol. Biotechnol.* 20: 483–486.

Perusek, L. and Maeda, T. (2013). Vitamin A derivatives as treatment options for retinal degenerative diseases. *Nutrients* 5: 2646–2666.

Pirastru, L., Darwish, M., Chu, F.L. et al. (2012). Carotenoid production and change of photosynthetic functions in *Scenedesmus* sp. exposed to nitrogen limitation and acetate treatment. *J. Appl. Phycol.* 24: 117–124.

Pisal, D.S. and Lele, S.S. (2005). Carotenoid production from microalga, *Dunaliella salina*. *Indian J. Biotechnol.* 4: 476–483.

Priyadarshani, I. and Rath, B. (2012). Commercial and industrial applications of micro algae – a review. *J. Algal. Biomass. Utln.* 3: 89–100.

Raja, R., Hemaiswarya, S., and Rengasamy, R. (2007). Exploitation of *Dunaliella* for β-carotene production. *Appl. Microbiol. Biotechnol.* 74: 517–523.

Saini, R.K. and Keum, Y.S. (2017). Progress in microbial carotenoids production. *Indian J. Microbiol.* 57: 129–130.

Saini, D.K., Chakdar, H., Pabbi, S., and Shukla, P. (2020). Enhancing production of microalgal biopigments through metabolic and genetic engineering. *Crit. Rev. Food Sci. Nutr.* 60: 391–405.

Sánchez, J.F., Fernández-Sevilla, J.M., Acién, F.G. et al. (2008). Biomass and lutein productivity of *Scenedesmus almeriensis:* influence of irradiance, dilution rate and temperature. *Appl. Microbiol. Biotechnol.* 79: 719–729.

Sandesh Kamath, B., Vidhyavathi, R., Sarada, R., and Ravishankar, G.A. (2008). Enhancement of carotenoids by mutation and stress induced carotenogenic genes in *Haematococcus pluvialis* mutants. *Bioresour. Technol.* 99: 8667–8673.

Sathasivam, R. and Ki, J.S. (2018). A review of the biological activities of microalgal carotenoids and their potential use in healthcare and cosmetic industries. *Mar. Drugs.* 16: 16677–16687.

Sathasivam, R., Kermanee, P., Roytrakul, S., and Juntawong, N. (2012). Isolation and molecular identification of β-carotene producing strains of *Dunaliella salina* and *Dunaliella bardawi*l from salt soil samples by using species-specific primers and internal transcribed spacer (ITS) primers. *Afr. J. Biotechnol.* 11: 16677–16687.

Sathasivam, R., Radhakrishnan, R., Hashem, A., and E.F., Abd_Allah (2019). Microalgae metabolites: a rich source for food and medicine. *Saudi. J. Biol. Sci.* 26: 709–722.

Shi, X.-M., Zhang, X.-W., and Chen, F. (2000). Heterotrophic production of biomass and lutein by *Chlorella protothecoides* on various nitrogen sources. *Enzym. Microb. Technol.* 27: 312–318.

Steinbrenner, J. and Sandmann, G. (2006). Transformation of the green alga *Haematococcus pluvialis* with a phytoene desaturase for accelerated astaxanthin biosynthesis. *Appl. Environ. Microbiol.* 72: 7477–7484.

Stephens, E., Wolf, J., Oey, M. et al. (2015). Genetic engineering for microalgae strain improvement in relation to biocrude production systems. In: *Biomass and Biofuels from Microalgae* (eds. N.R. Moheimani, M.P. McHenry, K. de Boer and P.A. Bahri), 191–249. Springer.

Sun, X.M., Ren, L.J., Zhao, Q.Y. et al. (2018). Microalgae for the production of lipid and carotenoids: a review with focus on stress regulation and adaptation. *Biotechnol. Biofuels* 11: 1–16.

Supamattaya, K., Kiriratnikom, S., Boonyaratpalin, M., and Borowitzka, L. (2005). Effect of a *Dunaliella* extract on growth performance, health condition, immune response and disease resistance in black tiger shrimp (Penaeus monodon). *Aquaculture* 248: 207–216.

Surai, A.P., Surai, P.F., Steinberg, W. et al. (2003). Effect of canthaxanthin content of the maternal diet on the antioxidant system of the developing chick. *Br. Poult. Sci.* 44: 612–619.

Takaichi, S. (2011). Carotenoids in algae: distributions, biosyntheses and functions. *Mar. Drugs* 9: 1101–1118.

Varela, J.C., Pereira, H., Vila, M., and León, R. (2015). Production of carotenoids by microalgae: achievements and challenges. *Photosynth. Res.* 125: 423–436.

Vickers, C.E., Bongers, M., Liu, Q. et al. (2014). Metabolic engineering of volatile isoprenoids in plants and microbes. *Plant Cell Environ.* 37: 1753–1775.

Wei, D., Chen, F., Chen, G. et al. (2008). Enhanced production of lutein in heterotrophic *Chlorella protothecoides* by oxidative stress. *Sci. China Ser. C Life Sci.* 51: 1088–1093.

Yaakob, Z., Ali, E., Zainal, A. et al. (2014). An overview: biomolecules from microalgae for animal feed and aquaculture. *J. Biol. Res.* 21: 6.

Ye, Z.-W., Jiang, J.-G., and Wu, G.-H. (2008). Biosynthesis and regulation of carotenoids in *Dunaliella*: progresses and prospects. *Biotechnol. Adv.* 26: 352–360.

Yi, Z., Xu, M., Magnusdottir, M. et al. (2015). Photo-oxidative stress-driven mutagenesis and adaptive evolution on the marine diatom *Phaeodactylum tricornutum* for enhanced carotenoid accumulation. *Mar. Drugs.* 13: 6138–6151.

Yuan, J.P., Chen, F., Liu, X., and Li, X.Z. (2002). Carotenoid composition in the green microalga *Chlorococcum*. *Food Chem.* 76: 319–325.

Zhan, J., Rong, J., and Wang, Q. (2017). Mixotrophic cultivation, a preferable microalgae cultivation mode for biomass/bioenergy production, and bioremediation, advances and prospect. *Int. J. Hydrog. Energy.* 42: 8505–8517.

Zhao, L., Chang, W., Xiao, Y. et al. (2013). Methylerythritol phosphate pathway of isoprenoid biosynthesis. *Annu. Rev. Biochem.* 82: 497–530.

Zhu, Q.-L., Zheng, J.-L., and Liu, J. (2020). Transcription activation of β-carotene biosynthetic genes at the initial stage of stresses as an indicator of the increased β-carotene accumulation in isolated *Dunaliella salina* strain GY-H13. *Aquat. Toxicol.* 222: 105472.

7

Microbial Xylanases: A Helping Module for the Enzyme Biorefinery Platform

Nisha Bhardwaj and Pradeep Verma

Bioprocess and Bioenergy Laboratory, Department of Microbiology, Central University of Rajasthan, Ajmer, Rajasthan, India

7.1 Introduction

Biorefining is a sustainable bioconversion of biomass (renewable resources) into the range of industrial products like chemical, foods, and feed, similarly bioenergy like electricity, heat, and fuel (Bhardwaj et al. 2021; De Jong et al. 2009; Kumar and Verma 2020a, b, c). Being a keystone of the bioeconomy, the aim of completely revealing the potential of biomass from lignocellulosic plants (agricultural and forestry) in the economic method remains undefinable. The continuous increase in the consumption of energy and the decrease in the supply of fossil fuels has increased the researcher's interest in developing sustainable methodologies for the production of biofuel (Yang et al. 2015). The biomass obtained from lignocellulosic plants is abundantly available in the environment and can be considered as a vital alternative of fossil fuels. The biomass can be found in the environment throughout the year in the bulk amount without being used in the form of agricultural and forestry waste/residues (Thomas et al. 2016). The residues of most biomass, e.g., rice and sugarcane after cultivation are burnt in the open fields, mostly in Asian countries causing environmental pollution (Thomas et al. 2016).

The composition of lignocellulosic plant biomass consists of three main components, i.e., lignin, hemicellulose, and cellulose, which together make the recalcitrant structure of plant biomass (Singh et al. 2017; Kumar and Verma 2020d, e). Due to this, the biorefinery process involves three major steps, such as pretreatment, saccharification, and hydrolysis, for the complete bioconversion (Bhardwaj et al. 2020). Other important aspects, such as the type of biomass to be used in the biorefinery process and biomass transportation, are also the matter of concern along with the structure recalcitrance of the biomass to expose valuable sugars to be utilized in the biorefinery process to fulfill the bioenergy requirement of the world (Hassan et al. 2019).

The microbial hydrolytic enzymes play an important role in the bioconversion of biomass by converting it into the fermentable sugar (Wei et al. 2012). Therefore, various strategies have been carried out to date, such as isolation of new microbes and various optimization studies to improve the production of enzymes (Attri and Garg 2014; Haitjema et al. 2014;

Bioenergy Research: Evaluating Strategies for Commercialization and Sustainability, First Edition.
Edited by Neha Srivastava and Manish Srivastava.

Nigam 2013). Enzymes are required in all the major steps of the biorefinery processes, e.g., in biological pretreatment methods, using laccase for the removal of lignin, which can help to reduce the recalcitrant nature of plant cell walls and making inner cellular parts, i.e., hemicellulose and cellulose, more accessible (Agrawal et al. 2019; Agrawal et al. 2020a, b; Bhardwaj et al. 2020; Kumar et al. 2020a). Hemicellulase, e.g., xylanases and cellulases, are required in the hydrolysis and saccharification of plant residues, which enhances the release of sugar molecules (Bala and Singh 2019a). These enzymes can either be used individually or as a cocktail (Bhardwaj et al. 2019a, b). Although the commercially available enzyme cocktails are costly and affect the economy of the process, microbial enzymes can be considered as the best alternative (Vaishnav et al. 2018). With the cost of the enzymes, another important factors to be considered are their required amount, as a wider suite of enzyme preparations are required to achieve the enhanced saccharification rate (Cunha et al. 2017). Also getting microorganisms that can produce an enzyme cocktail that can act on multiple agricultural residues is another option to improve the economic viability of the process (Thomas et al. 2016). With the availability of a huge range of cellulases, lignocellulases can be utilized to allow the adaptation of such cocktails (Ang et al. 2015). This can be achieved by xylanase supplementation; endoxylanase is known as one of the most suitable enzymes used in the hydrolysis process by breaking the internal glycosidic linkages present in the backbone of the complex structure of heteroxylan, resulting in the xylo-oligosaccharides formation (Thomas et al. 2014a, b). Later these xylo-oligosaccharides are converted into other fermentable sugars, such as trimers (xylotriose), dimers (xylobiose), and monomers (xylose) (Brienzo et al. 2012).

Therefore, considering the importance of enzymatic system in the field of biorefinery the main focus must be on finding new strains that can produce large amounts of xylanases along with other hydrolytic enzymes. Along with these, new methods should be found to enhance the production of fermentable sugar that can further be converted into biofuel. This chapter includes the brief overview of the process involved in the biorefinery system via microbial xylanases. A brief overview of biorefinery process has been shown in Figure 7.1.

7.2 Raw Material for Biorefinery

Lignocellulosic plant biomass has been recognized as an efficient raw material for the biorefinery processes, which can replace huge sections of fossil resources (Maiti et al. 2018). Biorefinery process can produce three main end-products, i.e., biofuels, bioenergy, and biochemicals. As compared to other renewable resources, such as sun, wind, and water, the use of lignocellulosic biomass has some advantages, as it contains carbon materials in addition to fossils (Pachapur et al. 2019). Biorefinery processes comprise a broad range of methods that can separate plant biomass resources, such as rice, wheat, wood, grass, and corn, etc., into their elements, such as carbohydrates, triglycerides, proteins, etc., which can further be converted into value-added end-products, such as biofuels and biochemicals (De Jong et al. 2009; Saba et al. 2015). Residues obtained from agricultural industries such as wheat straw and bran, rice straw and husk, sugarcane bagasse, and cotton stalk are some of the most abundant lignocellulose (Juodeikiene et al. 2011).

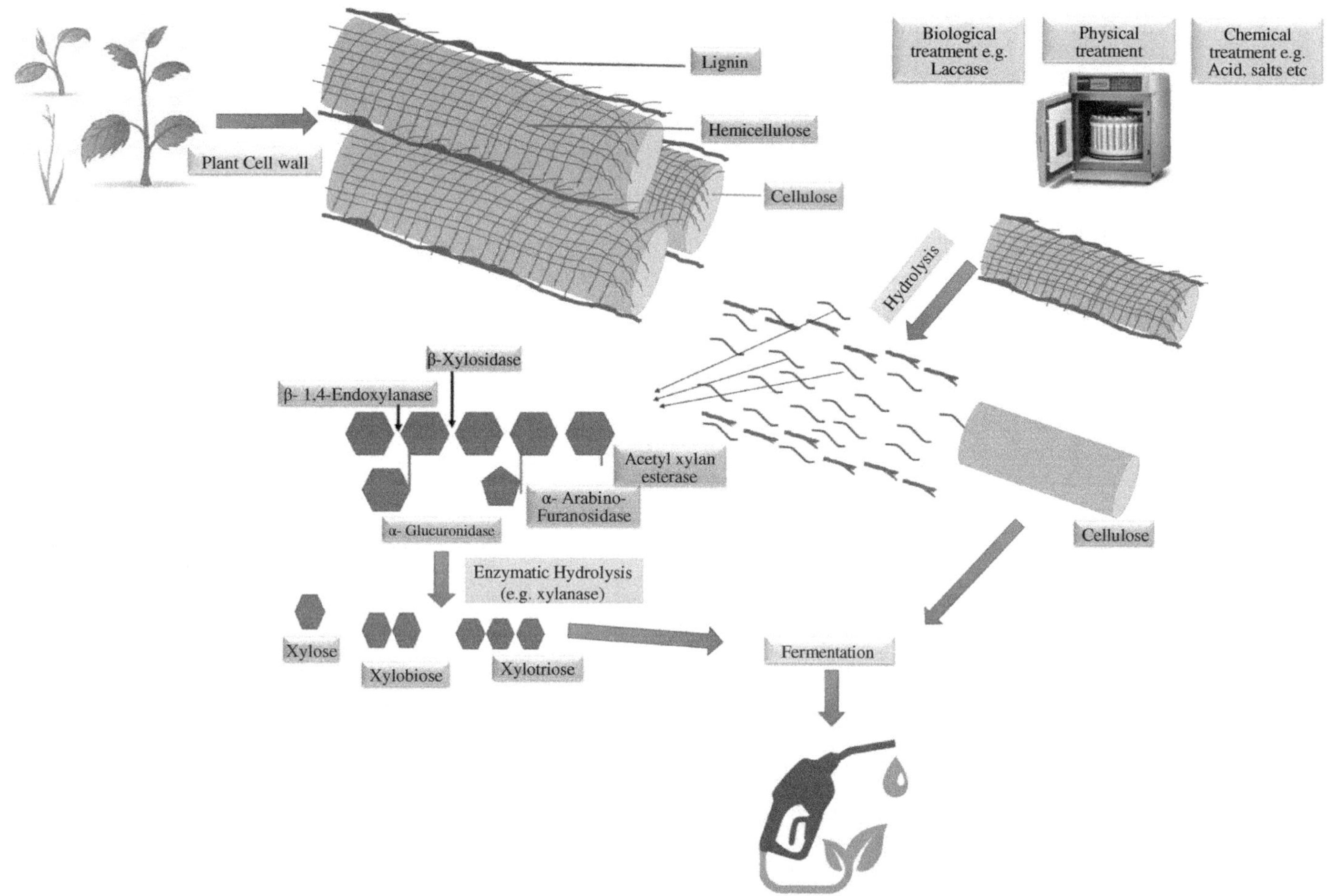

Figure 7.1 Brief overview of biorefinery process.

7.3 Structure of Lignocellulosic Plant Biomass

In the complete structure of plant cell wall cellulose is the principal component, which is present in a complex but systematic framework fibrous structure (Kumar et al. 2009) (Figure 7.2). This fibrous structure is made up of approximately 500–15 000 anhydrous glucose units linked with β-1,4-glycosidic linkages, which forms a linear homo-polysaccharide with the series of small cellobiose units. Extremely crystalline structures of cellulose comprising inter- and intra-molecular H-bonds are formed by β-1,4 arrangement of the glucoside bonds (Saini et al. 2015). Hemicellulose are found in the upper layer of cellulose and below the lignin in the plant cell wall (Saini et al. 2015) contains a short polypeptide chains of with 50–200 units of pentose and hexoses sugar, which is highly branched such as D-xylose, L-arabinose, and D-mannose-galactose-glucose, respectively. The hemicellulose part also have an acetate group, which is arranged randomly to the hydroxyl groups of the pentose sugar ring with ester linkages (Saini et al. 2015). Lignin is the third important component of the plant cell wall, which is highly crosslinked aromatic amorphous and heterogeneous polymer and comprises trans-coniferyl, -sinapyl, and -coumaryl alcohols. Lignin forms a complex matrix arranged covalently linked to side groups of other diverse hemicellulose and covers the cellulose microfibril. Lignin occupies the 2–40% plant cell wall in which C-C and C-O-C provides stability by protecting them from microbial attack (Mooney et al. 1998).

7.4 The Concept of Biorefinery

Biorefinery is classified in three different generations based on the use of different feedstock and the products (Azad et al. 2015). The raw material used for first-generation biorefinery is corn, barley, sunflower, etc. Biobased ethanol, diesel, biogas, methanol, and vegetable oils comes under this generation (Cherubini 2010). Due to the presence of high oil and sugar content, the bioconversion into biofuel is easy with this generation. Based on previous reports of life-cycle assessment analysis by Reinhardt et al. (2007) and Gasol et al. (2007), a remarkable decrease in the greenhouse gas emission, consumption of fossil energy as bioethanol and biodiesel has efficiently replaced the gasoline and diesel obtained from fossil resources. Apart from various benefits, this generation have the drawback of facing difficulties in feed and food industries as they use food resources and agricultural land (Cherubini 2010; Dutta et al. 2014).

In contrast, the second generation of biorefinery uses leftover residues from the food crops and cereals, which are known as lignocellulosic plant biomass, such as husks, bagasse, straws, animal fat, and municipal solid wastes, which can be used for biofuel production along with other value-added products (Azad et al. 2015; Geddes et al. 2011; Kumar et al 2020b; Zanuso et al. 2017). Based on various literature of life-cycle assessment analysis it was concluded that the second generation is more advantageous then the first, as it is more ecofriendly, economical, and more socially feasible as compared to food-based resources and requirements of agricultural land (Dutta et al. 2014), whereas in the third generation of biorefinery aquatic biomass, e.g., algae rice in proteins, oil, and carbohydrates, are used as feedstock for biofuel production (Martín and Grossmann 2012). Aquatic biomass consists of three groups: microalgae, cyanobacteria, and macroalgae. Although it is not a seasonal feedstock, with high oil productivity and high tolerance rate, its processing cost is very high due

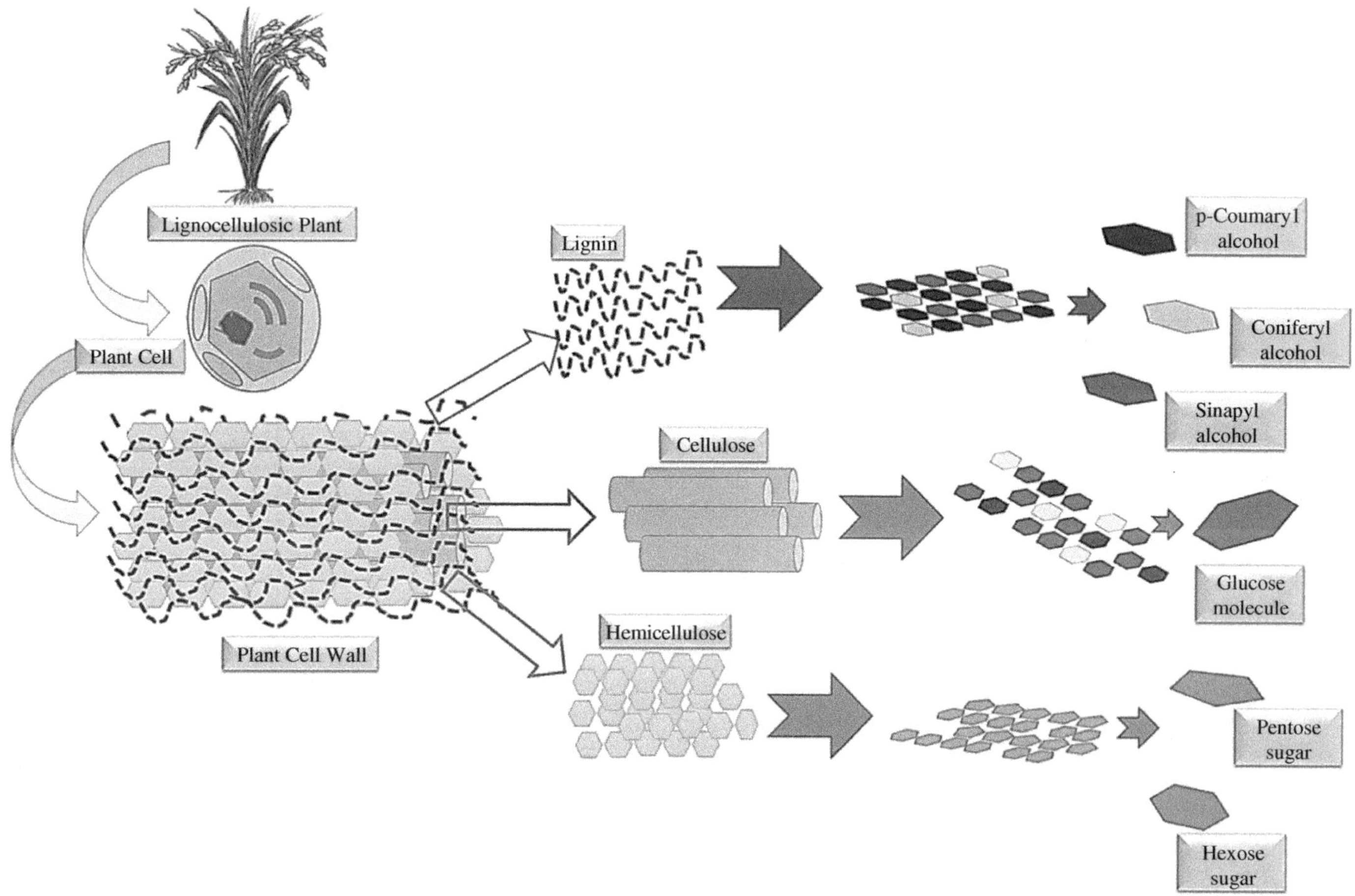

Figure 7.2 Structure of lignocellulosic plant biomass.

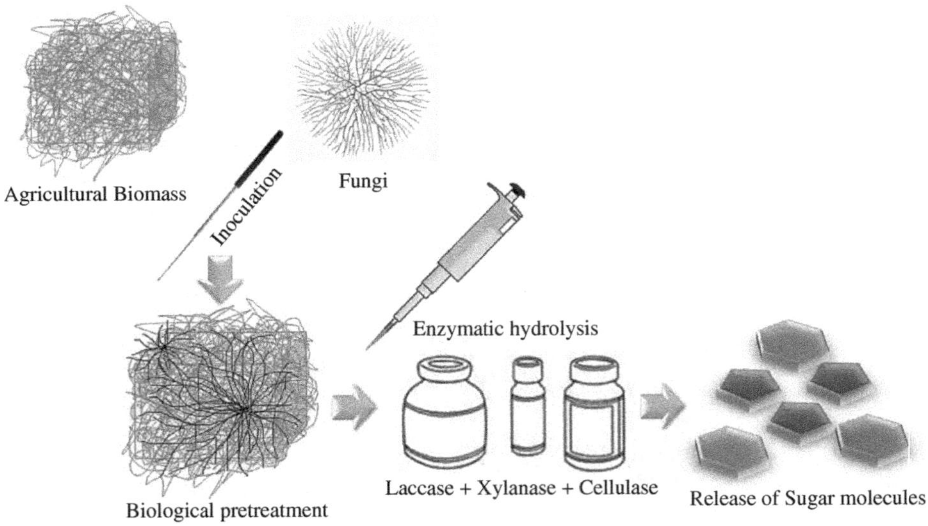

Figure 7.3 Role of enzymes in biological pretreatment and hydrolysis.

to high cultivation cost and energy input, which eventually affects the economic viability of the process (Cervantes-Cisneros et al. 2017). Among all the three-generation processes, the second generation has been considered more efficient, because the whole process can be considered economic from the use of waste products as resources until the production of value-added end-products.

7.5 Role of Enzymes in Biorefinery

7.5.1 In Biological Pretreatment

As discussed above, the biorefinery process involves three main steps of which pretreatment of biomass is but one of the important steps to enhance the production of fermentable sugar (Figure 7.3). Although pretreatment could be of three types: physical, chemical, or biological, biological pretreatment is preferred as it is ecofriendly and easy and safe to use, and involves the use of microbial enzymes and several microorganisms itself, e.g., white rot (*Myrothecium verrucaria*) and brown rot fungi (*Trametes versicolor, Pleurotus ostreatus*). It can be efficiently used in the delignification process without much requirement of energy (Kumar et al. 2009). Various enzymes, such as laccases, lignin peroxidases, manganese-dependent peroxidases, etc., are used for the delignification process (Agrawal et al. 2019). This process makes the inner hemicellulose and cellulose part much more accessible for the other hydrolytic enzymes, such as endoxylanases and cellulases respectively for the hydrolysis process (Bhardwaj et al. 2019). After this step, the accessibility of cellulose (carbon source) increases for efficient fermentation by microorganisms leading to the cost-effective enzyme production followed by hydrolysis of the same pretreated biomass. Therefore, it can be inferred that the rate of hydrolysis can be increased up to 90% after the pretreatment (Saini et al. 2015).

The pretreatment process via enzymes utilizes crude or purified enzymes or partially purified ligninolytic or hydrolytic enzymes. This may help to remove almost same amount of lignin similar to fungal pretreatment with a shorter time period (Plácido and Capareda 2015). Although the complete efficiency of enzymatic pretreatment process is not yet studied properly as compared to thermal and chemical pretreatment process, treatment of sugarcane using alkaline (NaOH) and crude *Anthracophyllum discolor* enzyme extracts for the production of bioethanol resulted in 48.7% and 33.6% lignin removal by NaOH enzymatic methods, i.e., 31% lower than the enzymatic process alone (Asgher et al. 2013). Although an increased cellulose hydrolysis of about 79% yield with enzymatic pretreated sugarcane bagasse was reported as compared to alkali pretreatment (Asgher et al. 2013). Hence, these results can be the example of continuing new researches on the use of both ligninolytic and cellulolytic enzymes to disrupt the structure of lignocellulosic plant biomass for the better saccharification and hydrolysis process (Asgher et al. 2013). Various reports in the enzymatic hydrolysis process, such as a microalgal pretreatment for the biomethane gas production, has been reported (Vanegas et al. 2015), production of biohydrogen (Mahdy et al. 2014), extraction of lipids for biodiesel generation (Fu et al. 2010), and production bioethanol (Kim et al. 2014). Similarly, by using manganese peroxidase in the *A. discolor* crude extract for the pretreatment of *Botryococcus braunii* has been for the production of biogas (Ciudad et al. 2014). Enzymatic pretreatment can be performed by using individual or cocktails of enzymes. Cocktail of enzymes are made by either using crude, partially, or purified enzymes. However, the use of a single enzymatic system has been reported with higher yield for the downstream processing of microalgal biomass (Vanegas et al. 2015), cocktails could be more hopeful for the hydrolysis of different biopolymers of plant biomass (Ehimen et al. 2013).

7.5.2 In Enzymatic Hydrolysis

For the economic generation of ethanol from cellulosic plant biomass, enzyme-based hydrolysis is an advantageous process as it is a very cost-effective method, with a probable vast yield when compared to chemical treatment (Figure 7.3). Long chain of carbohydrate present in the plant cell wall can be deconstructed by the hydrolysis method with the help of the enzyme catalysis process, and by forming a physical barrier, hemicellulose restricts the cellulase accessibility to cellulose (Zhang et al. 2012). Hence, supplying enzymes, such as xylanases, which can degrade them, can be the most suitable method to enhance the release of overall fermentable sugar from various pretreated lignocellulosic plant biomass (Kumar and Wyman 2009; Öhgren et al. 2007). Xylanases, e.g., *endo*-β-1,4-xylanases (EC 3.2.1.8) and β-xylosidase (EC 3.2.1.37), can act in main chains along with the side chain residues of complex structure of xylan. Endo-β-1,4-xylanase disrupts the long chain of xylan into smaller ones (Aditiya et al. 2016); similarly, xylopyranose are produced by β-xylosidase, which is a pyranose unit made up of xylose monomers that are formed by the continuous cleaving of oligosaccharide. Other xylanolytic accessory enzymes, such as feruloyl esterase (EC 3.1.139), and acetyl xylan esterase (EC 3.1.1.72) cleaves the outer chains (Aditiya et al. 2016). Due to a more amorphous nature, hemicelluloses are quite different from celluloses, and also hemicellulolytic enzymes are more complicated but with very particular actions. Hence it can be confirmed that destruction of xylan by enzymatic hydrolysis may remove the cellulose covering and it can also help in the improvement of cellulase performance (Zhang et al. 2011).

7.6 Enzyme Synergy: A Conceptual Strategy

Synergistic action of enzymes can be stated as the combination of pretreatment and hydrolysis steps to convert most of the polymeric components to fermentable sugar (Ang et al. 2015). In this process some attention must be taken that the process should not degrade or irreversibly transform the sugars, which will eventually lead to the loss in fermentable sugar. Hence, to avoid the degradation extent, less-severe pretreatment methodologies must be selected, e.g., biological pretreatment done by enzymes and microorganisms like fungi (Teter et al. 2014), because the slurries generated after the pretreatment may have some unwanted physical and chemical characteristics that may hinder the catalysis process of enzymatic proteins (Zhang et al. 2012).

In order to avoid the loss of fermentable sugar, all the three major steps, i.e., pretreatment, hydrolysis, and fermentation of biomass conversion, can be incorporated together, which will lead to a reduction in the multistep process. Hence, different enzymes can be mixed together in a sufficient ratio to prepare the suitable enzyme cocktail (Bhardwaj et al. 2019b). These enzymes will work synergistically and will lead to the enhanced biomass conversion and release of maximum sugar as compared to other physical and chemical methods (Chaturvedi and Verma 2013) (Figure 7.4). Later, the released sugar in the slurry can further be converted into bioethanol by the use of ethanologenic microorganisms, such as *Saccharomyces cerevisiae* (Bhardwaj et al. 2019b). Although biobased methods have various advantages, such as high specificity and no formation of toxic and inhibitory chemicals, expensive and sophisticated instruments are not required; they have some limitations also, like high enzyme cost, limited temperature, and pH stability (Bala and Singh 2019a).

A study has been reported on the use of thermo alkali stable ligno-hemicellulolytic enzyme laccase from *M. verrucaria* (Agrawal et al. 2019), xylanase from *Aspergillus oryzae* (Bhardwaj et al. 2017), and cellulase from *Schizophyllum commune* (Kumar et al. 2018) cocktails (crude, partially purified) in combination with *S. cerevisiae* MTCC-173, by using simultaneous delignification, saccharification, and fermentation (SDSF) in combination with *S. cerevisiae* MTCC-173 (Bhardwaj et al. 2019). Various forms of xylanase were produced by some thermophilic fungi, such as *Malbranchea cinnamomea* (Mahajan et al. 2014) *Pyrenophora phaeocomes* (Rastogi et al. 2016), *T. versicolor*, *P. ostreatus*, and *Piptoporus betulinus* (Valášková and Baldrian 2006). Similarly, thermophilic mold, such

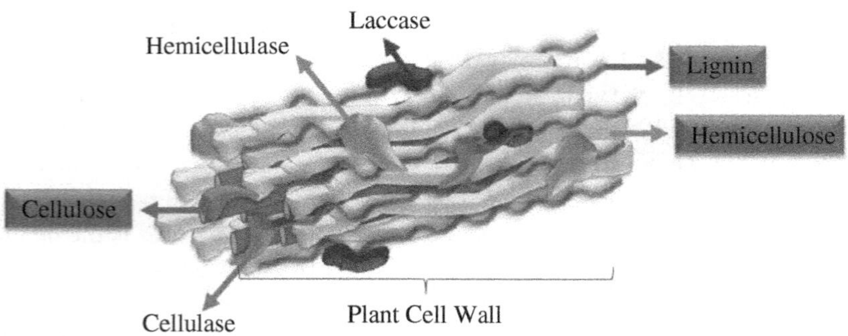

Figure 7.4 Synergistic action of enzymes.

as *Thermoascus aurantiacus,* was found capable of producing xylanase and cellulases by using an agricultural biomass (Jain et al. 2015).

Similarly, in the coculturing method, a combination of enzymes produced by *Aspergillus niger* and *Trichoderma reesei* resulted in a threefold higher hydrolysis rate of unwashed pretreated sugarcane bagasse with only 0.7 filter paper unit (FPU) activity/g glucan enzyme load when compared to 5–15 times enzyme loading (Florencio et al. 2016). Therefore, it can be stated that cocktails of various enzymes and coculture of microorganisms could be a better approach to enhance the fermentable sugar production (Kolasa et al. 2014).

7.7 Factors Affecting Biological Pretreatment

In order to get the highest yield via enzymatic pretreatment, it is required to understand the factors affecting the microbial growth and metabolism (Wan and Li 2012). The factors that affect the process may be nature, moisture content, and particle size of the biomass or substrates, microorganism type, and inoculum concentration, enzyme type, and conditions like time, pH, and temperature. Biomass surfaces contain internal and external areas where the particle size and shape is important for the maintenance biomass component capillary structure (Maurya et al. 2015). The particles with small sizes may be required for the increased digestibility and total yield by increasing the surface area, although use of small size particles is quite difficult in the downstream processing (Bolado-Rodríguez et al. 2016). However, the very small size of particles affects the efficiency of the pretreatment as it affects the proper microbial growth and metabolism by reducing the aeration rate (Sharma et al. 2019). In the case of larger particle size affecting the pretreatment process by reducing the penetration of microorganisms into the substrates and reducing the uniform air diffusion, similarly, time is another important factor, which varies according to the microorganism and microbial enzymes. A report of highest sugar yield with the rice straw hydrolysis, 60 days pretreated with *P. ostreatus* (Taniguchi et al. 2005), whereas less sugar concentration was reported with *Phanerochaete chrysosporium* pretreated wheat bran after 14 days (Salvachúa et al. 2011). An increased sugar yield was reported for the wood chips pretreatment by using *T. versicolor* (Hwang et al. 2008). Another important factor is biomass moisture content, as it is required in a specific amount for proper microbial growth and biodegradation (Gervais and Molin 2003), although this also varies on the basis of type of strain and biomass (Mustafa et al. 2016). Physical parameters, such as temperature, have also found to be another important parameter in the enzymatic pretreatment process, which is necessary for the optimum microbial growth and cell metabolic activities. Based on various microorganisms, the temperatures optima were also varied from 25 to 30 °C, and fungi from the ascomycetes group can grow at a higher temperature nearly up to 39 °C, whereas, in case of basidiomycetes, required temperature optima is 15 and 35 °C (Sindhu et al. 2016). This is because of the difference in the physiology of fungus substrate type and microbial strains (Isroi et al. 2011). The white rot fungi (WRF) metabolism in a solid-state system generates heat, which eventually enhances the bioreactors' gradient temperature (Wan and Li 2012), and plays an important challenge for the researchers while designing the bioreactor for the solid-state pretreatment application in a large scale. Similarly, pH in culture medium also affects the microbial growth, enzyme secretion, and hydrolysis (Sharma et al. 2019).

7.8 Advantages of Xylanases from Thermophilic Microorganisms in Biorefinery

Various thermophilic microorganisms have been reported for the production of different enzymes, such as hemicellulases, amylases, cellulases, phosphatases, proteases, laccases lipases, etc., which have various applications in different industries like food, textile, and detergent, dairy, pharmaceutical, and others (Singh 2016). The similarity of thermophilic microorganisms in their phylogenetic analysis and their enzymes showed a common origin with other mesophiles (Zeldes et al. 2015). Thus, cellulases and xylanases obtained from thermophilic origin and their mode of action was found to be similar, except only with some specific features that indicate their advantage at various industries. Thermophiles are found to be a good source of different enzymes, as they can produce thermostable enzymes. As compared to mesophilic enzymes, thermophiles have a high resistance for denaturing agents and high-pressure tolerance. Hence, they may be considered as the valuable domain for the production of biofuels at higher temperatures (Haki and Rakshit 2003), because high temperature may enhance the penetration of enzymes via the cell wall of lignocellulosic plant biomass and can behave as a physical factor for the disorganization of the cell wall of lignocellulosic biomass (Paës and O'Donohue 2006). Among various pretreatment methods, enzymatic degradation of lignocellulosic biomass using cellulase and xylanase is found to be the most suitable and specific with no other toxic effects or product formation and no loss of substrate. Thermostable xylanases and cellulases play a very important role in the pharmaceutical, pulp, chemical, food, and paper and pulp industries. Xylanases have been found to be an alternative of chlorine in paper and pulp due to its involvement in the leaching of xylan from carbohydrate-lignin complex. This way, xylanase can be useful in the replacement of chlorine and in the pulp-bleaching process and can reduce the environmental pollution caused by them. A thermostable xylanase obtained from *Myceliophthora thermophila* was found suitable as compared to a thermolabile xylanase obtained from *T. reesei* in the paper and pulp industry. A thermostable xylanase from *Bacillus* sp. NCIM5 was utilized in the bagasse pulp prebleaching by simultaneously reducing the demand of chlorine (Kulkarni and Rao 1996). Various bacterial strains, such as *Bacillus* sp. and *Dictyoglomus* sp. were successfully at a commercial scale (Rani and Nand 2000). Although for many xylanolytic and cellulolytic enzymes, the temperature and pH optima was found to be below 50 °C and acidic or neutral pH (Gessesse 1998). Various thermophilic fungi are found to be the good producers of xylanases and cellulases, which were successfully used in the lignocellulosic biomass saccharification (Kaur and Satyanarayana 2004) (Table 7.1).

7.9 The Products of Biorefinery

A list of some recent xylanases involved in the biorefinery process is shown in Table 7.2 and discussed as follows.

7.9.1 Bioethanol

Bioethanol produced from lignocellulosic plant biomass is a ecological process that can be enhanced by using suitable enzymes and microorganisms. Previous studies have reported that thermophilic microorganisms can produce more amounts of bioethanol via the SDSF process. Thermal stability has been found to be an important and desirable property

Table 7.1 Thermostable xylanases and their properties.

Xylanase source	Improved application	Improved property	References
Thermotoga thermarum DSM 5069	Enhances sugar release of pretreated lignocellulosic biomass	Enhanced hydrolysis then commercial HTec xylanase and five times lower xylanase amount	Long et al. (2018)
Camel rumen metagenome	Effective for the bioconversion of lignocellulosic biomass.	Demonstrated the power of in silico analysis to discover novel alkali-thermostable xylanases	Ariaeenejad et al. (2019)
Caldicellulosiruptor bescii	Exhibited high xylanase activity and thermo-/pH stability.	An insight into the structural basis for its substrate recognition mechanism	An et al. (2015)
Nonomuraea flexuosa (Nf Xyn, GH11) and *Thermoascus aurantiacus* (Ta Xyn, GH10)	Improved hydrolysis of hydrothermally pretreated wheat straw.	Better hydrolytic capacity of solubilizing xylan and acting synergistically with thermostable cellulases	Zhang et al. (2011)
Aspergillus fumigatus	Great potential in production of bioethanol from lignocellulose.	Sugars produced from hemicellulose except xylose show little effect on β-xylosidase.	Jin et al. (2020)
Thermomyces lanuginosus SSBP	Improved availability of fermentable sugars	The enzyme reduced phytate content in *Colocasia esculenta* starch.	Makolomakwa et al. (2017)
Streptomyces sp. SWU10	Two forms of a xylanase may have potential application in biofuel, food, textile industries and waste treatment.	Exhibited high pH and thermal stabilities.	Deesukon et al. (2011)
Chaetomium thermophilum	Relatively good thermostability	First detailed report on a novel bifunctional endoglucanase/xylanase enzyme from *C. thermophilum*.	Hua et al. (2018)
Kluyvera sp. *strain* OM3 and *Clostridium* sp. *strain* BOH3	High thermal and pH stability.	Application for butanol production directly from hemicelluloses.	Xin and He (2013)
Thermomonospora sp.	High yield of ethanol production	Simultaneous saccharification and fermentation (SSF) of oat spelt xylan (OSX) and wheat bran hemicellulose (WBH) using thermostable xylanase	Menon et al. (2010)

Table 7.2 Role of xylanases in the field of biorefinery.

Microorganisms	Agroresidues	Biorefinery product	References
Thermomyces lanuginosus, Trichoderma reesei	Rye, wheat	Bioethanol	Juodeikiene et al. (2011)
Aspergillus sp.	Rice straw	Bioethanol	Thomas et al. (2016)
Rhizopus oryzae	Sorghum stover	Bioethanol	Pandey et al. (2016)
Streptomyces variabilis (MAB3)	Rice straw	Bioethanol	Sanjivkumar et al. (2018)
Streptomyces thermovulgaris	Corn cob	Bioethanol	Boonchuay et al. (2018)
Aspergillus oryzae LC1	Rice straw	Bioethanol	Bhardwaj et al. (2019)
Penicillium chrysogenum	Sugar cane bagasse	Bioethanol	Terrone et al. (2018)
Aspergillus fumigatus	Kenaf (*Hibiscus cannabinus*)	Bioethanol	Damis et al. (2019)
Aspergillus terreus	Sugarcane bagasse	Bioethanol	Kamat et al. (2013)
Thermomyces lanuginosus	Wheat bran	Bioethanol	Wood et al. (2016)
Trichoderma atroviridae SS2	Sunflower oil sludge	Biobutanol	Sakthiselvan et al. (2015)
Trichoderma longibrachiatum	Barley straw	Acetone–Butanol–Ethanol	Yang et al. (2015)
Kluyvera species *OM3* *Clostridium* sp. strain BOH3	Xylan	Biobutanol	Xin and He (2013)
Methanocaldococcus sp. *Clostridium* sp.	Palm oil mill effluent	Biomethane	Prasertsan et al. (2017)
Acinetobacter johnsonii	Xylan	Ethanol	Xue et al. (2019)
Candida tropicalis MK-160	Xylan	Ethanol	Shariq and Sohail (2019)
Pennisetum sp.	Grasses	Bioethanol	Mohapatra et al. (2020)
Trichoderma longibrachiatum	Waste bamboo	Bioethanol	Song et al. (2020)
Penicillium chrysogenum	Sugarcane bagasse	Biorefinery	Terrone et al. (2018)
Aspergillus and *Trichoderma* sp.	Water hyacinth	Biofuel	Uday et al. (2016)
Peniciliium echinulatum	Sugar cane bagasse and elephant grass	Biofuel	Schneider et al. (2018)

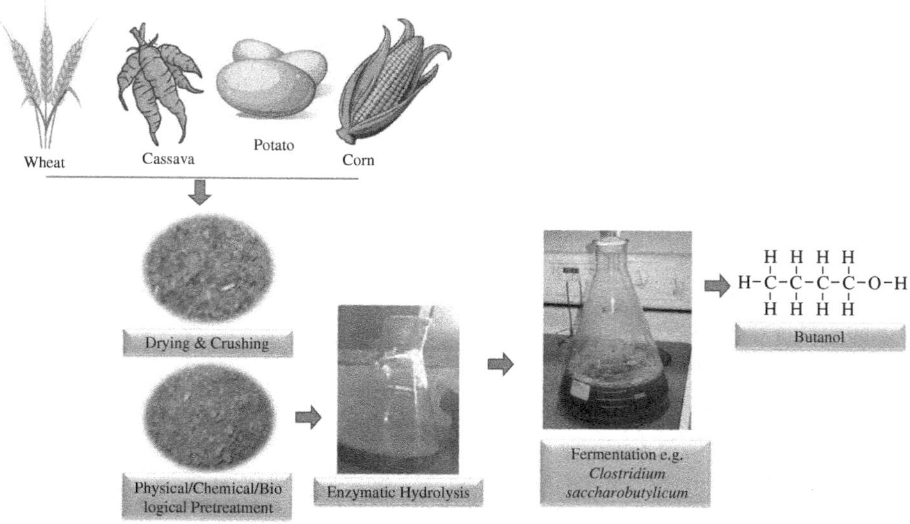

Figure 7.5 Generation of butanol using agricultural residues.

for cellulolytic and xylanolytic enzymes required for successful saccharification. The hydrolysis rate of *Trichoderma* is low, as it has less of a β-glucosidase level (Mohanram et al. 2013). Hence thermophilic fungi can serve as a suitable alternative of this. Various molds, e.g., *Sporotrichum thermophile*, *T. aurantiacus*, and *Scytalidium thermophillum* (Berka et al. 2011; Kaur et al. 2004), which are thermophilic in nature have shown sufficient enzymatic systems for the lignocellulosic plant biomass bioconversion process for enhanced bioethanol production. *S. cerevisiae* and *Pichia stipites* have been used for the production of bioethanol with high yield at 30 °C after 72 h (Bala and Singh 2019b). Similar reports with the rice straw and waste tea cup paper hydrolysis are there in the literature using partially purified cellulases and xylanase obtained from *S. thermophile* BJAMDU5, resulting in the high yield of reducing sugars (Bala and Singh 2016). Various thermophilic bacteria, such as *Clostridium*, *Caldanaerobacter*, and *Thermoanaerobacter*, were reported for high ethanol production (Taylor et al. 2009).

7.9.2 Biobutanol

Another product obtained from biorefinery that has attracted the attention of scientists as an efficient alternative for gasoline (Bhandiwad et al. 2014) (Figure 7.5) is biobutanol. Microorganisms, such as *Clostridium* spp., *Clostridium saccharoperbutylacetonicum*, *Clostridium acetobutylicum*, and *Clostridium beijerinckii*, were the suitable examples of biobutanol production by using the sugars from agricultural residues (Bhandiwad et al. 2014; Nakayama et al. 2011). Similarly *Thermo-anaerobacterium thermosaccharolyticum* showed 1.8–5.1 mM n-butanol production from the overexpression of *thl*, *hbd*, *crt*, *bcd*, *etfA*, and *etfB* genes of bcs operon required for butyryl CoA formation (Bhandiwad et al. 2014). Approximately 7.9 $g\,l^{-1}$ of n-butanol was produced by a coculture of *Clostridium thermocellum* and *C. saccharoperbutylacetonicum* (Nakayama et al. 2011). Approximatley 7.7 $g\,l^{-1}$ of acetoin and 14.5 $g\,l^{-1}$ of 2, 3-butanediol was reported from the *Geobacillus* strain XT15 from corn steep liquor at 55 °C (Yang et al. 2015).

7.9.3 Hydrogen

It is a carrier of energy having the high potential of being considered as an alternative for fossil fuel. As it is a clean fuel, it can be used as an internal fuel for combustion engines in combination with oxygen (Koskinen et al. 2008). Thermophilic microorganisms, e.g., *Pyrococcus furiosus*, *Thermococcus kodakaraensis*, and all *Thermotoga* and *Caldiecellulosiruptor* species have found to be the good producers of hydrogen with only water vapor emission (Verhaart et al. 2010). adhE and aldH genes are not present in these microorganisms, and therefore do not produce ethanol, and hence, due to hydrogenase, hydrogen production increases. However, *Clostridium uzonii* strain AK15 and *Thermoanaero bacterium aciditolerans* AK17 isolated from Iceland in geothermal springs showed good hydrogen production along with bioethanol (Koskinen et al. 2008).

7.10 Molecular Aspects of Enzymes in Biorefinery

The advances of effective hydrolysis enzymes with advanced properties, e.g., better interaction with cheap substrates, higher specific activity, and higher stability are important factors for the industrial production of biofuel (Table 7.3). As discussed above, lignocellulose plant biomass degradation into their monomeric sugars comprises two important constituents, i.e., hemicellulose and cellulose (Balat 2011; Pareek et al. 2013; Ulaganathan et al. 2017), and the composite hemicellulose structure needs the synergistic action of different enzymes, and endo-1,4-β-xylanase plays an important role in order to degrade the complex polymer of xylan into oligosaccharides and other monomeric sugars (Madadi et al. 2017). Also hemicellulolytic enzymes are not sufficient for the complete hydrolysis of recalcitrance lignocellulosic biomass (Himmel et al. 2007). Hence, there is a requirement of enzymes and they are commercially expensive, which will eventually lead to product loss (Visser et al. 2015). The only solution for this problem is that the efficiency of the enzymes should be increased (Morone and Pandey 2014) along with the exploitation of accessory enzymes, e.g., xylanase and β-glucosidase, which can be synergistically acted with cellulases (Berlin et al. 2005). Recently, various reports are found in the literature based on the improvements of hydrolytic enzymes, which has only considered the cellulase and their synergy with hemicellulases (Diogo et al. 2015; Quiñones et al. 2015; Yang et al. 2018), but very few reports are there focusing on xylanases individually. Molecular biology aspects that include directed evolution, library construction strategies, mutagenesis, and gene recombination have gained researchers' interest to improve the genetic variations on enzymes (McLachlan et al. 2009). The increased hydrolysis of pretreated sugarcane bagasse was reported with xylanase (Ribeiro et al. 2014). Two xylanase genes (GH10 and GH11) from *M. cinnamomea*, i.e., XYN10A_MALCI and XYN11A_MALCI, respectively, expressed in *Pichia pastoris* X33 showed improved hydrolysis of substituted arabinoxylan and unsubstituted xylan. The synergistic action of recombinant xylanase with commercial cellulase resulted in better hydrolysis of acid- and alkali-treated rice straw (Basotra et al. 2018). Similarly, *Geobacillus thermodenitrificans* JK1 showed the production of isoforms of xylanase, i.e., XynA1 and XynA2, acted synergistically with β-xylosidases and arabinofuranosidase for the improved birch wood xylan hydrolysis (Huang et al. 2017).

Table 7.3 Genetically modified xylanases in the field of biorefinery.

Xylanase sources	Host microorganisms	Applications	References
Fusarium oxysporum	*Agrobacterium tumefaciens*	Bioethanol	Anasontzis et al. (2011)
Cellulomonas flavigena KCTC 9104	*Pichia pastoris* X-33	Biofuel	Kim et al. (2018)
Pleurotus ostreatus HAUCC 162 and *Irpex lacteus* CD2	*Escherichia coli* BL21 (DE3)	Saccharification, oligosaccharides	Zhuo et al. (2018)
Thermotoga neapolitana	*E. coli* TG1	Xylan hydrolysis	Velikodvorskaya et al. (1997)
Bispora sp. MEY-1	*Pichia pastoris.* xyl11B	Xylan hydrolysis	Luo et al. (2009)
Penicillium citrinum	*Yarrowia lipolytica*	Xylan hydrolysis	Ouephanit et al. (2019)
Intein-modified xylanases (iXyn GH10, iXyn GH11*)*	*Panicum virgatum*	Bioethanol production	Chen et al. (2017)
Talaromyces thermophilus F1208	*E. coli* BL21 (DE3)	Modulating hydrolysis and transglycosylation activity	Li et al. (2017)
Thermobacillus xylanilyticus (Tx-Xyn11)	*E. coli* BL21 (DE3)	Improved adsorption onto lignin and thermostabilty	Rakotoarivonina et al. (2015)

7.11 Conclusion

Advancement in enzyme efficiency and effective hydrolysis is required in the world of biorefinery so that scientists can focus on the economic and ecofriendly processes. Xylanase plays a key role in the biorefinery process, hence, its production, and hydrolytic efficiency must be enhanced by finding new microorganisms that can produce isoforms of xylanases. Overexpression of new genes from novel xylanases from different microorganisms can be explored for future applications. Hence, using the advantage of gene editing and synthetic biological techniques in the future, with improved characteristics like thermostability, can be a fruitful contribution toward the high demand of biorefinery.

References

Aditiya, H., Mahlia, T.M.I., Chong, W. et al. (2016). Second generation bioethanol production: a critical review. *Renew. Sustain. Energy Rev.* 66: 631–653.

Agrawal, K., Bhardwaj, N., Kumar, B. et al. (2019). Process optimization, purification and characterization of alkaline stable white laccase from *Myrothecium verrucaria* ITCC-8447 and its application in delignification of agroresidues. *Int. J. Biol. Macromol.* 125: 1042–1055.

Agrawal, K. and Verma, P. (2020a). Production optimization of yellow laccase from Stropharia sp. ITCC 8422 and enzyme-mediated depolymerization and hydrolysis of lignocellulosic biomass for biorefinery application. *Biomass Convers. Bioref.* 1–20.

Agrawal, K. and Verma, P. (2020b). Laccase-mediated synthesis of bio-material using agro-residues. In: *Biotechnological Applications in Human Health*, 87–93. Springer, Singapore.

An, J., Xie, Y., Zhang, Y. et al. (2015). Characterization of a thermostable, specific GH10 xylanase from *Caldicellulosiruptor bescii* with high catalytic activity. *J. Mol. Catal. B: Enzym.* 117: 13–20.

Anasontzis, G.E., Zerva, A., Stathopoulou, P.M. et al. (2011). Homologous overexpression of xylanase in *Fusarium oxysporum* increases ethanol productivity during consolidated bioprocessing (CBP) of lignocellulosics. *J. Biotechnol.* 152 (1–2): 16–23.

Ang, S.K., Abd-Aziz, S., and Madihah, M.S. (2015). Potential uses of xylanase-rich lignocellulolytic enzymes cocktail for oil palm trunk (OPT) degradation and lignocellulosic ethanol production. *Energy Fuels* 29 (8): 5103–5116.

Ariaeenejad, S., Maleki, M., Hosseini, E. et al. (2019). Mining of camel rumen metagenome to identify novel alkali-thermostable xylanase capable of enhancing the recalcitrant lignocellulosic biomass conversion. *Bioresour. Technol.* 281: 343–350.

Asgher, M., Ahmad, Z., and Iqbal, H.M.N. (2013). Alkali and enzymatic delignification of sugarcane bagasse to expose cellulose polymers for saccharification and bio-ethanol production. *Ind. Crops Prod.* 44: 488–495.

Attri, S. and Garg, G. (2014). Isolation of microorganisms simultaneously producing xylanase, pectinase and cellulase enzymes using cost effective substrates. *J. Innov. Biol.* 1 (1): 45–50.

Azad, A.K., Rasul, M., Khan, M.M.K. et al. (2015). Prospect of biofuels as an alternative transport fuel in Australia. *Renewable Sustainable Energy Rev.* 43: 331–351.

Bala, A. and Singh, B. (2016). Cost-effective production of biotechnologically important hydrolytic enzymes by *Sporotrichum thermophile*. *Bioprocess. Biosyst. Eng.* 39 (1): 181–191.

Bala, A. and Singh, B. (2019a). Cellulolytic and xylanolytic enzymes of thermophiles for the production of renewable biofuels. *Renewable Energy* 136: 1231–1244.

Bala, A. and Singh, B. (2019b). Development of an environmental-benign process for efficient pretreatment and saccharification of *Saccharum* biomasses for bioethanol production. *Renewable Energy* 130: 12–24.

Balat, M. (2011). Production of bioethanol from lignocellulosic materials via the biochemical pathway: a review. *Energy Convers. Manage.* 52 (2): 858–875.

Basotra, N., Joshi, S., Satyanarayana, T. et al. (2018). Expression of catalytically efficient xylanases from thermophilic fungus *Malbranchea cinnamomea* for synergistically enhancing hydrolysis of lignocellulosics. *Int. J. Biol. Macromol.* 108: 185–192.

Berka, R.M., Grigoriev, I.V., Otillar, R. et al. (2011). Comparative genomic analysis of the thermophilic biomass-degrading fungi *Myceliophthora thermophila* and *Thielavia terrestris*. *Nat. Biotechnol.* 29 (10): 922–927.

Berlin, A., Gilkes, N., Kilburn, D. et al. (2005). Evaluation of novel fungal cellulase preparations for ability to hydrolyze softwood substrates–evidence for the role of accessory enzymes. *Enzyme Microb. Technol.* 37 (2): 175–184.

Bhandiwad, A., Shaw, A.J., Guss, A. et al. (2014). Metabolic engineering of *Thermoanaerobacterium saccharolyticum* for n-butanol production. *Metab. Eng.* 21: 7–25.

Bhardwaj, N., Chanda, K., Kumar, B. et al. (2017). Statistical optimization of nutritional and physical parameters for xylanase production from newly isolated *Aspergillus oryzae* LC1 and its application in the hydrolysis of lignocellulosic agro-residues. *BioResources* 12 (4): 8519–8538.

Bhardwaj, N., Kumar, B., Agrawal, K., and Verma, P. (2019a). Bioconversion of rice straw by synergistic effect of in-house produced ligno-hemicellulolytic enzymes for enhanced bioethanol production. *Bioresour. Technol. Rep.* 10 (2020): 100352.

Bhardwaj, N., Kumar, B., and Verma, P. (2019b). A detailed overview of xylanases: an emerging biomolecule for current and future prospective. *Bioresour. Bioprocess.* 6: 1–36.

Bhardwaj, N., Kumar, B., and Verma, P. (2020). Microwave-assisted pretreatment using alkali metal salt in combination with orthophosphoric acid for generation of enhanced sugar and bioethanol. *Biomass Convers. Biorefin.*: 1–8.

Bhardwaj, N., Agrawal, K., Kumar, B., and Verma, P. (2021). Role of enzymes in deconstruction of waste biomass for sustainable generation of value-added products. In: *Bioprospecting of Enzymes in Industry, Healthcare and Sustainable Environment*, 219–250. Springer, Singapore.

Bhardwaj, N. and Verma, P. (2020). Extraction of fungal xylanase using ATPS-PEG/sulphate and its application in hydrolysis of agricultural residues. In: *Biotechnological Applications in Human Health*, 95–105. Springer, Singapore.

Bolado-Rodríguez, S., Toquero, C., Martín-Juárez, J. et al. (2016). Effect of thermal, acid, alkaline and alkaline-peroxide pretreatments on the biochemical methane potential and kinetics of the anaerobic digestion of wheat straw and sugarcane bagasse. *Bioresour. Technol.* 201: 182–190.

Boonchuay, P., Techapun, C., Leksawasdi, N. et al. (2018). An integrated process for xylooligosaccharide and bioethanol production from corncob. *Bioresour. Technol.* 256: 399–407.

Brienzo, M., Monte, J., and Milagres, A.M.F. (2012). Induction of cellulase and hemicellulase activities of *Thermoascus aurantiacus* by xylan hydrolyzed products. *World J. Microbiol. Biotechnol.* 28 (1): 113–119.

Cervantes-Cisneros, D.E., Arguello-Esparza, D., Cabello-Galindo, A. et al. (2017). Hydrothermal processes for extraction of macroalgae high value-added compounds. In: *Hydrothermal Processing in Biorefineries* (eds. H. Ruiz, T.M. Hedegaard and H. Trajano), 461–481. Springer.

Chaturvedi, V. and Verma, P. (2013). An overview of key pretreatment processes employed for bioconversion of lignocellulosic biomass into biofuels and value added products. *3 Biotech* 3 (5): 415–431.

Chen, M.-H., Kim, S.M., Raab, R.M. et al. (2017). Heterologous expression of thermoregulated xylanases in switchgrass reduces the amount of exogenous enzyme required for saccharification. *Biomass Bioenergy* 107: 305–310.

Cherubini, F. (2010). The biorefinery concept: using biomass instead of oil for producing energy and chemicals. *Energy Convers. Manage.* 51 (7): 1412–1421.

Ciudad, G., Rubilar, O., Azócar, L. et al. (2014). Performance of an enzymatic extract in *Botryococcus braunii* cell wall disruption. *J. Biosci. Bioeng.* 117 (1): 75–80.

Cunha, F.M., Badino, A.C., and Farinas, C.S. (2017). Effect of a novel method for in-house cellulase production on 2G ethanol yields. *Biocatal. Agric. Biotechnol.* 9: 224–229.

Damis, S.I.R., Murad, A.M.A., Bakar, F.D.A. et al. (2019). Protein engineering of GH11 xylanase from *Aspergillus fumigatus* RT-1 for catalytic efficiency improvement on kenaf biomass hydrolysis. *Enzyme Microb. Technol.* 131: 109383.

De Jong, E., van Ree, R., and Kwant, I.K. (2009). Biorefineries: adding value to the sustainable utilisation of biomass. *IEA Bioenergy* 1: 1–16.

Deesukon, W., Nishimura, Y., Harada, N. et al. (2011). Purification, characterization and gene cloning of two forms of a thermostable endo-xylanase from *Streptomyces* sp. SWU10. *Process Biochem.* 46 (12): 2255–2262.

Diogo, J.A., Hoffmam, Z.B., Zanphorlin, L.M. et al. (2015). Development of a chimeric hemicellulase to enhance the xylose production and thermotolerance. *Enzyme Microb. Technol.* 69: 31–37.

Dutta, K., Daverey, A., and Lin, J.-G. (2014). Evolution retrospective for alternative fuels: first to fourth generation. *Renewable Energy* 69: 114–122.

Ehimen, E.A., Holm-Nielsen, J.-B., Poulsen, M., and Boelsmand, J. (2013). Influence of different pre-treatment routes on the anaerobic digestion of a filamentous algae. *Renewable Energy* 50: 476–480.

Florencio, C., Cunha, F.M., Badino, A.C. et al. (2016). Secretome analysis of *Trichoderma reesei* and *Aspergillus niger* cultivated by submerged and sequential fermentation processes: enzyme production for sugarcane bagasse hydrolysis. *Enzyme Microb. Technol.* 90: 53–60.

Fu, C.-C., Hung, T.-C., Chen, J.-Y. et al. (2010). Hydrolysis of microalgae cell walls for production of reducing sugar and lipid extraction. *Bioresour. Technol.* 101 (22): 8750–8754.

Gasol, C.M., Gabarrell, X., Anton, A. et al. (2007). Life cycle assessment of a *Brassica carinata* bioenergy cropping system in southern Europe. *Biomass Bioenergy* 31 (8): 543–555.

Geddes, C.C., Nieves, I.U., and Ingram, L.O. (2011). Advances in ethanol production. *Curr. Opin. Biotechnol.* 22 (3): 312–319.

Gervais, P. and Molin, P. (2003). The role of water in solid-state fermentation. *Biochem. Eng. J.* 13 (2–3): 85–101.

Gessesse, A. (1998). Purification and properties of two thermostable alkaline xylanases from an alkaliphilic *Bacillus* sp. *Appl. Environ. Microbiol.* 64 (9): 3533–3535.

Haitjema, C.H., Solomon, K.V., Henske, J.K. et al. (2014). Anaerobic gut fungi: advances in isolation, culture, and cellulolytic enzyme discovery for biofuel production. *Biotechnol. Bioeng.* 111 (8): 1471–1482.

Haki, G. and Rakshit, S. (2003). Developments in industrially important thermostable enzymes: a review. *Bioresour. Technol.* 89 (1): 17–34.

Hassan, S.S., Williams, G.A., and Jaiswal, A.K. (2019). Moving towards the second generation of lignocellulosic biorefineries in the EU: drivers, challenges, and opportunities. *Renewable Sustainable Energy Rev.* 101: 590–599.

Himmel, M.E., Ding, S.-Y., Johnson, D.K. et al. (2007). Biomass recalcitrance: engineering plants and enzymes for biofuels production. *Science* 315 (5813): 804–807.

Hua, C., Li, W., Han, W. et al. (2018). Characterization of a novel thermostable GH7 endoglucanase from *Chaetomium thermophilum* capable of xylan hydrolysis. *Int. J. Biol. Macromol.* 117: 342–349.

Huang, D., Liu, J., Qi, Y. et al. (2017). Synergistic hydrolysis of xylan using novel xylanases, β-xylosidases, and an α-l-arabinofuranosidase from *Geobacillus thermodenitrificans* NG80-2. *Appl. Microbiol. Biotechnol.* 101 (15): 6023–6037.

Hwang, S.-S., Lee, S.-J., Kim, H.-K. et al. (2008). Biodegradation and saccharification of wood chips of *Pinus strobus* and *Liriodendron tulipifera* by white rot fungi. *J. Microbiol. Biotechnol.* 18: 1819–1825.

Isroi, I., Millati, R., Niklasson, C. et al. (2011). Biological treatment of lignocelluloses with white-rot funghi and its applications. *BioResources* 6 (4): 5224–5259.

Jain, K.K., Dey, T.B., Kumar, S., and Kuhad, R.C. (2015). Production of thermostable hydrolases (cellulases and xylanase) from *Thermoascus aurantiacus* RCKK: a potential fungus. *Bioprocess. Biosyst. Eng.* 38 (4): 787–796.

Jin, X., Song, J., Ma, J., and Liu, G.-Q. (2020). Thermostable β-xylosidase from *Aspergillus fumigatus*: purification, characterization and potential application in lignocellulose bioethanol production. *Renewable Energy* 155: 1425–1431.

Juodeikiene, G., Basinskiene, L., Vidmantiene, D. et al. (2011). The use of β-xylanase for increasing the efficiency of biocatalytic conversion of crop residues to bioethanol. *Catal. Today* 167 (1): 113–121.

Kamat, S., Khot, M., Zinjarde, S. et al. (2013). Coupled production of single cell oil as biodiesel feedstock, xylitol and xylanase from sugarcane bagasse in a biorefinery concept using fungi from the tropical mangrove wetlands. *Bioresour. Technol.* 135: 246–253.

Kaur, G. and Satyanarayana, T. (2004). Production of extracellular pectinolytic, cellulolytic and xylanoytic enzymes by thermophilic mould *Sporotrichum thermophile* Apinis in solid state fermentation. *Indian J. Biotechnol.* 3: 552–557.

Kaur, G., Kumar, S., and Satyanarayana, T. (2004). Production, characterization and application of a thermostable polygalacturonase of a thermophilic mould *Sporotrichum thermophile* Apinis. *Bioresour. Technol.* 94 (3): 239–243.

Kim, K.H., Choi, I.S., Kim, H.M. et al. (2014). Bioethanol production from the nutrient stress-induced microalga *Chlorella vulgaris* by enzymatic hydrolysis and immobilized yeast fermentation. *Bioresour. Technol.* 153: 47–54.

Kim, C.K., Choi, H.S., Lee, S.J. et al. (2018). Production of xylanase from a novel engineered *Pichia pastoris* and application to enzymatic hydrolysis process for biorefinery. *Process Biochem.* 65: 130–135.

Kolasa, M., Ahring, B.K., Lübeck, P.S., and Lübeck, M. (2014). Co-cultivation of *Trichoderma reesei* RutC30 with three black *Aspergillus* strains facilitates efficient hydrolysis of pretreated wheat straw and shows promises for on-site enzyme production. *Bioresour. Technol.* 169: 143–148.

Koskinen, P.E., Beck, S.R., Örlygsson, J., and Puhakka, J.A. (2008). Ethanol and hydrogen production by two thermophilic, anaerobic bacteria isolated from Icelandic geothermal areas. *Biotechnol. Bioeng.* 101 (4): 679–690.

Kulkarni, N. and Rao, M. (1996). Application of xylanase from alkaliphilic thermophilic *Bacillus* sp. NCIM 59 in biobleaching of bagasse pulp. *J. Biotechnol.* 51 (2): 167–173.

Kumar, R. and Wyman, C.E. (2009). Effect of xylanase supplementation of cellulase on digestion of corn stover solids prepared by leading pretreatment technologies. *Bioresource technology* 100 (18): 4203–4213.

Kumar, P., Barrett, D.M., Delwiche, M.J., and Stroeve, P. (2009). Methods for pretreatment of lignocellulosic biomass for efficient hydrolysis and biofuel production. *Ind. Eng. Chem. Res.* 48 (8): 3713–3729.

Kumar, B., Bhardwaj, N., Alam, A. et al. (2018). Production, purification and characterization of an acid/alkali and thermo tolerant cellulase from *Schizophyllum commune* NAIMCC-F-03379 and its application in hydrolysis of lignocellulosic wastes. *AMB Express* 8 (1): 173.

Kumar, B. and Verma, P. (2020a). Biomass-based biorefineries: an important architype towards a circular economy. *Fuel*, 119622.

Kumar, B. and Verma, P. (2020b). Techno-economic assessment of biomass-based integrated biorefinery for energy and value-added product. In: *Biorefineries: A Step Towards Renewable and Clean Energy*, 581–616. Springer, Singapore.

Kumar, B. and Verma, P. (2020c). Life cycle assessment: blazing a trail for bioresources management. *Energy Convers. Manage. X*: 100063.

Kumar, B. and Verma, P. (2020d). Enzyme mediated multi-product process: a concept of bio-based refinery. *Ind. Crops Prod.* 154: 112607.

Kumar, B. and Verma, P. (2020e). Application of hydrolytic enzymes in biorefinery and its future prospects. In: *Microbial Strategies for Techno-economic Biofuel Production*, 59–83. *Springer*, Singapore.

Kumar, B., Bhardwaj, N., Agrawal, K., Chaturvedi, V., and Verma, P. (2020a). Current perspective on pretreatment technologies using lignocellulosic biomass: an emerging biorefinery concept. *Fuel Process. Technol.* 199: 106244.

Kumar, B., Bhardwaj, N., Agrawal, K., and Verma, P. (2020b). Bioethanol production: generation-based comparative status measurements. In: *Biofuel Production Technologies: Critical Analysis for Sustainability*, 155–201. Springer, Singapore.

Li, Q., Sun, B., Xiong, K. et al. (2017). Improving special hydrolysis characterization into *Talaromyces thermophilus* F1208 xylanase by engineering of N-terminal extension and site-directed mutagenesis in C-terminal. *Int. J. Biol. Macromol.* 96: 451–458.

Long, L., Tian, D., Zhai, R. et al. (2018). Thermostable xylanase-aided two-stage hydrolysis approach enhances sugar release of pretreated lignocellulosic biomass. *Bioresour. Technol.* 257: 334–338.

Luo, H., Wang, Y., Li, J. et al. (2009). Cloning, expression and characterization of a novel acidic xylanase, XYL11B, from the acidophilic fungus *Bispora* sp. MEY-1. *Enzyme Microb. Technol.* 45 (2): 126–133.

Madadi, M., Tu, Y., and Abbas, A. (2017). Recent status on enzymatic saccharification of lignocellulosic biomass for bioethanol production. *Electron J. Biol.* 13 (2): 135–143.

Mahajan, C., Chadha, B., Nain, L., and Kaur, A. (2014). Evaluation of glycosyl hydrolases from thermophilic fungi for their potential in bioconversion of alkali and biologically treated *Parthenium hysterophorus* weed and rice straw into ethanol. *Bioresour. Technol.* 163: 300–307.

Mahdy, A., Mendez, L., Ballesteros, M., and González-Fernández, C. (2014). Enhanced methane production of *Chlorella vulgaris* and *Chlamydomonas reinhardtii* by hydrolytic enzymes addition. *Energy Convers. Manage.* 85: 551–557.

Maiti, S., Gallastegui, G., Suresh, G. et al. (2018). Microwave-assisted one-pot conversion of agro-industrial wastes into levulinic acid: an alternate approach. *Bioresour. Technol.* 265: 471–479.

Makolomakwa, M., Puri, A.K., Permaul, K., and Singh, S. (2017). Thermo-acid-stable phytase-mediated enhancement of bioethanol production using *Colocasia esculenta*. *Bioresour. Technol.* 235: 396–404.

Martín, M. and Grossmann, I.E. (2012). Optimal synthesis of sustainable biorefineries. In: *Integrated Biorefineries Design Analysis and Optimization* (eds. P.R. Stuart and M.M. El-Halwagi), 325–347. USA: CRC.

Maurya, D.P., Singla, A., and Negi, S. (2015). An overview of key pretreatment processes for biological conversion of lignocellulosic biomass to bioethanol. *3 Biotech* 5 (5): 597–609.

McLachlan, M.J., Sullivan, R.P., and Zhao, H. (2009). Directed enzyme evolution and high throughput screening. In: *Biocatalysis for the Pharmaceutical Industry-Discovery, Development, and Manufacturing* (eds. J.(.A.). Tao, G.-Q. Lin and A. Liese), 45–64. Singapore: Wiley.

Menon, V., Prakash, G., Prabhune, A., and Rao, M. (2010). Biocatalytic approach for the utilization of hemicellulose for ethanol production from agricultural residue using thermostable xylanase and thermotolerant yeast. *Bioresour. Technol.* 101 (14): 5366–5373.

Mohanram, S., Amat, D., Choudhary, J. et al. (2013). Novel perspectives for evolving enzyme cocktails for lignocellulose hydrolysis in biorefineries. *Sustainable Chem. Processes* 1 (1): 15.

Mohapatra, S., Jena, S., Jena, P.K. et al. (2020). Partial consolidated bioprocessing of pretreated *Pennisetum* sp. by anaerobic thermophiles for enhanced bioethanol production. *Chemosphere* 256: 127126.

Mooney, C.A., Mansfield, S.D., Touhy, M.G., and Saddler, J.N. (1998). The effect of initial pore volume and lignin content on the enzymatic hydrolysis of softwoods. *Bioresour. Technol.* 64 (2): 113–119.

Morone, A. and Pandey, R. (2014). Lignocellulosic biobutanol production: gridlocks and potential remedies. *Renewable Sustainable Energy Rev.* 37: 21–35.

Mustafa, A.M., Poulsen, T.G., and Sheng, K. (2016). Fungal pretreatment of rice straw with *Pleurotus ostreatus* and *Trichoderma reesei* to enhance methane production under solid-state anaerobic digestion. *Appl. Energy* 180: 661–671.

Nakayama, S., Kiyoshi, K., Kadokura, T., and Nakazato, A. (2011). Butanol production from crystalline cellulose by cocultured *Clostridium thermocellum* and *Clostridium saccharoperbutylacetonicum* N1-4. *Appl. Environ. Microbiol.* 77 (18): 6470–6475.

Nigam, P.S. (2013). Microbial enzymes with special characteristics for biotechnological applications. *Biomolecules* 3 (3): 597–611.

Öhgren, K., Bura, R., Saddler, J., and Zacchi, G. (2007). Effect of hemicellulose and lignin removal on enzymatic hydrolysis of steam pretreated corn stover. *Bioresource Tech.* 98 (13): 2503–2510.

Ouephanit, C., Boonvitthya, N., Theerachat, M. et al. (2019). Efficient expression and secretion of endo-1, 4-β-xylanase from *Penicillium citrinum* in non-conventional yeast *Yarrowia lipolytica* directed by the native and the preproLIP2 signal peptides. *Protein Expression Purif.* 160: 1–6.

Pachapur, V.L., Brar, S.K., and Le Bihan, Y. (2019). Wood hydrolysate biorefinery: improvements and tailor-made two-step strategies on hydrolysis techniques. *Bioresour. Technol.* 299: 122632.

Paës, G. and O'Donohue, M.J. (2006). Engineering increased thermostability in the thermostable GH-11 xylanase from *Thermobacillus xylanilyticus*. *J. Biotechnol.* 125 (3): 338–350.

Pandey, A.K., Edgard, G., and Negi, S. (2016). Optimization of concomitant production of cellulase and xylanase from *Rhizopus oryzae* SN5 through EVOP-factorial design technique and application in Sorghum Stover based bioethanol production. *Renewable Energy* 98: 51–56.

Pareek, N., Gillgren, T., and Jönsson, L.J. (2013). Adsorption of proteins involved in hydrolysis of lignocellulose on lignins and hemicelluloses. *Bioresour. Technol.* 148: 70–77.

Plácido, J. and Capareda, S. (2015). Ligninolytic enzymes: a biotechnological alternative for bioethanol production. *Bioresour. Bioprocess.* 2 (1): 23.

Prasertsan, P., Khangkhachit, W., Duangsuwan, W. et al. (2017). Direct hydrolysis of palm oil mill effluent by xylanase enzyme to enhance biogas production using two-steps thermophilic fermentation under non-sterile condition. *Int. J. Hydrogen Energy* 42 (45): 27759–27766.

Quiñones, T.S., Retter, A., Hobbs, P.J. et al. (2015). Production of xylooligosaccharides from renewable agricultural lignocellulose biomass. *Biofuels* 6 (3–4): 147–155.

Rakotoarivonina, H., Hermant, B., Aubry, N., and Rémond, C. (2015). Engineering the hydrophobic residues of a GH11 xylanase impacts its adsorption onto lignin and its thermostability. *Enzyme Microb. Technol.* 81: 47–55.

Rani, D.S. and Nand, K. (2000). Production of thermostable cellulase-free xylanase by *Clostridium absonum* CFR-702. *Process Biochem.* 36 (4): 355–362.

Rastogi, S., Soni, R., Kaur, J., and Soni, S.K. (2016). Unravelling the capability of *Pyrenophora phaeocomes* S-1 for the production of ligno-hemicellulolytic enzyme cocktail and simultaneous bio-delignification of rice straw for enhanced enzymatic saccharification. *Bioresour. Technol.* 222: 458–469.

Reinhardt, G.A., Rettenmaier, N., Gärtner, S.O., Pastowski, A. (2007). Rain forest for biodiesel? Ecological effects of using palm oil as a source of energy. WWF GERMANY, 50.

Ribeiro, L.F., De Lucas, R.C., Vitcosque, G.L. et al. (2014). A novel thermostable xylanase GH10 from *Malbranchea pulchella* expressed in *Aspergillus nidulans* with potential applications in biotechnology. *Biotechnol. Biofuels* 7 (1): 115.

Saba, N., Jawaid, M., Hakeem, K. et al. (2015). Potential of bioenergy production from industrial kenaf (*Hibiscus cannabinus* L.) based on Malaysian perspective. *Renewable Sustainable Energy Rev.* 42: 446–459.

Saini, J.K., Saini, R., and Tewari, L. (2015). Lignocellulosic agriculture wastes as biomass feedstocks for second-generation bioethanol production: concepts and recent developments. *3 Biotech* 5 (4): 337–353.

Sakthiselvan, P., Madhumathi, R., and Partha, N. (2015). Eco friendly bio-butanol from sunflower oil sludge with production of xylanase. *Eng. Agric. Environ. Food* 8 (4): 212–221.

Salvachúa, D., Prieto, A., López-Abelairas, M. et al. (2011). Fungal pretreatment: an alternative in second-generation ethanol from wheat straw. *Bioresour. Technol.* 102 (16): 7500–7506.

Sanjivkumar, M., Silambarasan, T., Balagurunathan, R., and Immanuel, G. (2018). Biosynthesis, molecular modeling and statistical optimization of xylanase from a mangrove associated actinobacterium *Streptomyces variabilis* (MAB3) using Box-Behnken design with its bioconversion efficacy. *Int. J. Biol. Macromol.* 118: 195–208.

Schneider, W.D.H., dos Reis, L., Fontana, R.C. et al. (2018). Exploring strategies for the use of glycerol in the production of cellulases and xylanases, and the use of these enzymes in the hydrolysis of lignocellulosic biomass. *Ind. Crops Prod.* 122: 114–118.

Shariq, M. and Sohail, M. (2019). Application of *Candida tropicalis* MK-160 for the production of xylanase and ethanol. *J. King Saud Univ. Sci.* 31 (4): 1189–1194.

Sharma, H.K., Xu, C., and Qin, W. (2019). Biological pretreatment of lignocellulosic biomass for biofuels and bioproducts: an overview. *Waste Biomass Valorization* 10 (2): 235–251.

Sindhu, R., Binod, P., and Pandey, A. (2016). Biological pretreatment of lignocellulosic biomass–an overview. *Bioresour. Technol.* 199: 76–82.

Singh, B. (2016). *Myceliophthora thermophila* syn. *Sporotrichum thermophile*: a thermophilic mould of biotechnological potential. *Crit. Rev. Biotechnol.* 36 (1): 59–69.

Singh, A., Patel, A.K., Adsul, M. et al. (2017). Genetic modification: a tool for enhancing cellulase secretion. *Biofuel Res. J.* 4 (2): 600–610.

Song, Y., Lee, Y.G., Cho, E.J., and Bae, H.-J. (2020). Production of xylose, xylulose, xylitol, and bioethanol from waste bamboo using hydrogen peroxicde-acetic acid pretreatment. *Fuel* 278: 118247.

Taniguchi, M., Suzuki, H., Watanabe, D. et al. (2005). Evaluation of pretreatment with *Pleurotus ostreatus* for enzymatic hydrolysis of rice straw. *J. Biosci. Bioeng.* 100 (6): 637–643.

Taylor, M.P., Eley, K.L., Martin, S. et al. (2009). Thermophilic ethanologenesis: future prospects for second-generation bioethanol production. *Trends Biotechnol.* 27 (7): 398–405.

Terrone, C.C., de Freitas, C., Terrasan, C.R.F. et al. (2018). Agroindustrial biomass for xylanase production by *Penicillium chrysogenum*: purification, biochemical properties and hydrolysis of hemicelluloses. *Electron. J. Biotechnol.* 33: 39–45.

Teter, S., Sutton, K.B., and Emme, B. (2014). Enzymatic processes and enzyme development in biorefining. In: *Advances in Biorefineries* (ed. K. Waldron), 199–233. Elsevier.

Thomas, L., Joseph, A., and Gottumukkala, L.D. (2014a). Xylanase and cellulase systems of *Clostridium* sp.: an insight on molecular approaches for strain improvement. *Bioresour. Technol.* 158: 343–350.

Thomas, L., Ushasree, M.V., and Pandey, A. (2014b). An alkali-thermostable xylanase from *Bacillus pumilus* functionally expressed in Kluyveromyces lactis and evaluation of its deinking efficiency. *Bioresour. Technol.* 165: 309–313.

Thomas, L., Parameswaran, B., and Pandey, A. (2016). Hydrolysis of pretreated rice straw by an enzyme cocktail comprising acidic xylanase from *Aspergillus* sp. for bioethanol production. *Renewable Energy* 98: 9–15.

Uday, U.S.P., Choudhury, P., Bandyopadhyay, T.K., and Bhunia, B. (2016). Classification, mode of action and production strategy of xylanase and its application for biofuel production from water hyacinth. *Int. J. Biol. Macromol.* 82: 1041–1054.

Ulaganathan, K., Goud, S., Reddy, M., and Kayalvili, U. (2017). Genome engineering for breaking barriers in lignocellulosic bioethanol production. *Renewable Sustainable Energy Rev.* 74: 1080–1107.

Vaishnav, N., Singh, A., Adsul, M. et al. (2018). Penicillium: the next emerging champion for cellulase production. *Bioresour. Technol. Rep* 2: 131–140.

Valášková, V. and Baldrian, P. (2006). Estimation of bound and free fractions of lignocellulose-degrading enzymes of wood-rotting fungi *Pleurotus ostreatus*, *Trametes versicolor* and *Piptoporus betulinus*. *Res. Microbiol.* 157 (2): 119–124.

Vanegas, C., Hernon, A., and Bartlett, J. (2015). Enzymatic and organic acid pretreatment of seaweed: effect on reducing sugars production and on biogas inhibition. *Int. J. Ambient Energy* 36 (1): 2–7.

Velikodvorskaya, T., Volkov, I., Vasilevko, V. et al. (1997). Purification and some properties of *Thermotoga neapolitana* thermostable xylanase B expressed in *E. coli* cells. *Biochemistry. Biokhimiia* 62 (1): 66–70.

Verhaart, M.R., Bielen, A.A., Oost, J.v.d. et al. (2010). Hydrogen production by hyperthermophilic and extremely thermophilic bacteria and archaea: mechanisms for reductant disposal. *Environ. Technol.* 31 (8–9): 993–1003.

Visser, E.M., Leal, T.F., de Almeida, M.N., and Guimarães, V.M. (2015). Increased enzymatic hydrolysis of sugarcane bagasse from enzyme recycling. *Biotechnol. Biofuels* 8 (1): 5.

Wan, C. and Li, Y. (2012). Fungal pretreatment of lignocellulosic biomass. *Biotechnol. Adv.* 30 (6): 1447–1457.

Wei, W., Wu, S., and Liu, L. (2012). Enzymatic saccharification of dilute acid pretreated eucalyptus chips for fermentable sugar production. *Bioresour. Technol.* 110: 302–307.

Wood, I.P., Cook, N.M., Wilson, D.R. et al. (2016). Ethanol from a biorefinery waste stream: saccharification of amylase, protease and xylanase treated wheat bran. *Food Chem.* 198: 125–131.

Xin, F. and He, J. (2013). Characterization of a thermostable xylanase from a newly isolated *Kluyvera* species and its application for biobutanol production. *Bioresour. Technol.* 135: 309–315.

Xue, D., Zeng, X., Lin, D., and Yao, S. (2019). Thermostable ethanol tolerant xylanase from a cold-adapted marine species *Acinetobacter johnsonii*. *Chin. J. Chem. Eng.* 27 (5): 1166–1170.

Yang, M., Zhang, J., Kuittinen, S. et al. (2015). Enhanced sugar production from pretreated barley straw by additive xylanase and surfactants in enzymatic hydrolysis for acetone–butanol–ethanol fermentation. *Bioresour. Technol.* 189: 131–137.

Yang, Y., Yang, J., Liu, J. et al. (2018). The composition of accessory enzymes of *Penicillium chrysogenum* P33 revealed by secretome and synergistic effects with commercial cellulase on lignocellulose hydrolysis. *Bioresour. Technol.* 257: 54–61.

Zanuso, E., Lara-Flores, A.A., Aguilar, D.L. et al. (2017). Kinetic modeling, operational conditions, and biorefinery products from hemicellulose: depolymerization and solubilization during hydrothermal processing. In: *Hydrothermal Processing in Biorefineries* (eds. H. Ruiz, T.M. Hedegaard and H. Trajano), 141–160. Springer.

Zeldes, B.M., Keller, M.W., Loder, A.J. et al. (2015). Extremely thermophilic microorganisms as metabolic engineering platforms for production of fuels and industrial chemicals. *Front. Microbiol.* 6: 1209.

Zhang, J., Tuomainen, P., Siika-Aho, M., and Viikari, L. (2011). Comparison of the synergistic action of two thermostable xylanases from GH families 10 and 11 with thermostable cellulases in lignocellulose hydrolysis. *Bioresour. Technol.* 102 (19): 9090–9095.

Zhang, J., Tang, M., and Viikari, L. (2012). Xylans inhibit enzymatic hydrolysis of lignocellulosic materials by cellulases. *Bioresour. Technol.* 121: 8–12.

Zhuo, R., Yu, H., Qin, X. et al. (2018). Heterologous expression and characterization of a xylanase and xylosidase from white rot fungi and their application in synergistic hydrolysis of lignocellulose. *Chemosphere* 212: 24–33.

8

Microbial Cellulolytic-Based Biofuel Production

S.M. Bhatt

Agriculture Department, ACET Amritsar, Amritsar Group of Colleges, Punjab, India

8.1 Introduction

The main driving force for biofuel production is the need for blending due to a depleted fossiliferous fuel, which has a permissible limit of 10–20% and for a country like India, which has a lot of biowaste or agriculture waste produced during alternating seasons and is burned in almost every state in India without exploring any possibilities of conversion into biofuel ethanol and without caring much about increasing risk of GHG emissions today.

Competitive resources like the uses of starch against lignocellulosic-based biofuel may be an economical option, but feeding humans is of prime importance, even more than biofuel productions. Therefore, the biofuel cost crisis is a major issue when we are talking about lignocellulosic-based biofuel, because of the enzymatic conversion involved, which is costly.

Therefore, lignocellulosic-based biofuel is not very economical today because bioprocessing is costlier. As we all know, fossil fuel is depleting rapidly in the country and around 50% of the economy is wasted buying fossil-based fuel. In this chapter, we have focused on the type of biofuels, their requirements, and diversity, as well as the application of various cellulase enzyme and economical productions.

8.2 Biofuel Classifications

8.2.1 Generations of Biofuel

As mentioned in Figure 8.1, based on food sources, biofuel can be categorized into first, second, and third generations. Sometimes a fourth generation also involved. First-generation examples include bioethanol biodiesel, while second generations involve biomasses, such as lignocellulosic biomass. Third generations involve bioethanol production from algae.

First-generation biofuel includes food sources like starches and vegetables are often used for production of first-generation biofuel. There is easy procedure for bioethanol production and with direct bioprocessing, it is possible after hydrolysis using yeast microbes for ethanol using food sources and can be obtained after hydrolysis.

Bioenergy Research: Evaluating Strategies for Commercialization and Sustainability, First Edition.
Edited by Neha Srivastava and Manish Srivastava.

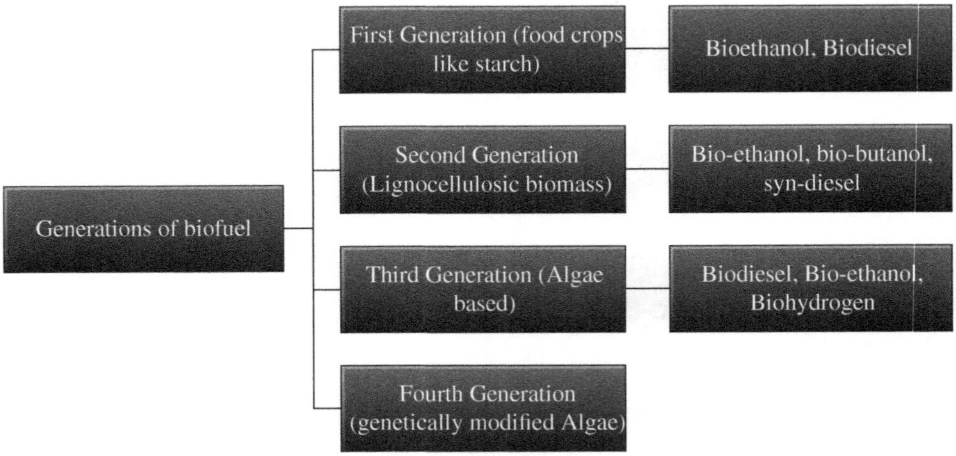

Figure 8.1 Biofuel generation.

A good example of the first-generation biofuel is biodiesel production and bioethanol production. Transesterification is the main process used for synthesis of biodiesel where cooking oil is used that has rich transfats. Biodiesel can be directly used in the engine or after mixing up with the diesel. So for this intention, mostly vegetable oil is utilized, which is also used for cooking purposes.

Biogas can be produced after anaerobic digestion of organic materials under anaerobic conditions are involved in the biodegradation of waste material in a correctly designed anaerobic digester. They have a dual benefit for creating biofertilizer as well as biogas.

Another form of first-generation use is *bio-alcohol and biodiesel fuel,* which can be obtained by direct fermentation of starch and sugar. Mostly ethanol butanol is produced and in some cases, propanol is also obtained.

Biobutanol is not known to be better than fossil fuel, because this can be directly used in place of petrol in the engine.

Biobutanol production can be produced by ABE fermentation method. Some workers believe that biobutanol is more energy-efficient for direct use in the petrol engine as compared to ethanol.

The last first-generation biofuel is syngas. Gasification method is used to produce syngas where oxygen and organic matter gets pyrolysis after combustion. The carbon monoxide helps in converting gas.

8.2.2 Bioethanol Production Using Lignocellulose

Bioethanol's main source in the first generation is mostly starch materials (e.g. corn, cassava, potato, barley, wheat sugar beet, sweet sorghum, and sugarcane) (Ho et al. 2014).

Fermentation of C6 sugars yields a high amount of ethanol, where yeast *Saccharomyces cerevisiae* can be used to convert glucose into ethanol (Turner et al. 2018).

The process is $C_6H_{12}O_6 \leftrightarrow 2\,C_2H_5OH + 2\,CO_2$.

Largest Biological Source Available on Earth Lignocellulose can be Harnessed for Bioethanol Production

S. No.	Biowaste used	Process followed/yield	References
1.	Coir waste substrate	SSF/submerged fermentation (SmF) Cellulase production Cellulase in SSF 14.6 fold> SmF *Apergillus niger* were shown to be at pH 6, temperature 30 °C.	Mrudula and Murugammal (2011)
2.	Wheat bran	*Trichoderma reesei* NCIM 992 cellulase production by wheat bran SSF mode, cellulase production was 2.63 U ml^{-1}	Maurya et al. (2012)
3.	Rice straw + wheat bran	*Apergillus niger and Trichoderma reesei,* respectively. Mixed-culture SSF using rice straw + wheat bran in 3 : 2 high endoglucanase (CMCase) FP cellulase, β-glucosidase, and xylanase activities	Dhillon et al. (2011)
4.	Various agro waste	Submerged fermentation (SmF)	Srivastava et al. (2018)
5.	Various agro waste	SSF > SmF	Imran et al. (2016)
6.			
7.	Sorghum stover	*Trichoderma reesei* RUT C-30 + solid-state fermentation (SSF)	Idris et al. (2017)
8.	Sugarcane bagasse	*Pycnoporus sanguineus* +	Yoon et al. (2013)
9.	Kinnow mandarin (*Citrus reticulata*)	SSF central composite design	Oberoi et al. (2011)
10.	Olive mill wastes	SSF sonication xylanase with 3.6-fold increase; cellulase with 1.2-fold increase	Leite et al. (2016)
11.	Sugarcane bagasse	CG 104NH strain	Florencio et al. (2012)
12	Sugarcane bagasse	*Trichoderma reesei* NRRL 11460	Singhania et al. (2006)
13	Sugarcane bagasse	SSF MODE *Penicillium funiculosum*	Maeda et al. (2013)
14	Sugarcane bagasse	SSF + CBP (SSF) simultaneous saccharification and fermentation (CBP) consolidated bioprocessing	Zanchetta et al. (2018)
15	Sugarcane bagasse	*Geobacillus* sp. HTA426 thermophilic cellulase	Potprommanee et al. (2017)
16	Sugarcane bagasse	*Kluyveromyces marxianus* IMB3, SSF MODE, Cellulase 30 FPU cellulase g^{-1}	Silva et al. (2015)

S. No.	Biowaste used	Process followed/yield	References
17	Sugarcane bagasse	*Myceliophthora thermophile*	Mendes et al. (2017)
18	Sugarcane bagasse	Xylnase, *Streptomyces viridosporus* T7A	Alberton et al. (2009)
19	Sugarcane bagasse	*Trichoderma harzaianum*	Asgher et al. (2016)
20	Sugarcane bagasse	*Pycnoporus sanguineus* PF-2	Falkoski et al. (2012)
21	Sugarcane bagasse	*Aspergillus tubingensis* NKBP-55	Prajapati et al. (2018)
22	Sugarcane bagasse	*Bacillus* sp. SMIA-2	Ladeira et al. (2015)
23	Sugarcane bagasse	Two yeast strains, MK-157 and MK-118	Qadir et al. (2018)
24	Wheat straw	SSF mode/SSF optimal conditions 40 °C, 120 h, cellulase loading 40 FPU/(g wheat straw)	Chen et al. (2008)
25	Wheat straw	Recombinant bacterium	Saha et al. (2015)
26	Wheat straw	*Bacillus licheniformis* A99	Archana and Satyanarayana (1997)
27	Wheat straw	*Saccharomyces cerevisiae*	Pierre et al. (2011)
28	Wheat straw	Lactobacillus plantarum	Moraïs et al. (2013)
29	Wheat straw	cellulase	Coimbra et al. (2016)
30	Wheat straw	*Trametes versicolor and Pleurotus ostreatus*	Shrivastava et al. (2011)
31	Wheat straw	*Trametes versicolor* m4D and *Phanerochaete chrysosporium*, *Trametes versicolor 52J*	Gao et al. (2017)
32	Wheat straw	*Aspergillus niger* NS-2	Bansal et al. (2012)
33	Wheat straw	*Saccharomyces cerevisiae* and *Fusarium oxysporum*	Paschos et al. (2015)
34	Wheat straw	*Aspergillus niger* NS-2	Bansal et al. (2012)
35	Wheat straw	M2n[TLG1-SFA1], and *Saccharomyces cerevisiae* MEL2[TLG1-SFA1]	Cripwell et al. (2015)
36	Wheat straw	*Aspergillus tubingensis* JP-1 xylanase	Pandya and Gupte (2012)
37	Wheat straw	*Bacillus* sp. H1666	Harshvardhan et al. (2013)
38	Wheat straw	*Trichoderma reesei*	Grewal et al. (2017)
39	Wheat straw	Soybean hulls *Trichoderma reesei* and *Aspergillus oryzae*	Brijwani et al. (2010)
40	Wheat straw	*Thermoascus aurantiacus* RCKK	Jain et al. (2017)
41	Wheat straw	*Trichoderma reesei* 3EMS35 mutant,	Khokhar et al. (2014)
42	Wheat straw	*Kluyveromyces marxianus* DBTIOC-35	Saini et al. (2015)

8.2.2.1 Polymeric Lignocellulosic Composition

Cellulosic biopolymer have β1,4-glycosidic bonds between various joined units of D-glucose. The breaking β-1,4-glycosidic bonds are difficult because of highly crystalline nature, and thus resist while in enzymatic hydrolysis. It needs various enzymes, such as endoglucanase, exoglucanase, and beta-glucosidase, to finally convert cellulose first into cellobiose and then to glucose (Chen and Chen 2014).

Plant cell walls are often rich in C5 sugar. Hemicellulose mainly contains xylans made up of xylose residues cross-linking cellulose microfibrils in a plant secondary wall and lignin are impregnated in between, as shown in Figure 8.4 via ferulic acid components (Taherzadeh and Karimi 2008).

Hemicellulose is made up of pentose sugar l-arabinose and d-xylose subunit xylose, and mannose is one of the most abundant sugars among hemicellulose-rich biomass, which usually has been rich in corn or any other herbs (Chen and Chen 2014).

Enzymatic hydrolysis of lignocellulose for ethanol production is usually done (Ahring et al. 1996; Margeot et al. 2009). Lignocellulose is composed of cellulose and hemicellulose, and upon hydrolysis, gives rise to hexoses (C_6) and pentoses (C_5) sugars (Rubin 2008). A perfect blend of pretreatment strategies and fermentation of C_6 and C_5 sugars can enhance commercial cellulosic ethanol productions.

Lignocellulosic materials from various wastes, e.g., bagasse or wheat stalk or rice straw, can be a suitable feedstock source for ethanol production and many bioprocessing techniques have been found to be useful such as enzymatic bioprocessing or microbial consortia (Liu et al. 2020).

8.3 Bioprocessing of Bagasse for Bioethanol Production

Sugarcane bagasse has been largely used for bioethanol production, and is reported to be rich in glucan and xylan. Bagasse is a common feedstock for biofuel production in Brazil and other European countries and is producing the highest ethanol based on bagasse.

The flow chart of the complete process is shown in Figure 8.2, which is simple to perform for production of ethanol, but needs the following steps: (i) Crushing to remove sucrose using water; (ii) Purification of sugar, removing impurities before starting fermentations.

So to use for bioethanol production from bagasse, pretreatment is required. During pretreatment (which is the first part of processing), cellulosic biomass is hydrolyzed into the crystalline cellulosic structure and then further into glucose.

In Figure 8.2, the method to obtain the cellulosic ethanol is depicted. Obtaining ethanol from sugarcane bagasse is a challenge since it contains lignocellulose biomass and has not been adopted commercially yet. Ethanol production can be divided into prehydrolysis and posthydrolysis process.

Prehydrolysis requires the cleaning of biomass before the pretreatment steps. Then pretreatment is done in order to remove lignin from cellulose and hemicellulose, which often includes laborious techniques and releases a large number of inhibitors along with C5 and C6 sugars. Table 8.1 shows the percentage of ethanol produced using various feedstock.

Some authors recommended and applied various techniques such as milling (Da Silva et al. 2010). According to Saha and colleagues, sugarcane bagasse contains 40% hemicellulose with additional pretreatment methods (Saha 2003).

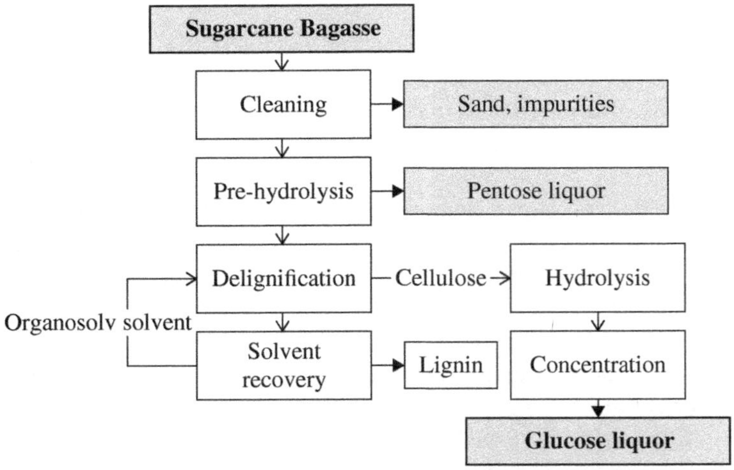

Figure 8.2 Flow chart of separating sugar from bagasse. *Source:* From Dias et al. (2009). © 2009, Elsevier.

Table 8.1 Feedstock and ethanol production shown in (%).

Feedstock	Ethanol produced
Sawdust	0.70%
Switchgrass	0.37%
Cardboard	1.67%
Corn stalks	0.60%

For bioconversion of biomass into ethanol, the first step is to hydrolyze lignocellulosic biomass, which then gets hydrolyzed into fermentable sugars, and that is why enzymatic-based pretreatment is excogitated in order to get sub-components separated into C5 and C6 sugars using endoxylanase, or β-xylosidase. Sometimes, α-L-arabinofuranosidase is recommended for hemicellulose bioconversion. Hemicellulose is beneficial in that it can be processed into a diversity of valuable by-products, such as hemicellulose, for conversion of various useful biofuels, such as 2,3-butanediol or ethanol.

Advantages of pretreatment can be noted as follows:

1. Crystallinity is reduced
2. Cellulose is separated from lignin and hemicellulose
3. It improves cellulase hydrolysis
4. With organosolv and acid pretreatment, it is possible to separate C5 and C6 sugars with an alkaline hydrogen peroxide pretreatment and it does have better digestibility and separation of cellulose and hemicellulose (Rabelo et al. 2014).
   With CO_2 subcritical pretreatment followed by enzymatic pretreatment. up to a 93% glucose yield can be obtained (the highest glucose yield = 38.5 g/100 g sugarcane bagasse after 72 hours).

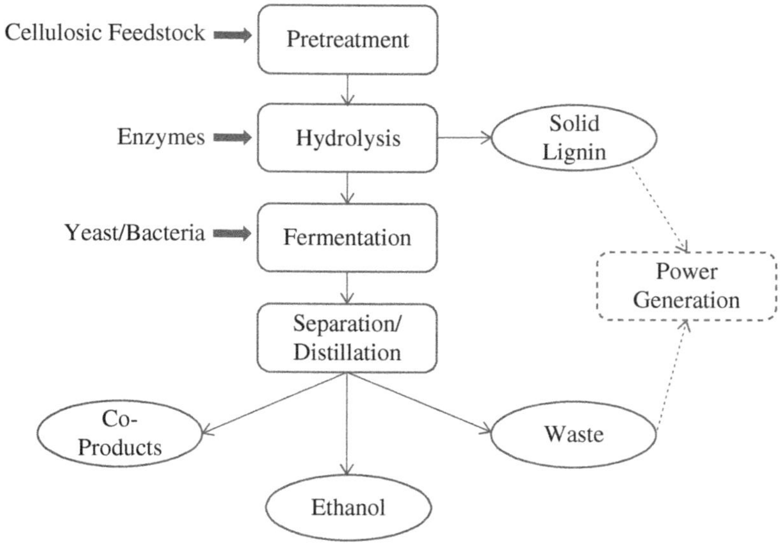

Figure 8.3 Common procedure to obtain the ethanol using cellulosic feedstock. *Source:* Based on Mesa et al. (2010).

5. The organosolv pretreatment followed by enzymatic hydrolysis (Mesa et al. 2010), along with xylanase can be very useful because of high fermentable sugars obtained in the form of glucose, which could yield high levels of ethanol.

A common procedure to obtain the cellulosic feedstock–based ethanol is presented in Figure 8.3. The first step is the milling, then the pretreatment, and then separation of the cellulose and hemicellulose out of lignin and other carbon materials. After separation of lignin, the second step is the enzymatic hydrolysis of cellulose and hemicellulose at a large scale. The third and the last step is fermentation by yeast or bacteria to obtain ethanol. Ethanol is distilled after the fermentation and another by-product is also recovered.

8.3.1 Enzymatic Hydrolysis and Cellulose Structure

8.3.1.1 Cellulolytic Microbes

There are various known microbes involved in cellulolytic production but only in specific conditions. Various thermophilic microbes are involved in productions of ethanol bacteria (Chang and Yao 2011). Various fungus are now identified as industrial strains: *Trichoderma,* viz. *T. reesei* or mutants can best produce high extracellular cellulase (Ryu and Mandels 1980), whereas insects have their own cellulase secreting (Watanabe and Tokuda 2010) and are known to hydrolyze the β-1, 4-glycosidic linkages present within the polymeric cellulose. Some bacteria fermented to produce butanol and yeast can be engineered to produce more butanol instead of ethanol. The combined action of exocellulases, endocellulases, glucanohydrolases, and beta-glucosidases are required to convert cellulosic-based feedstock into glucose (Li and Papageorgiou 2019; Li et al. 2011).

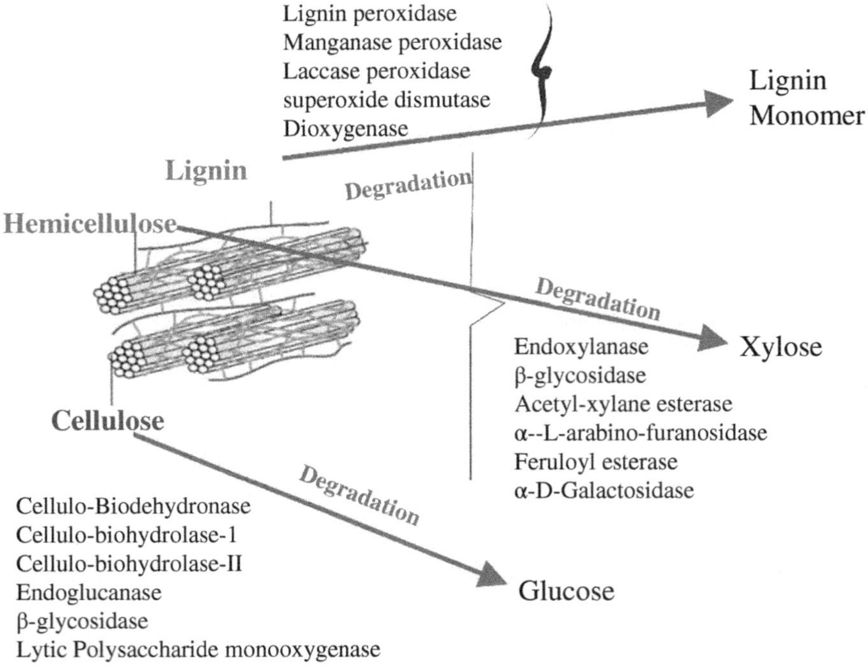

Figure 8.4 Diversity of enzymes required for lignocellulosic hydrolysis adapted from (Champreda et al. 2019). *Source:* Adapted from Champreda et al. (2019).

Enzyme used in lignocellulosic-based ethanol are also depicted in Figure 8.4.

1. Cellulase
2. Hemicellulase
3. Lignase
4. Pectinase
5. Xylanse

8.4 Microbial Cellulase

There are various microbes that produces cellulase enzyme (Roth et al. 2020). Such microbes have been listed here and are mostly from symbiotic fungi, some bacteria, and some termites harboring special microbes (Juturu and Wu 2014). Table 8.2 shows genetically modified microbes for high cellulase production.

Glycoside hydrolases (EC 3.2.1.-) family

1. Endoglucanase (E.C. 3.2.1.4),
2. Exoglucanase (E.C. 3.2.1.176) and (E.C. 3.2.1.91)
3. β-glucosidase (E.C. 3.2.1.21).

Attached to the cell wall of bacteria, are cellulosomes and cellulose-binding domain (CBD) with a unique catalytic domain (CD) that play significant roles on binding and conversion of biomass. Cellulosomes are found to be a highly active component found in

Table 8.2 List of microbes genetically modified (Bischof et al. 2016).

S. No	Species	Type of exoglucanases	References
1	*Trichoderma reesei*	produces two exoglucanases (CBH I and CBH II) eight endoglucanases (EGI-EGVIII), and seven ß-glucosidases (BG I-BG VII)	Meenu et al. (2014)
2	*Trichoderma reesei* and *Aspergillus niger*	moderately tolerant to temperatures above 50 °C	Biswas et al. (2014)

Source: Based on Bischof et al. (2016).

Table 8.3 Biomass detail composition can be found in Zhang (2019).

Biomass commonly used for biofuel productions
Corn stover, sugarcane bagasse
Corn, switchgrass obs, grasses
Wheat straw, oat straw, rice straw, rice husk
Hardwood Black locust Hybrid poplar Eucalyptus
Newspaper, Softwood-pine, Hardwood stems Nut shells

Source: Based on Zhang (2019).

many microbes and are involved in the hydrolysis of cellulose and are also evident from avicel hydrolysis (You et al. 2012).

According to environmental temperature, they may be classified as:

1. Thermophilic cellulases
2. Mesophilic cellulases
3. Psychrophilic cellulases

There are various improvements in the action of cellulase enzyme due to genetic engineering improvements in order to obtain novel cellulase enzyme (Druzhinina and Kubicek 2017). Improvement started with *Trichoderma* where *reesei* species were improved tremendously. Sometimes cocktails of the enzyme are required, owing to the low catalytic activity, in order to improved hydrolysis and cellulosome is mostly present in anaerobic microbes, and mostly *T. reesei* is used for the industrial production of cellulase (Figure 8.5; Tables 8.3 and 8.4).

8.5 Mode of Economical Production of Enzyme

Cellulase and endoxylanase are distinguished components known for quick hydrolysis and dissolution into cellodextrin (Figure 8.3).

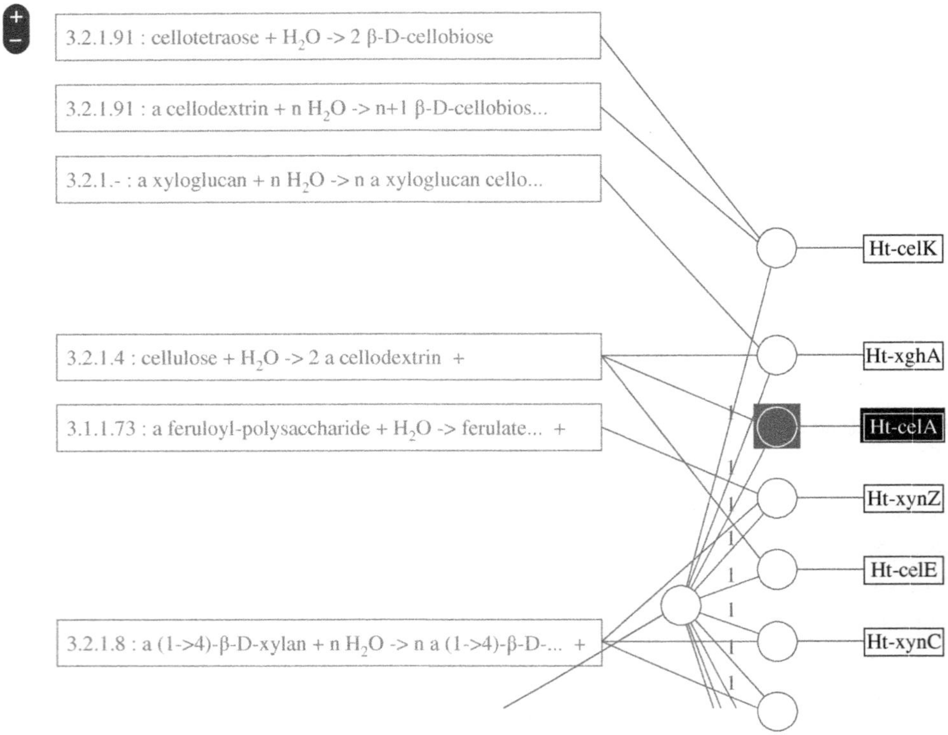

Figure 8.5 Different reaction by cellulase. Source: From Caspi R., Gene-Reaction Schematic. © 2011, SRI International.

For cost-effective biowaste management, there is a need for economic process developments of pilot-scale production of cellulase and hemicellulose. SSF is the key bioreactor system where cellulase and xylanase can be produced using a perfect strain of fungi and using solid substrates, such as saw dust or any other fine cellulose particles. The growth of fungi requires a fine solid substrate with a high moisture content at optimum pH and temperature. For improved enzyme production, one needs to isolate high strains. Rapid enzymatic hydrolysis depends on a suitable fermenter condition, such as the right pH and the right temperature, which is called optimum temperature, salt combinations, and moisture.

In the traditional methods, an ethanol-producing strain, such as *Saccharomyces cerevisiae* can be used, but has problems of sustainability in high concentrations of ethanol, which will be there and which can be solved by some genetic modifications and by yeast, which can now tolerate inhibitory compounds and ethanol.

However, pentoses, such as xylose and arabinose, are difficult to metabolize by *S. cerevisiae* as well as the development of a novel yeast strain with capacity to C6 and C5 hydrolysis sugars.

$$C_{12}H_{22}0_{11} + H_20- \rightarrow 4CH_3CH_2OH + 4\,CO_2$$

The cellulase action is because of the following reasons: (i) increased β-glucosidase activity and (ii) improved cellulose degradation. Therefore, more action of the enzyme is required, such as the addition of xylanase and also LMPO. The LMPO addition has

Table 8.4 Cellulosome microbes.

S. No	Cellulosome Microbes[a]	References
1	• *Clostridium thermocellum*	Gefen et al. (2012)
2	• *Ruminococcus flavefaciens*	Kirby et al. (1997)
3	• *Ruminococcus albus*	Vijayarani et al. (2000)
4	• *Acetivibrio cellulolyticus*	MacKenzie and Bilous (1982)
5	• *Bacteroides cellulosolvens*	Xu et al. (2004)
6	• *Clostridium acetobutylicum*	Fierobe et al. (2012)
7	• *Clostridium papyrosolvens*	Rani et al. (2004)
8.	• *Clostridium josui*	Ohmiya (2005)
9.	• *Clostridium clariflavum*	Paye et al. (2016)
10	• *Clostridium cellulovorans*	Ciolacu et al. (2010)
11	• *Clostridium cellulolyticum*	Ravachol et al. (2014)

a) *T. reesei* genome database (http://genome.jgi.doe.gov/Trire2/Trire2.home.html).
Source: Based on CELLULOSOME MICROBES http://genome.jgi.doe.gov/Trire2/Trire2.home.html

improved cellulose production in a much more efficient way (Bischof et al. 2016). Lytic polysaccharide monoxygenases (LPMO) mainly solubilize lignin, and thus, are more accessible to cellulose.

8.6 Structure of Cellulase

Typically, cellulase structure of two components are cellobiohydrolases and endocellulases and their structure consists of four different domains or regions as shown in Figure 8.6. A CBD acts as an anchor with feedstock. A linker region, which is rich in tri-amino acid, such as Ser, Thr, and Pro residues, and a CD region, are involved in feedstock hydrolysis The mature proteins are *O*- and *N*-glycosylated in the hinge region and the CDs, respectively. Glycosylation sites are present in the hinge region, but its role is not clear yet (Li and Papageorgiou 2019; Li et al. 2011).

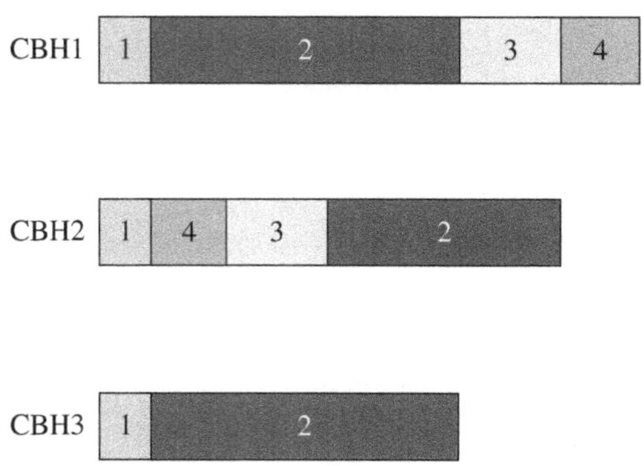

Figure 8.6 DOMAIN structure of CBH1, CBH2, and CBH3.

Table 8.5 Mode of cellulase production (SSF SMF).

S. No.	Feedstock	SSF/SMF	Cellulase	References
1	Lignocellulosic biomass	SSF	Factors are genetic modification, co-culture of different fungal strains	Yoon et al. (2014)
2			*Aspergillus niger*	Journal (2011)
3			*Bacillus coagulans*	Ou et al. (2009)
4		The sugar cane bagasse	*Penicillium funiculosum*	Maeda et al. (2013)
5		Agricultural Wastes		Imran et al. (2016)

8.6.1 CBH1 Structure

This domain belongs to cellobiohydrolase from uniport-annotated function is hydrolase activity, hydrolyzing O-glycosyl compounds, and cellulose binding a domain.

8.6.2 Thermophilic Cellulase Enzyme

GH5 cellulase are most abundant in fungi, as shown in Table 8.5.

8.7 Family Classification

Swollenins is an expansin-like protein, which is reported in Trichoderma, Aspergillus, and Penicillium species. These are helping to loosen the crystalline structure of cellulose by disrupting H-bonds, and have domains similar to G-45-like domains. A cellulose-digesting enzyme has been grouped in a database called CAZymes, which has been classified into glycoside hydrolases (GH), polysaccharides lyases (PL), glycosyltransferases (GT), carbohydrate esterases (CE), and the auxiliary activity (AA), based on the amino acid sequence (www.cazy.org).

Modified enzyme for various carbohydrate-digesting enzymes including the following:

a. **Glycoside hydrolases (GHs)**: Attacks glycosidic bonds
b. **Glycosyltransferases (GTs)**: Formation of glycosidic bonds
c. **Polysaccharide lyases (PLs):** Nonhydrolytic cleavage of glycosidic bonds
d. **Carbohydrate esterases (CEs):** Hydrolysis of carbohydrate esters
e. **Auxiliary activities (AAs):** Redox enzymes that act in conjunction with CAZymes

CBMs are classified as CBDs, which are bound to cellulose (Tomme et al. 1988), and amino acid similarity has been the base for the classification of CBDs (Alvarez et al. 2013), but now, classification has been introduced using roman numerals (e.g. type I or type II CBDs).

A list of the CBM-containing proteins in the CAZy database in the CBM family. CBM from different families can be grouped into superfamilies or clans (Cantarel et al. 2009; Lombard et al. 2013).

Most cellulase enzymes are grouped in various names, such as GH5, 7, and 12 in the family of GH6 and 45. The nonreducing end of carbohydrates or the reducing end of the cellulose chain cleaves cellobiose into glucose and smaller subcomponents.

8.8 Consortia-Based Cellulase Production

Although *T. reesei* cellulases have been proclaimed to be "the industry standard" for enzymatic lignocellulose hydrolysis, that is why the economical way of high production using a pilot scale has been done via developing consortia (Cerda et al. 2017) in reactors. Consortia-based technology works best in SSF mode for the production of cellulase and hemicellulose. Before introducing in 50 l bioreactor, a microbial community was selected and grown and then productivity was measured at high temperatures. Results were shown to produce high cellulase and hemicellulose specially, and that is why consortia is required. *T. reesei* is good only for production of cellulase and some parts of hemicellulose but lacks β-galactosidase at the pilot scale (it produces only a negligible amount); therefore, keep in mind that complete digestion hydrolysis and conversion of cellulose requires the blending of various enzymes (Srivastava et al. 2015). Thus, supplementation meets complementation, and that is why the consortia model is more likely to be industrially fit for biofuel production. For this, *Aspergillus niger* is supplemented, along with *T. reesei* to complement the enzyme production (Chauve et al. 2010; Narasimha et al. 2016). *Aspergillus niger*, an addition was reported to alleviate additional problems of inhibition by the by-product cellobiose and glucose, which were found to be decreased by the cellulolytic hydrolysis rate, as well as the product yield in cellulose hydrolysis (Aliyah et al. 2017).

Using biological consortia, biogas production can be done (Tantayotai et al. 2017).

The designing of consortia needs genomic tools in order to understand their natural and ecological interactions, as well as the perfect environment to grow (McCarty and Ledesma-Amaro 2019). A very good review was recently written about synthetic consortia design using genetic engineering approaches. In actuality, it uses synthetic biology tools to understand the physiology of microbes, so that we can say that this is the most evolving technology of the future and where, based on optimal designs of system biology approaches, scientists can predict the most powerful microbial consortia for a high cellulase system (Qian et al. 2020). Unique interactions among the diverse group of microbes makes them a robust future tool in various biotechnological applications, such as bioremediations and biofuel production.

8.9 Cellulase Production SSF Mode

SSF has the benefit at the industrial scale, because end-product inhibition is minimized, and this enhanced production of cellulase can be done along with a number of other benefits. The main benefit of SSF is that they are more economical when compared to other techniques (Journal 2011) when using the trichoderma species (Maurya et al. 2012; Srivastava et al. 2018). As per one report, enhanced production in SSF lies in various factors, such as substrate size, fungal species used, and activation of cellulosome (Imran et al. 2016). Various optimization methods are also helpful in the high production of cellulase, for example, SSF, and optimizing media, by using response surface methodology (RSM)(Kumar et al. 2011). Some other methods of optimization are working fine and can be adapted to optimize various nutrient conditions for the optimal growth of microbes. Since enzyme is protein, suitable nutrient media must also be included. Many combinations of nutrients have carbon source nitrogen phosphorous and minerals to enhance growth. Other olive pomace agrowastes (Filipe et al. 2020) have been used beside rice waste, wheat straw, and water hyacinth has been used (Deka et al. 2018). Some authors experimented with a mixture of cellulosic substrate and it was found to have an enhanced result in one particular experiment conducted recently (Masutti et al. 2015). The author has intensely mixed a number of substrates, such as grape stalks, grape seeds, and wheat bran to enhance the production of cellulase in SSF mode. This often meets with several delimitations, such as oxygen transfer, mass transfer, and lack of proper aerations under normal conditions required for the proper growth of microbes. Recently, it was reported that various edible mushrooms can be used for high enzymes, for instance, mushroom-spent compost gives a better yield of cellulase by using Trichoderma species (Grujić et al. 2015). Some authors have studied cellulase enzyme production using mushroom-Pleurotus sajor-caju over sugarcane pressmud (Pandit and Maheshwari 2012). Once such novel strain has been isolated from mushroom compost and purified for a high production of cellulase (Sharma and Bajaj 2018).

8.10 Concluding Remarks

Cellulase production depends on various factors required to produce large-scale biofuel production. Genetic engineering has improved the production of cellulase by a lot. Mostly, experiments have been done on *T. reesei* modifications. So, it is expected that in the future, the price can be lowered due to modification.

Declarations

Author declares that there is no conflict of interested with any agencies.

Acknowledgment

Author acknowledges the support of Ms. Shilpa Bhatt in handling the article.

References

Ahring, B.K., Jensen, K., Nielsen, P. et al. (1996). Pretreatment of wheat straw and conversion of xylose and xylan to ethanol by thermophilic anaerobic bacteria. *Bioresour. Technol.* 58 (2): 107–113.

Alberton, L.R., Vandenberghe, L.P.d.S., Assmann, R. et al. (2009). Xylanase production by *Streptomyces viridosporus* T7A in submerged and solid-state fermentation using agro-industrial residues. *Braz. Arch. Biol. Technol.* https://doi.org/10.1590/S1516-89132009000700022.

Aliyah, A., Alamsyah, G., Ramadhani, R., and Hermansyah, H. (2017). Production of α-amylase and β-glucosidase from *Aspergillus niger* by solid state fermentation method on biomass waste substrates from rice husk, bagasse and corn cob. *Energy Proc.* https://doi.org/10.1016/j.egypro.2017.10.269.

Alvarez, T.M., Paiva, J.H., Ruiz, D.M. et al. (2013). Structure and function of a novel cellulase 5 from sugarcane soil metagenome. *PLoS One.* https://doi.org/10.1371/journal.pone.0083635.

Archana, A. and Satyanarayana, T. (1997). Xylanase production by thermophilic *Bacillus licheniformis* A99 in solid-state fermentation. *Enzyme Microb. Technol.* https://doi.org/10.1016/S0141-0229(96)00207-4.

Asgher, M., Wahab, A., Bilal, M., and Nasir Iqbal, H.M. (2016). Lignocellulose degradation and production of lignin modifying enzymes by *Schizophyllum* commune IBL-06 in solid-state fermentation. *Biocatal. Agric. Biotechnol.* https://doi.org/10.1016/j.bcab.2016.04.003.

Bansal, N., Tewari, R., Soni, R., and Soni, S.K. (2012). Production of cellulases from *Aspergillus niger* NS-2 in solid state fermentation on agricultural and kitchen waste residues. *Waste Manage. (Oxford)* 32: 1341–1346. https://doi.org/10.1016/j.wasman.2012.03.006.

Bischof, R.H., Ramoni, J., and Seiboth, B. (2016). Cellulases and beyond: the first 70 years of the enzyme producer *Trichoderma reesei*. *Microb. Cell Fact.* https://doi.org/10.1186/s12934-016-0507-6.

Biswas, R., Persad, A., and Bisaria, V.S. (2014). Production of cellulolytic enzymes. In: *Bioprocessing of Renewable Resources to Commodity Bioproducts.* https://doi.org/10.1002/9781118845394.ch5.

Brijwani, K., Oberoi, H.S., and Vadlani, P.V. (2010). Production of a cellulolytic enzyme system in mixed-culture solid-state fermentation of soybean hulls supplemented with wheat bran. *Process Biochem.* https://doi.org/10.1016/j.procbio.2009.08.015.

Cantarel, B.I., Coutinho, P.M., Rancurel, C. et al. (2009). The carbohydrate-active EnZymes database (CAZy): an expert resource for glycogenomics. *Nucleic Acids Res.* https://doi.org/10.1093/nar/gkn663.

Cerda, A., Mejías, L., Gea, T., and Sánchez, A. (2017). Cellulase and xylanase production at pilot scale by solid-state fermentation from coffee husk using specialized consortia: the consistency of the process and the microbial communities involved. *Bioresour. Technol.* https://doi.org/10.1016/j.biortech.2017.07.076.

Champreda, V., Mhuantong, W., Lekakarn, H. et al. (2019). Designing cellulolytic enzyme systems for biorefinery: from nature to application. *J. Biosci. Bioeng.* https://doi.org/10.1016/j.jbiosc.2019.05.007.

Chang, T. and Yao, S. (2011). Thermophilic, lignocellulolytic bacteria for ethanol production: current state and perspectives. *Appl. Microbiol. Biotechnol.* https://doi.org/10.1007/s00253-011-3456-3.

Chauve, M., Mathis, H., Huc, D. et al. (2010). Comparative kinetic analysis of two fungal β-glucosidases. *Biotechnol. Biofuels.* https://doi.org/10.1186/1754-6834-3-3.

Chen, H. and Chen, H. (2014). Chemical composition and structure of natural lignocellulose. In: *Biotechnology of Lignocellulose.* https://doi.org/10.1007/978-94-007-6898-7_2.

Chen, H., Han, Y., and Xu, J. (2008). Simultaneous saccharification and fermentation of steam exploded wheat straw pretreated with alkaline peroxide. *Process Biochem.* 43: 462–1466. https://doi.org/10.1016/j.procbio.2008.07.003.

Ciolacu, D., Kovac, J., and Kokol, V. (2010). The effect of the cellulose-binding domain from *Clostridium cellulovorans* on the supramolecular structure of cellulose fibers. *Carbohydr. Res.* https://doi.org/10.1016/j.carres.2009.12.023.

Coimbra, M.C., Duque, A., Saéz, F. et al. (2016). Sugar production from wheat straw biomass by alkaline extrusion and enzymatic hydrolysis. *Renewable Energy* https://doi.org/10.1016/j.renene.2015.09.026.

Cripwell, R., Favaro, L., Rose, S.H. et al. (2015). Utilisation of wheat bran as a substrate for bioethanol production using recombinant cellulases and amylolytic yeast. *Appl. Energy* https://doi.org/10.1016/j.apenergy.2015.09.062.

Da Silva, A.S.A., Inoue, H., Endo, T. et al. (2010). Milling pretreatment of sugarcane bagasse and straw for enzymatic hydrolysis and ethanol fermentation. *Bioresour. Technol.* 101: 7402–7409. https://doi.org/10.1016/j.biortech.2010.05.008.

Deka, D., Das, S.P., Ravindran, R., Jawed, M., and Goyal, A. (2018). Water hyacinth as a potential source of biofuel for sustainable development. doi: https://doi.org/10.1007/978-3-319-74494-0_27

Dhillon, G.S., Oberoi, H.S., Kaur, S. et al. (2011). Value-addition of agricultural wastes for augmented cellulase and xylanase production through solid-state tray fermentation employing mixed-culture of fungi. *Ind. Crops Prod.* https://doi.org/10.1016/j.indcrop.2011.04.001.

Dias, M.O.S., Ensinas, A.V., Nebra, S.A. et al. (2009). Production of bioethanol and other bio-based materials from sugarcane bagasse: integration to conventional bioethanol production process. *Chem. Eng. Res. Des.* 87: 1206–1216. https://doi.org/10.1016/j.cherd.2009.06.020.

Druzhinina, I.S. and Kubicek, C.P. (2017). Genetic engineering of *Trichoderma reesei* cellulases and their production. *Microb. Biotechnol.* https://doi.org/10.1111/1751-7915.12726.

Falkoski, D.L., Guimarães, V.M., De Almeida, M.N. et al. (2012). Characterization of cellulolytic extract from *Pycnoporus sanguineus* PF-2 and its application in biomass saccharification. *Appl. Biochem. Biotechnol.* https://doi.org/10.1007/s12010-012-9565-3.

Fierobe, H.P., Mingardon, F., and Chanal, A. (2012). Engineering cellulase activity into *Clostridium acetobutylicum*. *Method. Enzymol.* https://doi.org/10.1016/B978-0-12-415931-0.00016-1.

Filipe, D., Fernandes, H., Castro, C. et al. (2020). Improved lignocellulolytic enzyme production and antioxidant extraction using solid-state fermentation of olive pomace mixed with winery waste. *Biofuels, Bioprod. Biorefin.* https://doi.org/10.1002/bbb.2073.

Florencio, C., Couri, S., and Farinas, C.S. (2012). Correlation between agar plate screening and solid-state fermentation for the prediction of cellulase production by trichoderma strains. *Enzyme Res.* https://doi.org/10.1155/2012/793708.

Gao, W., Tabil, L.G., Dumonceaux, T. et al. (2017). Optimization of biological pretreatment to enhance the quality of wheat straw pellets. *Biomass Bioenergy* https://doi.org/10.1016/j.biombioe.2016.12.012.

Gefen, G., Anbar, M., Morag, E. et al. (2012). Enhanced cellulose degradation by targeted integration of a cohesin-fused β-glucosidase into the *Clostridium thermocellum* cellulosome. *Proc. Natl. Acad. Sci. U.S.A.* https://doi.org/10.1073/pnas.1202747109.

Grewal, J., Ahmad, R., and Khare, S.K. (2017). Development of cellulase-nanoconjugates with enhanced ionic liquid and thermal stability for in situ lignocellulose saccharification. *Bioresour. Technol.* https://doi.org/10.1016/j.biortech.2017.04.007.

Grujić, M., Dojnov, B., Potočnik, I. et al. (2015). Spent mushroom compost as substrate for the production of industrially important hydrolytic enzymes by fungi *Trichoderma* spp. and *Aspergillus niger* in solid state fermentation. *Int. Biodeterior. Biodegrad.* https://doi.org/10.1016/j.ibiod.2015.04.029.

Harshvardhan, K., Mishra, A., and Jha, B. (2013). Purification and characterization of cellulase from a marine *Bacillus* sp. H1666: a potential agent for single step saccharification of seaweed biomass. *J. Mol. Catal. B: Enzym.* https://doi.org/10.1016/j.molcatb.2013.04.009.

Ho, D.P., Ngo, H.H., and Guo, W. (2014). A mini review on renewable sources for biofuel. *Bioresour. Technol.* https://doi.org/10.1016/j.biortech.2014.07.022.

Idris, A.S.O., Pandey, A., Rao, S.S., and Sukumaran, R.K. (2017). Cellulase production through solid-state tray fermentation, and its use for bioethanol from sorghum stover. *Bioresour. Technol.* https://doi.org/10.1016/j.biortech.2017.03.092.

Imran, M., Anwar, Z., Irshad, M. et al. (2016). Cellulase production from species of fungi and bacteria from agricultural wastes and its utilization in industry: a review. *Adv. Enzyme Res.* https://doi.org/10.4236/aer.2016.42005.

Jain, K.K., Kumar, S., Deswal, D., and Kuhad, R.C. (2017). Improved production of thermostable cellulase from *Thermoascus aurantiacus* RCKK by fermentation bioprocessing and its application in the hydrolysis of office waste paper, algal pulp, and biologically treated wheat straw. *Appl. Biochem. Biotechnol.* https://doi.org/10.1007/s12010-016-2249-7.

Journal, B. (2011). Production of cellulase by. *Braz. J. Microbiol.* 42: 1119–1127.

Juturu, V. and Wu, J.C. (2014). Microbial cellulases: engineering, production and applications. *Renewable Sustainable Energy Rev.* https://doi.org/10.1016/j.rser.2014.01.077.

Khokhar, Z.U., Syed, Q.u.A., Wu, J., and Athar, M.A. (2014). On-site cellulase production by Trichoderma reesei 3EMS35 mutant and same vessel saccharification and fermentation of acid treated wheat straw for ethanol production. *EXCLI J.* https://doi.org/10.17877/DE290R-15480.

Kirby, J., Martin, J.C., Daniel, A.S., and Flint, H.J. (1997). Dockerin-like sequences in cellulases and xylanases from the rumen cellulolytic bacterium *Ruminococcus flavefaciens. FEMS Microbiol. Lett.* https://doi.org/10.1016/S0378-1097(97)00078-5.

Kumar, S., Sharma, H.K., and Sarkar, B.C. (2011). Effect of substrate and fermentation conditions on pectinase and cellulase production by *Aspergillus niger* NCIM 548 in submerged (SmF) and solid state fermentation (SSF). *Food Sci. Biotechnol.* https://doi.org/10.1007/s10068-011-0178-3.

Ladeira, S.A., Cruz, E., Delatorre, A.B. et al. (2015). Cellulase production by thermophilic *Bacillus* sp. SMIA-2 and its detergent compatibility. *Electron. J. Biotechnol.* https://doi.org/10.1016/j.ejbt.2014.12.008.

Leite, P., Salgado, J.M., Venâncio, A. et al. (2016). Ultrasounds pretreatment of olive pomace to improve xylanase and cellulase production by solid-state fermentation. *Bioresour. Technol.* https://doi.org/10.1016/j.biortech.2016.05.028.

Li, D.-C. and Papageorgiou, A.C. (2019). Cellulases from thermophilic fungi: recent insights and biotechnological potential. In: *Fungi in Extreme Environments: Ecological Role and Biotechnological Significance*. https://doi.org/10.1007/978-3-030-19030-9_20.

Li, D.C., Li, A.N., and Papageorgiou, A.C. (2011). Cellulases from thermophilic fungi: recent insights and biotechnological potential. *Enzyme Res.* https://doi.org/10.4061/2011/308730.

Liu, Y.-J., Li, B., Feng, Y., and Cui, Q. (2020). Consolidated bio-saccharification: leading lignocellulose bioconversion into the real world. *Biotechnol. Adv.*: 107535.

Lombard, V., Golaconda Ramulu, H., Drula, E. et al. (2013). The carbohydrate-active EnZymes database (CAZy): an expert resource for glycogenomics. *Nucleic Acids Res.* https://doi.org/10.1186/1754-6834-6-41.

MacKenzie, C.R. and Bilous, D. (1982). Location and kinetic properties of the cellulase system of Acetivibrio cellulolyticus. *Can. J. Microbiol.* https://doi.org/10.1139/m82-171.

Maeda, R.N., Barcelos, C.A., Anna, L.M.M.S., and Pereira, N. (2013). Cellulase production by *Penicillium funiculosum* and its application in the hydrolysis of sugar cane bagasse for second generation ethanol production by fed batch operation. *J. Biotechnol.* https://doi.org/10.1016/j.jbiotec.2012.10.014.

Margeot, A., Hahn-Hagerdal, B., Edlund, M. et al. (2009). New improvements for lignocellulosic ethanol. *Curr. Opin. Biotechnol.* 20 (3): 372–380.

Masutti, D.C., Borgognone, A., Scardovi, F. et al. (2015). Effects on the enzymes production from different mixes of agro-food wastes. *Chem. Eng. Trans.* https://doi.org/10.3303/CET1543082.

Maurya, D.P., Singh, D., Pratap, D., and Maurya, J.P. (2012). Optimization of solid state fermentation conditions for the production of cellulase by *Trichoderma reesei*. *J. Environ. Biol.*

McCarty, N.S. and Ledesma-Amaro, R. (2019). Synthetic biology tools to engineer microbial communities for biotechnology. *Trends Biotechnol.* https://doi.org/10.1016/j.tibtech.2018.11.002.

Meenu, K., Singh, G., and Vishwakarma, R.A. (2014). Molecular mechanism of Cellulase production systems in trichoderma. In: Biotechnology and Biology of Trichoderma. https://doi.org/10.1016/B978-0-444-59576-8.00022-9.

Mendes, F., Dias, M., Ferraz, A., Milagres, A.M.F., Santos, J.C., & Bonomi, A. (2017). Impact of varied compositional profiles of sugarcane experimental hybrids on a biorefinery producing 1G-2G ethanol and electricity. In: *39th Symposium on Biotechnology for Fuels and Chemicals*.

Mesa, L., González, E., Cara, C. et al. (2010). An approach to optimization of enzymatic hydrolysis from sugarcane bagasse based on organosolv pretreatment. *J. Chem. Technol. Biotechnol.* https://doi.org/10.1002/jctb.2404.

Moraïs, S., Shterzer, N., Grinberg, I.R. et al. (2013). Establishment of a simple lactobacillus plantarum cell consortium for cellulase-xylanase synergistic interactions. *Appl. Environ. Microbiol.* https://doi.org/10.1128/AEM.01211-13.

Mrudula, S. and Murugammal, R. (2011). Production of cellulase by *Aspergillus niger* under submerged and solid state fermentation using coir waste as a substrate. *Braz. J. Microbiol.* 42 (3): 1119–1127.

Narasimha, G., Sridevi, A., Ramanjaneyulu, G., and Rajasekhar Reddy, B. (2016). Purification and characterization of β-glucosidase from *Aspergillus niger*. *Int. J. Food Prop.* https://doi.org/10.1080/10942912.2015.1023398.

Oberoi, H.S., Vadlani, P.V., Nanjundaswamy, A. et al. (2011). Enhanced ethanol production from Kinnow mandarin (*Citrus reticulata*) waste via a statistically optimized simultaneous saccharification and fermentation process. *Bioresour. Technol.* 102: 1593–1601. https://doi.org/10.1016/j.biortech.2010.08.111.

Ohmiya, K. (2005). Effective utilization of biomass by using anaerobic bacterial functions. *Seibutsu Kogaku Kaishi.*

Ou, M.S., Mohammed, N., Ingram, L.O., and Shanmugam, K.T. (2009). Thermophilic bacillus coagulans requires less cellulases for simultaneous saccharification and fermentation of cellulose to products than mesophilic microbial biocatalysts. *Appl. Biochem. Biotechnol.* https://doi.org/10.1007/s12010-008-8509-4.

Pandit, N.P. and Maheshwari, S.K. (2012). Optimization of cellulase enzyme production from sugarcane pressmud using oyster mushroom - Pleurotus Sajor-Caju by solid state fermentation. *J. Bioremed. Biodegrad.* https://doi.org/10.4172/2155-6199.1000140.

Pandya, J.J. and Gupte, A. (2012). Production of xylanase under solid-state fermentation by *Aspergillus tubingensis* JP-1 and its application. *Bioprocess. Biosyst. Eng.* https://doi.org/10.1007/s00449-011-0657-1.

Paschos, T., Xiros, C., and Christakopoulos, P. (2015). Simultaneous saccharification and fermentation by co-cultures of *Fusarium oxysporum* and *Saccharomyces cerevisiae* enhances ethanol production from liquefied wheat straw at high solid content. *Ind. Crops Prod.* https://doi.org/10.1016/j.indcrop.2015.07.061.

Paye, J.M.D., Guseva, A., Hammer, S.K. et al. (2016). Biological lignocellulose solubilization: comparative evaluation of biocatalysts and enhancement via cotreatment. *Biotechnol. Biofuels* https://doi.org/10.1186/s13068-015-0412-y.

Pierre, G., Maache-Rezzoug, Z., Sannier, F. et al. (2011). High-performance hydrolysis of wheat straw using cellulase and thermomechanical pretreatment. *Process Biochem.* https://doi.org/10.1016/j.procbio.2011.09.002.

Potprommanee, L., Wang, X.Q., Han, Y.J. et al. (2017). Characterization of a thermophilic cellulase from *Geobacillus* sp. HTA426, an efficient cellulase-producer on alkali pretreated of lignocellulosic biomass. *PLoS One.* https://doi.org/10.1371/journal.pone.0175004.

Prajapati, B.P., Kumar Suryawanshi, R., Agrawal, S. et al. (2018). Characterization of cellulase from *Aspergillus tubingensis* NKBP-55 for generation of fermentable sugars from agricultural residues. *Bioresour. Technol.* https://doi.org/10.1016/j.biortech.2017.11.099.

Qadir, F., Shariq, M., Ahmed, A., and Sohail, M. (2018). Evaluation of a yeast co-culture for cellulase and xylanase production under solid state fermentation of sugarcane bagasse using multivariate approach. *Ind. Crops Prod.* https://doi.org/10.1016/j.indcrop.2018.07.021.

Qian, X., Chen, L., Sui, Y. et al. (2020). Biotechnological potential and applications of microbial consortia. *Biotechnol. Adv.* https://doi.org/10.1016/j.biotechadv.2019.107500.

Rabelo, S.C., Andrade, R.R., Maciel Filho, R., and Costa, A.C. (2014). Alkaline hydrogen peroxide pretreatment, enzymatic hydrolysis and fermentation of sugarcane bagasse to ethanol. *Fuel.* https://doi.org/10.1016/j.fuel.2014.07.033.

Rani, D.S., Thirumale, S., and Nand, K. (2004). Production of cellulase by *Clostridium papyrosolvens* CFR-703. *World J. Microbiol. Biotechnol.* https://doi.org/10.1023/B:WIBI.0000043193.39486.32.

Ravachol, J., Borne, R., Tardif, C. et al. (2014). Characterization of all family-9 glycoside hydrolases synthesized by the cellulosome-producing bacterium *Clostridium cellulolyticum*. *J. Biol. Chem.* https://doi.org/10.1074/jbc.M113.545046.

Roth, J.C.G., Hoeltz, M., and Benitez, L.B. (2020). Current approaches and trends in the production of microbial cellulases using residual lignocellulosic biomass: a bibliometric analysis of the last 10 years. *Arch. Microbiol.* (0123456789). doi: https://doi.org/10.1007/s00203-019-01796-9.

Rubin, E.M. (2008). Genomics of cellulosic biofuels. *Nature* 454 (7206): 841–845.

Ryu, D.D.Y. and Mandels, M. (1980). Cellulases: biosynthesis and applications. *Enzyme Microb. Technol.* https://doi.org/10.1016/0141-0229(80)90063-0.

Saha, B.C. (2003). Hemicellulose bioconversion. *J. Ind. Microbiol. Biotechnol.* 30: 279–291. https://doi.org/10.1007/s10295-003-0049-x.

Saha, B.C., Nichols, N.N., Qureshi, N. et al. (2015). Pilot scale conversion of wheat straw to ethanol via simultaneous saccharification and fermentation. *Bioresour. Technol.* https://doi.org/10.1016/j.biortech.2014.10.060.

Saini, J.K., Agrawal, R., Satlewal, A. et al. (2015). Second generation bioethanol production at high gravity of pilot-scale pretreated wheat straw employing newly isolated thermotolerant yeast *Kluyveromyces marxianus* DBTIOC-35. *RSC Adv.* https://doi.org/10.1039/c5ra05792b.

Sharma, S. and Bajaj, B.K. (2018). Xylanase production from a new strain of *Aspergillus terreus* S9 and its application for saccharification of rice straw using combinatorial approach. *Environ. Prog. Sustainable Energy* https://doi.org/10.1002/ep.12779.

Shrivastava, B., Thakur, S., Khasa, Y.P. et al. (2011). White-rot fungal conversion of wheat straw to energy rich cattle feed. *Biodegradation.* https://doi.org/10.1007/s10532-010-9408-2.

Silva, G.M., Giordano, R.L.C., Cruz, A.J.G. et al. (2015). Ethanol production from sugarcane bagasse using SSF process and thermotolerant yeast. *Trans. ASABE* https://doi.org/10.13031/trans.58.11024.

Singhania, R.R., Sukumaran, R.K., Pillai, A., Prema, P., Szakacs, G., and Pandey, A. (2006). Solid-state fermentation of lignocellulosic substrates for cellulase production by *Trichoderma reesei*. NRRL 11460. 5(July):332–336.

Srivastava, N., Srivastava, M., Mishra, P.K. et al. (2015). Application of cellulases in biofuels industries: an overview. *J. Biofuel. Bioenergy* https://doi.org/10.5958/2454-8618.2015.00007.3.

Srivastava, N., Srivastava, M., Mishra, P.K. et al. (2018). Applications of fungal cellulases in biofuel production: advances and limitations. *Renewable Sustainable Energy Rev.* https://doi.org/10.1016/j.rser.2017.08.074.

Taherzadeh, M.J. and Karimi, K. (2008). Pretreatment of lignocellulosic wastes to improve ethanol and biogas production: a review. *Int. J. Mol. Sci.* https://doi.org/10.3390/ijms9091621.

Tantayotai, P., Pornwongthong, P., Muenmuang, C. et al. (2017). Effect of cellulase-producing microbial consortium on biogas production from lignocellulosic biomass. *Energy Proc.* 141: 180–183. https://doi.org/10.1016/j.egypro.2017.11.034.

Tomme, P., Van Tilbeurgh, H., Pettersson, G. et al. (1988). Studies of the cellulolytic system of *Trichoderma reesei* QM 9414: analysis of domain function in two cellobiohydrolases by limited proteolysis. *Eur. J. Biochem.* https://doi.org/10.1111/j.1432-1033.1988.tb13736.x.

Turner, T.L., Kim, H., Kong, I.I. et al. (2018). Engineering and evolution of *Saccharomyces cerevisiae* to produce biofuels and chemicals. *Adv. Biochem. Eng. Biotechnol.* https://doi.org/10.1007/10_2016_22.

Vijayarani, K., Thirumurugan, G., Nachimuthu, K., and Padmanaban, V.D. (2000). Cloning and expression of cellulase genes from ruminococcus albus. *Ind. Vet J.*

Watanabe, H. and Tokuda, G. (2010). Cellulolytic systems in insects. *Annu. Rev. Entomol.* https://doi.org/10.1146/annurev-ento-112408-085319.

Xu, Q., Bayer, E.A., Goldman, M. et al. (2004). Architecture of the *Bacteroides cellulosolvens* cellulosome: description of a cell surface-anchoring scaffoldin and a family 48 cellulase. *J. Bacteriol.* https://doi.org/10.1128/JB.186.4.968-977.2004.

Yoon, L.W., Ngoh, G.C., and Chua, A.S.M. (2013). Simultaneous production of cellulase and reducing sugar through modification of compositional and structural characteristic of sugarcane bagasse. *Enzyme Microb. Technol.* https://doi.org/10.1016/j.enzmictec.2013.05.005.

Yoon, L.W., Ang, T.N., Ngoh, G.C., and Chua, A.S.M. (2014). Fungal solid-state fermentation and various methods of enhancement in cellulase production. *Biomass Bioenergy.* https://doi.org/10.1016/j.biombioe.2014.05.013.

You, C., Zhang, X.Z., Sathitsuksanoh, N. et al. (2012). Enhanced microbial utilization of recalcitrant cellulose by an ex vivo cellulosome-microbe complex. *Appl. Environ. Microbiol.* https://doi.org/10.1128/AEM.07138-11.

Zanchetta, A., dos Santos, A.C.F., Ximenes, E. et al. (2018). Temperature dependent cellulase adsorption on lignin from sugarcane bagasse. *Bioresour. Technol.* https://doi.org/10.1016/j.biortech.2017.12.061.

Zhang, C. (2019). Lignocellulosic ethanol: technology and economics. In: *Alcohol Fuels-Current Technologies and Future Prospect*. IntechOpen.

9

Recent Developments of Bioethanol Production

Arla Sai Kumar[1], Sana Siva Sankar[2], S K Godlaveeti[1], Dinesh Kumar[3], S Dheiver[4], Ram Prasad[5], Chandrasekhar Nb[6], Thi Hong Chuong Nguyen[7,8] and Quyet Van Le[8]

[1] *Department of Materials Science and Nanotechnology, Yogi Vemana University, Kadapa, Andhra Pradesh, India*
[2] *School of Chemical Engineering and Technology, North University of China, Taiyuan, China*
[3] *Centre for Nanoscience and Nanotechnology, Sathayabama Institute of Science and Technology, Chennai*
[4] *Institute of Computing, Federal University of Alagoas (UFAL), Maceio, Brazil*
[5] *Department of Botany, Mahatma Gandhi Central University, Motihari, Bihar, India*
[6] *Research and Development Center Department of Biotechnology, Shridevi Institute of Engineering and Technology, Tumakuru, India*
[7] *Institute of Research and Development, Duy Tan University, Da Nang, Vietnam*
[8] *Faculty of Natural sciences, Duy Tan University, Danang, Vietnam*

9.1 Introduction

Bioethanol is otherwise known as ethanol, which has formula CH_3CH_2OH, which is used in beverages. This is one of the most commonly used biofuels across the globe (Ruan et al. 2019). Usually, they are synthesized by fermenting various raw materials rich with sugar or carbohydrate. Available feedstocks and substrates are categorized into four orders, which are shown in Table 9.1 (Gupta et al. 2014, Li et al. 2018, Sadeghinezhad et al. 2014, Taghizadeh-Alisaraei et al. 2017). The first generation is usually edible crops like potato, corn, vegetable oil, and so on. This has contributed to the production of bioethanol, but this has a large number of demerits. The food supply and land usage is the main problem, which can make the production less efficient. So the output from good source becomes a problem, so it is required to synthesize from nonedible raw materials. Then came the use of second-generation biomass. which includes lignocellulosic substances, etc. This is better as compared to the former one. The advantage is that they can grow with the least amount of water as well as manure. However, the first-generation feedstock has more amounts of sugar content compared to the later. The main thing is that the second-generation feedstock requires more capital cost as it needs more complicated equipment.

The second-generation biomass releases less amounts of greenhouse gas emissions (GHGs), and hence, the pollution level can be controlled. In comparison with other feedstocks, this is the best method (Halder et al. 2019). By considering the limitation of this first- and second-generation biofuel, scientists have developed other feedstock, which is much better as compared to the existing ones. This came into the beginning of third-generation biomass, which is by using the algal source (Figure 9.1) (Pandiyan et al.

Bioenergy Research: Evaluating Strategies for Commercialization and Sustainability, First Edition.
Edited by Neha Srivastava and Manish Srivastava.

Table 9.1 The generations of bioethanol.

Generation	Substrates	Feedstocks
1st	Sugar, grain, vegetable oil	Potato, corn, wheat, palm, maize, soybean, sugarcane, rapeseed, oil crops, sugar beet
2nd	Lignocellulosic	Agricultural crops (wheat, corn, sugarcane, etc.), forest residues, grass
3rd	Lipids and carbohydrates	Algae, microbes, yeast
4th	Industrial waste carbon dioxide	Vegetable oil, biodiesel

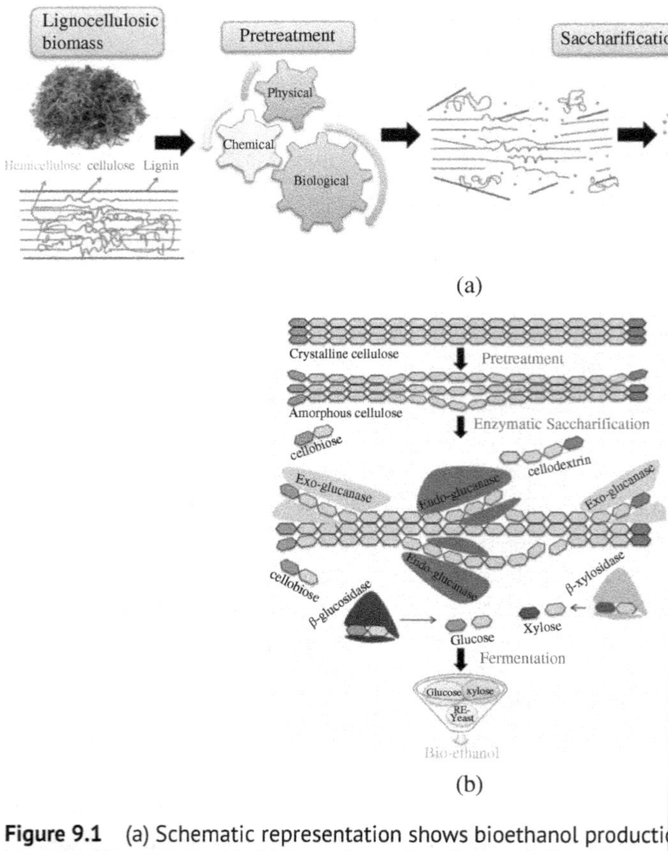

Figure 9.1 (a) Schematic representation shows bioethanol production from second-generation biomass (lignocellulosic), (b) Preliminary description of changes in pretreatment and enzymatic saccharification of lignocellulosic biomass. Source: Adapted with permission from Ref. Pandiyan et al. (2018).

2018), ie., microalgae, as well as microbes, in this case. These have brought a positive outcome and have more advantages compared to the other two. These microalgae can be grown in marginal land in the most efficient way. There is a maximum energy density produced as well as highest conversion efficiency, but this has a disadvantage that the bioethanol synthesized is the least stable, comparatively (Pandiyan et al. 2018).

Other chemical methods are available by collecting carbon dioxide with the help of techniques like electrochemical, electrolysis, petro-hydro processes, and so on, which is the origin of fourth-generation biofuel. They release only some amount of compounds containing carbon; hence as the GHG will. These methods are in their developing stages. The demand for biofuel is expected to increase every year, so in order to solve this issue, new techniques and processes are introduced. These biofuels are mainly used as transportation fuel as they are more versatile in nature. Bioethanol, as well as biodiesel, are mainly produced from renewable sources (Izaskun et al. 2019). The four generations of biofuel are described in Figure 9.2.

The most commonly used biofuel is bioethanol, which is produced from sugarcane or corn grain initially. This lignocellulosic feedstock is said to be an attractive one as it is readily available and easy to supply. Many modern techniques are developed in order to make the process more beneficial. Many methods are developed to make the method efficient and for faster conversion of this biomass into fuel (Rajiv and Rintu 2020). These include thermochemical treatments, new advanced fermentation techniques, and so on. Emerging techniques are there in order to obtain maximum yield and make the process more effective (Gray et al. 2006). Use of these sources makes the generation more efficient, and with this, the required pretreatment should be carried out for better conversion (Rastogi and

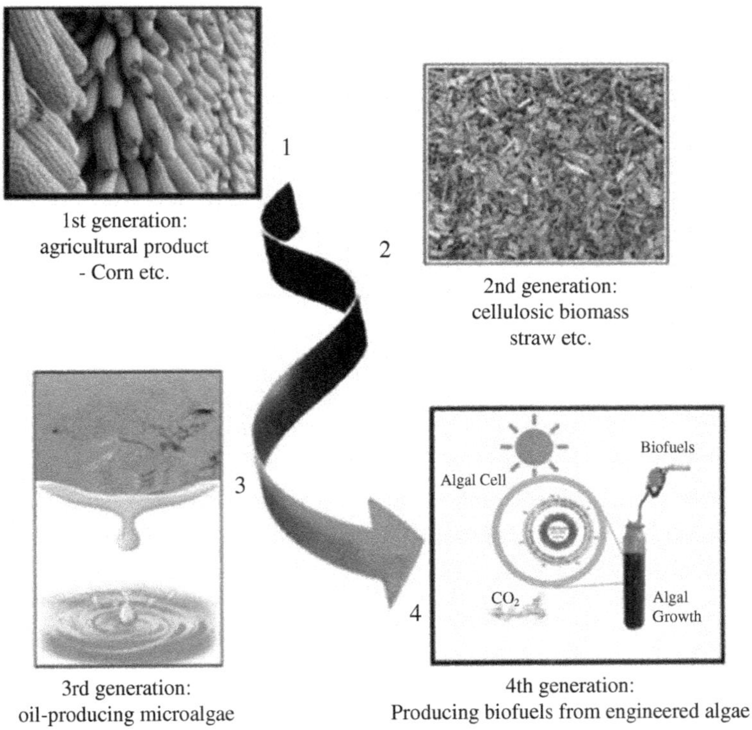

Figure 9.2 Four generations of biofuel. Source: Adapted with permission from Ref. Lü et al. (2011). Copyright (2011) Royal Society of Chemistry.

Shrivastava 2017). Bioethanol synthesis from these biomass sources can reduce the release of GHG as well as fossil fuel consumption can be minimized. From past technologies, a large number of by-products are developed, which becomes one of the major drawbacks (Gavahian et al. 2019). The various advancements in bioethanol production are discussed in this chapter.

9.2 Emerging Techniques in Bioethanol Production

The worldwide population will increase by approximately 8 billion by 2040, so this will ultimately increase the requirement of biofuel by 30% compared to current projections. Nowadays, the price of crude oil is rising because the source is getting depleted. Along with this, by using this type of source, the amount of GHG to the atmosphere also increases. This forced the energy sector to make changes in production methods and to find sustainable procedures. As a result of this, more focus was made to use an advanced technology that could produce maximum energy in the most ecofriendly way. In this case, in the transport sector, this biofuel holds great importance due to its advantage in replenishment as well as storage capacity. In some countries like Brazil biofuel is produced and sold at a competitive price. Biofuels also have other merits like less carbon dioxide emission, safer for the environment, better lubricant, produce less pollutant, and so on (Catarina et al. 2018). In addition, this emerging biofuel technology is expected to be of further assistance to the agricultural sector as well as to new domestic employment opportunities. In addition, bioethanol is a good option that can be used as an alternative for existing fuel. This is synthesized in many countries, and research has been done to commercialize the biofuel most efficiently. This is renewable energy, which is mainly used in the transport sector. Many nations have encouraged production, as this can reduce the overall pollution level, reduce GHG emissions, help the farmers, and provide employment for the local community (Sawatdeenarunat et al. 2019).

Usually, the distillation is carried out in a way that consumes lots of energy, so another method for bioethanol production is that integrated techniques are carried out. The main thing is the recovery of the bioethanol at low power can use standard water. When ethanol is consumed, energy conservation and sustainability have to be considered. All these are identified because of the properties of ethanol and water combination, which can make the process simple (Pei-Yao et al. 2019). Another parameter is the production of residue, which is produced after the process. Nowadays, this is recycled to make the process more efficient. This reduces the cost as well as increases the potential activity. Various energy-saving techniques during fermentation have been identified, as well as the effectiveness of environmental restoration methods through the linking of potential biomass with bioactivity (Gavahian et al. 2019). Bioethanol production can be performed in a variety of processes, including batch and fed-batch (Jan et al. 2015). Various advanced bioconversion processes for bioethanol production are shown in Figure 9.3.

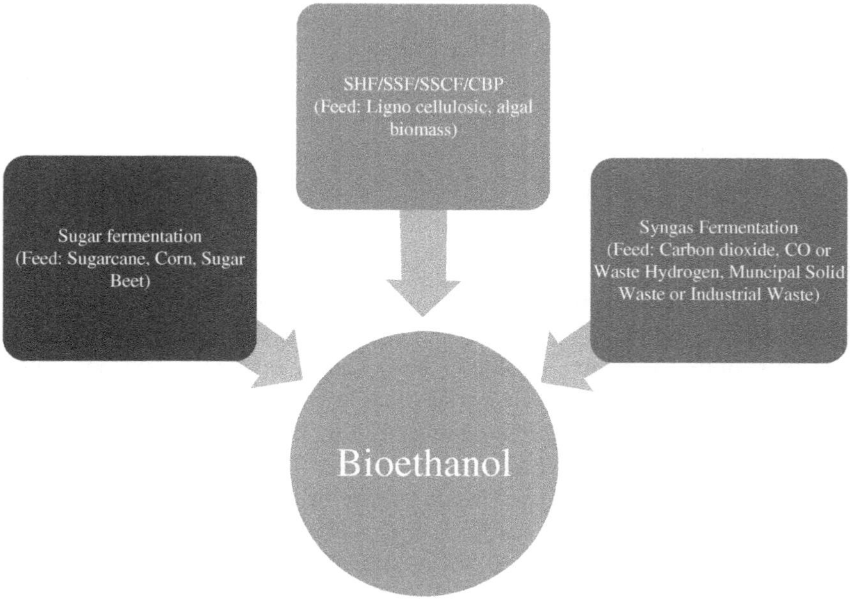

Figure 9.3 Various biological conversion process for bioethanol production from different feedstock. Source: Lü, et al. (2011). © 2011, Royal Society of Chemistry.

9.3 Advancement in Distillation and Waste-Valorization Techniques

9.3.1 Heat Integrated Distillation

This system helps in conservation of energy during a distillation process. The heat that is usually wasted is consumed and makes it productive. This method can reduce the amount of energy utilized by biofuel and make it more effective. The consumption of biofuel is reduced by about 40% (Gavahian et al. 2020). Now the research is on to reduce the amount of energy utilization in different processes. This process is considered as one of the most beneficial ones, but great scientific and experimental progress is still needed to maximize its potential (Dias et al. 2011; Dias et al. 2012). They found conventional disease rehabilitation procedures. They took the bagasse as well as sugar and used their mixture to produce the bioethanol. By using 10% of the stock, about 26% is found to be increased. The impact of reduced combustion planning to minimize the need for bioethanol production capacity from sugarcane Diaz and Tost (2016).

The principle of heat-integrated distillation includes improving the distillation system's first use of energy. Depending on the strategy, the system structure to accomplish better energy-saving distillation differs. This technique has been reported to reduce energy consumption during biofuel generation as far as 40% (Gavahian et al. 2016). Investigators explored the possibility of using different devices to energy consumption (e.g. heat pumps, heat exchangers, adiabatic columns, and multi-effect columns). Even though this method drastically decreases the need for fossil fuels and the effect of manufacturing operations in the environment, there is a need to increase its significant theoretical and experimental advances to make it more feasible (Jana 2010).

9.3.2 Membrane Technology

This technique is used in the recovery of ethanol from the aqueous solution by the principle of difference in the partial pressure of ethanol with water (Liu et al. 2015). This leads to the elution of ethanol from the broth. This process can be accelerated by decreasing the flow of gas or its pressure in the membrane side when this condition is supplied due to the concentration gradient. This process is called as pervaporation. The initial aim is to reduce the amount of water from the ethanol solution (Ueno et al. 2017). The structure of the membrane plays an important role in this separation from this broth or solution. When considering this, the fermentation broth is more complex to elute the ethanol separately, so most of them tried glucose as enrichment for microbes to convert these sugars more efficiently (Song et al. 2017). This obtained ethanol cannot be easily purified as it forms azeotropes with water, so the membrane is used in this case. The ZSM-5 zeolites/polydimethylsiloxane (PDMS) hybrid membranes are typical examples of this (Figure 9.4) (Liu et al. 2015). The excessive partial pressure of ethanol with regard to water is the mechanism involved in ethanol recovery by membrane technology. The transfer of ethanol from the fermented broth to membrane pores is favored by this difference. Acceleration of the process to recover ethanol is achieved by pressure reduction of gas flow in the membrane's permeate side. During this state, pervaporation occurs, which is the transfer of target compound by concentration gradient across the membrane. The removal of water from concentrated ethanolic solutions is the focus of primary industrial applications, owing to the kinetic diameter of water and ethanol (0.32 vs. 0.57 nm, respectively). In the separation of ethanol from binary solutions and fermented broths (whether it is glucose solution or fermentable sugars derived from naturally occurring sources), membrane composition plays a crucial role (Wei et al. 2014).

9.3.2.1 Membrane-Assisted Vapor Stripping

This method in which the vapor is generated from the broth is then separated into two. Ethanol was recovered with the help of the hydrophilic membrane and stripping column (Wang et al. 2016), and then it produced the output with 80% yield of ethanol. Vapor stripping, a procedure that involves the passage of vapor to volatilize ethanol, generally vapor, the water content present in ethanolic solution from the fermented broth. This vapor flow is set apart into two courses – retentate and permeate – which contains either higher or lower ethanol concentration depending on the membrane's properties. There are only limited configurations proposed hitherto: a batch system and a continuous system (Vane et al. 2013).

9.3.2.2 Combining Extractive and Azeotropic Distillation

These techniques, in which the mixtures are having a close boiling point, as well as azeotropes, are separated. The primary mechanism is that volatile compounds are

Figure 9.4 (a) SEM image of original ZSM-5 zeolites, (b) SEM image of ZSM-5 zeolites filled PDMS/PVDF composite membrane, (c) SEM image of cross-section of ZSM-5 zeolites filled PDMS/PVDF composite membrane. Source: Reprinted with permission from Ref. Liu et al. (2015). Copyright (2015) Elesevier B.V.

separated with the help of miscible liquids (Vane et al. 2013), thus, a simple distillation can be carried out. In this case, the difference is that it is entrained, which acts as an azeotrope, which can be obtained at the top as vapor form. In the case of the posterior-conventional column, this separation can be done quickly.

Li and Bai recently proposed a new system, extractive distillation, for ethanol-water separation (Li and Bai 2012). As the authors state, the significance of identifying energy efficient sequences of ethanol-water separation is now linked to the development of bioethanol. The most energy-saving process is the thermally combined ternary extractive distillation process (combination of two extractive distillation columns) along with mixed trainer, and also shows evident gains in cost-effective and environmental aspects (Yongteng et al. 2018). The key explanation is that it is relieved of remixing effect, and the good mixed trainer performance increases the relative volatility of the original variable. In addition, a comparison between the traditional ternary extractive distillation process one and the thermally coupled ternary extractive distillation process two (combination of extractive distillation column and trainer-recovery column) shows that a thermally coupled extractive distillation series with a side rectifier did not every time produce the desired result owing to remixing that still occured. While comparing to the one bar an existing process for the extractive distillation column, which attributes mainly the flow rate of the trainer. Decreasing from 200.020%, extractive pressure-swinging distillation with 4 bar for the optimized extractive distillation column is reduced in regard to the total annual cost, CO_2 emissions, and energy loss of by 44.09, 44.16, and 41.54%, respectively (Ao et al. 2019).

9.3.2.3 Feed-Splitting

This is done to make the process more efficient. The heat exchanger is utilized in which the temperature will be high at the top and very low at the bottom. The process simulation is carried out in this case (Kaur et al. 2020). A feed-splitting system reduces the utilization of energy. This can reduce the usage of energy consumption by 27%. The splitting of feed is a technique for increasing energy efficiency in distillation systems. Feed flow is heated by a heat exchanger, and the bottom product flow is cooled (Soave and Feliu 2002). As far as we know, process simulation for the segregation of azeotropic ethanol is investigated only in a single study. The energy utilization in the retrieval process of azeotropic mixtures like the fermented broth from biofuel production was put forth by Tavan and Shahhosseini (Tavan and Shahhosseini 2016). The investigators used the hysys process simulation tools and detailed optimization analysis in this context. This suggested system reduced energy utilization by 27% as opposed to the traditional system. A conclusion was drawn that the suggested feed-splitting method may be taken into account as an alternate energy-saving choice to conventional azeotropic ethanol-water cycle.

9.3.2.4 Ohmic-Assisted Hydro Distillation (OADH)

This system mainly contains an ohmic heater as well as a condenser. Initially, the heat is generated, and then it is collected and cooled to produce vapor. The major components are the electrodes and a nonelectroconductive enclosure, and usually, the integrated system is much beneficial (Mohsen and Asgar 2018), thus the safety system, transformers, etc., were added. This was first used as an energy-saving tool, which has more applications. Later, it was used for extracting ethanol concentration. One of the common applications is oil extraction. This method was used to extract ethanol from raw corn material and increase production from 10 to 47%. This can conserve the electric current more than 70%. Laterm it is proved that this can improve the concentration and yield of bioethanol produced (Mohsen et al. 2016a,b).

9.4 Green Extraction of Bioactive Products

When the bioethanol is produced, a large number of by-products formed, which contain valuable substances, such new technologies like green and innovative, which takes only less time and requires low temperature as well as a large amount of solvent (Al-Hilphy et al. 2020). This process can also reduce the GHG emission as well as energy consumption. Common methods include electron technologies ultrasound, microwaves, etc., used for the "green" process. The advantages and disadvantages of the green extraction methods are presented in Table 9.2 (Boussetta and Vorobiev 2014, Deng et al. 2018, Gavahian et al. 2020, Herrero et al. 2006, Li et al. 2019, Liu et al. 2011, Martínez et al. 2020).

Table 9.2 Advantages and disadvantages of green extraction methods.

No.	Method	Advantages	Disadvantages
1	Pulsed electric fields (PFE)	- Short-time operation - High selectivity - Low cost - Scaling easily - No thermal effects	- Expensive equipment - Dependence on conductivity
2	High voltage electrical discharges	- Effective in increasing yields - Use low temperatures	- Consume a lot of energy - Affects organic molecules - Requires high technology - Restrictions on operating machinery (operation only in batch mode)
3	Enzyme-assisted extraction	- High selectivity - Nondestructive conditions	- High price - Difficulty in restoring enzymes - Need a long time
4	Ultrasound-assisted extraction	Mixing effect	- Difficulty in scaling up - High price - Influence the quality of thermo-labile compounds and extraction materials. - Not selective
5	Microwave-assisted extraction	- Simple in scaling - Effective in operating	- Consume a lot of energy - Influence the quality of thermo-labile compounds
6	Subcritical fluid extraction	- High selectivity - Easily performed - Short extraction time - Compatible with the environment	- High operational costs - High labor safety requirements
7	Ohmic-assisted extraction	- Simple in operation	- Pollution of waste by metal ions - Negative effect on plant extracts

Sources: From Boussetta and Vorobiev (2014); Deng et al. (2018); Gavahian et al. (2020); Herrero et al. (2006); Li et al. (2019); Liu et al. (2011); Martínez et al. (2020).

9.4.1 Pulsed Electric Fields (PFE)

This electric field method can improve the recovery of bioethanol. The mechanism involves a delay of the natural-dipole moment between molecules that trap bioactive substances. As a result, the distance in between increases, so the electron transport occurs. In this similar method, the plant cells can be brokered by using such an electric field, which can lead to the discharge of intracellular substances. The conditions must be optimized to obtain the maximum yield (Cristina et al. 2018). Improvement in the yield of extraction depend on the control of processing conditions. In this regard, the intensity of field strength, selecting the number of pulses, and energy input, temperature, and time of treatment are of great value for the extraction of intended compounds from a compound matrix (Puértolas and Barba 2016). Even though limited information is available regarding the pulsed electric field application as the technique for extraction, a research carried out by Almohammed et al. on the exaction of fermentable sugars for ethanol fermentation (Almohammed et al. 2016) provided promising results. The presence of high amounts of sugar (3.75 folds added sugar than the exact procured by a solid-liquid exaction) by pulsed electric field treatment (intensity of 450 V cm^{-1} for 10 minutes) suggests the influence of sugar beet matrix by pulsed electric fields. Several publications for the application of the PFE method in bioethanol production are shown in Table 9.3.

9.4.2 High-Voltage Electrical Discharges

Two electrodes are placed, and the electric field is applied for a short period. Samples should be maintained in a wet condition, and the electric field causes the release of ions into the solution and energy gets transferred into the matrix. The process starts with the vapor formed at the electrode. As a result, the conductivity increases, and thus, voltage reduces (Prasoulas et al. 2020). This process takes only a few seconds. This process is dependent upon many factors like electrical discharge intensity, temperature, time taken, and so on. The mass transfer also occurs due to the formation of matrix fragments (Mohsen and Asgar 2018).

Electrical discharges operate with high voltage by positioning samples within electrodes and implementing an electrical discharge right away for a brief time. Specimens are customarily immersed for extraction in an aqueous solution (Puértolas and Barba 2016). Even though high-voltage electrical discharge mechanisms in liquid mediums remain partially transparent, there is an agreement that the ionization of components leads to multiplied charging carriers and subsequent transportation of energy to the sample matrices. The development of vapor channels in the middle of electrodes initiates the process. The conductivity increases with the development of vapor channels. This state heightens the current and lowers the tension in between electrodes. The entire procedure takes place in just a couple of nanoseconds. For example, the influence of high-voltage electrical discharge on the structure depends on the intensity of the electrical discharge, frequency, operating time, and temperature. Membrane integrity and complexity of structures are weakened, leading to matrix fragmentation and increased mass transfer (Almohammed et al. 2017).

9.4.3 Enzyme-Assisted Extraction

This method is used to recover active biocompounds in which the enzymatic degradation of the complex matrix occurs; as a result, the required compounds are removed. The usual enzyme used is cellulose, pectinase, and other types, and one of the common examples is substances with antioxidant activity (Kirupa et al. 2018). The activity is controlled by various parameters like pH, temperature, etc., which has to be optimized (Koel et al. 2017).

Sea buckthorn seeds and pomace are processed using supercritical carbondioxide (SFE-CO_2), pressurized ethanol (PLE-EtOH), and enzyme-assisted (EAE) extaction via three-step fractionation. SFE-CO_2 yields 146 and 135 g kg^{-1} of pomace and seed lipophilic fractions, respectively. SFE-CO_2 remnants PLE-EtOH retrieved 158 g kg^{-1} of polar components from the pomace and five times less from the seeds. Eventually, the treatment

Table 9.3 The publications for the application of the PFE method in bioethanol production.

Matrix	Extracted compounds	Conditions	References
Blueberries	Polyphenols Anthocyanins	Electrical, 1–5 kV; 10 Hz; 1–10 kJ kg^{-1}; – pulses (20 μs); Temperature, 23 °C	Bobinaitė et al. (2015)
Potato peels	Steroidal alkaloids	Electrical, 10 Hz; 0.75 kV cm^{-1}; 200 pulses (3 μs); Temperature, 16 °C	Hossain et al. (2015)
Grape pomace	Polyphenols Anthocyanins	Electrical, 40 kV; 0.5 Hz; 13.3 kV cm^{-1}; 0–564 kJ kg^{-1}; – pulses (10 μs); Temperature, 22 °C; Solvent, water	Barba et al. (2015)
Spearmints	Polyphenols	Electrical, 20 mV; 100 kHz; 3 kV cm^{-1}; 99 pulses (10 μs); Mannitol solution	Fincan (2015)
Microalgae *Nannochloropsis*	Polyphenols Proteins Carotenoids Carbohydrates	Electrical, 40 kV; 20 kV cm^{-1}; 400 pulses (10 μs); Temperature, 20–30 °C; Solvent, water	Parniakov et al. (2015)
Rapeseed	Polyphenols Proteins	Electrical, 400 V; 0.5 kHz; 5 kV cm^{-1} (20 kV cm^{-1} for proteins); 200 pulses (10 μs); Time, 20 min; Temperature, 50 °C for polyphenols, 20 °C for proteins; Solvent, 75% ethanol for polyphenols, water for proteins	Yu et al. (2015)
Button mushroom	Polysaccharide Polyphenols Proteins	Electrical, 30 kV; 1 Hz; 38.4 kV cm^{-1}; 136 pulses (2 μs); Temperature, 20 °C; Solvent, water	Xue et al. (2015)
Sesame cake	Polyphenols Proteins Lignans	Electrical, 40 kV; 0.5 Hz; 13.3 kV cm^{-1}; 83 kJ kg^{-1}; 100 pulses (10 μs); Time, 20 min; Temperature, 60 °C (40 °C for proteins); Solvent, 10% ethanol (50% ethanol for lignans)	Sarkis et al. (2015)
Norway spruce Bark	Polyphenols	Electrical, 40 kV; 0.5 Hz; 20 kV cm^{-1}; 200 pulses (10 μs); Time, 10 min; Temperature, 20 °C; Solvent, water +0.01 M NaOH	Bouras et al. (2016)

(Continued)

Table 9.3 (Continued)

Matrix	Extracted compounds	Conditions	References
Red prickly pear	Colorants	Electrical, 40 kV; 0.5 Hz; 20 kV cm^{-1}; 50 pulses (10 μs); Temperature, 20 °C; Solvent, water	Koubaa et al. (2016)
Red beet	Betanine	Electrical, 1 Hz; 6 kV cm^{-1}; 50 pulses (3 μs)	Luengo et al. (2016)
Tea	Polyphenols	Electrical, 1.1 kV cm^{-1}; 50 pulses (100 μs); Solvent, water	Zderic et al. 2016
Red wine	Polyphenols	Electrical, 10 kV; 200 Hz; 5 kV cm^{-1}; 3.5 kJ kg^{-1}; 1 pulses (100 μs); Solvent, ethanol 30%	Saldaña et al. (2017)
Microalgae *Chlorella vulgaris*	Proteins	Electrical, 20 kV cm^{-1}; – pulses (2 μs); Temperature, 20 °C; Solvent, water	Postma et al. (2017)
Tomato	Carotenoids	Electrical, 3.8 kV; 0.33 Hz; 600 pulses (350 μs of total pulses); Temperature, 45 °C	Bot et al. (2018)
Microalgae *P. cruentum*	β-pycoerythrin	Electrical, 2–10 kV cm^{-1}, 0.5 Hz, 50 pulses (3 μs); Temperature, 22–30 °C	Martínez et al. (2019)
Microalgae *Chlorella vulgaris*	Proteins	Electrical, 40 kV cm^{-1}, 4.5 Hz, – pulses (1 μs); Temperature, 21–38 °C	Scherer et al. (2019)

of cellulolytic and xylanolytic enzyme PLE-EtOH residue (Viscozyme, CeluStar XL) preparation increases the quantity of soluble components by 24–80% relative to samples not treated with enzymes (Vaida et al. 2017).

The biomass fill up and preheating time effect on the fractional hydrolysis procedure was investigated in conjunction with different approaches until 30% H_2SO_4 (v/v) to minimize the numerous acid use stages. Of the total sugar reduction existing in kans grass biomass, 84.88% was extricated as distinct fractions of hexose and pentose sugar with trace toxic substances. During co-culture fermentation, *Scheffersomyces shehatae* and *Zymomonas mobilis* were used. The production of ethanol (25.0 $g\,l^{-1}$) from the kans grass biomass hydrolysate used 92.13% of the sugar from the xylose-rich fraction (initial sugar: 21.87 $g\,l^{-1}$) and 96.32% of the glucose-rich fraction (initial sugar: 40.32 $g\,l^{-1}$); around 78.6% of the full theoretical ethanol prod was produced and the mean yield is 0.435 Archana and Sanjoy (2020).

Kirupa et al. co-immobilized laccase, cellulose, and β-glucosidase to instigate a novel one-pot pretreatment technology to serve as a tri-enzyme biocatalyst to assess the bioethanol fabrication potential of four feasible lignocellulosic biomasses, viz., *Saccharum arundinaceum, Arundo donax, Typha angustifolia,* and *Ipomoea carnea*. Comparing to other biomass in the course of enzymatic saccharification, *Saccharum arundinaceum* exhibited a notable total reduction in sugar of $205 \pm 3.73\,mg\,g^{-1}$. The prime percentage of bioethanol yield was obtained from *Ipomoea carnea* of $63.43 \pm 9.35\%$.

Distillers gried grains (DDG) explored the impact of oil extaction aids on ethanol yield, skim, and insoluble fiber, oil retrieval, and quality, as well as oil partition. Two hydrolyzing fiber enzymes (cellulase and pectinase), an acid protease, and a surfactant (Tween 80) was assessed. The inclusion of skim, skim, and insoluble fiber mixture or Fermgen® to maize fermentation results in a reduced fermentation time of approximately 32 hours. In addition to soy coproducts, the oil partitioning in thin stillage without added enzyme or surfactant treatment increased by approximately 10–28%. Insoluble fiber addition solely resulted in a reduction of approximately 19% in the thin stillage partition of solids. Full free oil recovery was achieved with a combination treatment of enzymes (cellulase, pectinase, and Fermgen®) and surfactant (Tween 80) from corn-insoluble fiber thin stillage. Maximum recovery of extractable oil, $67 \pm 3.2\%$, was achieved with the treatment of the enzyme solely. DDG of corn-soy has about 11% elevated protein content, about 2% lesser fiber content, and about 2% lesser fat content relative to DDG corn content. After enzyme treatment, the fiber content was additionally lessened to approximately 2 %. This study illustrates the effective utilization of soy EAEP coproducts and enzymes to optimize oil partitioning in thin stillage, and produces a high-quality DDG corn-soy (Jasreen et al. 2018).

9.4.4 Ultrasound-Assisted Extraction

When the ultrasound is allowed to propagate through the liquid, a phenomenon called cavitation occurs, which can reduce the amount of pressure. The name is given because due to this pressure, vapor cavities are formed, and when it collapses they form micro jetting. This, when subjected to ultrasound treatment, the interaction in between the solvent and the valuable substances increases (Nagappan and Nakkeeran 2020). The main method involves capillarity, detexturation, erosion, fragmentation, as well as sonoporation (Arup et al. 2019).

9.4.5 Microwave-Assisted Extraction

This technique is mainly used to obtain bioactive substances, especially pectin. The main principle behind this is ionic conduction as well as the dipole rotation of the solvent when the electromagnetic field is applied. These lead to the migration of ions, which causes dipole movement. The solution should be selected accordingly so that the energy can be dissociated easily (Ma et al. 2020). Water is most commonly used, as it has a high dielectric constant when compared with other solvents. There are three methods: the first one is by adding the solvent that can absorb the microwave energy, the second one is by using a mixture that has the capacity to absorb energy, and the third one, which has the weak capacity, is a solvent that does not absorb energy (Harish and Senthil 2020).

9.4.6 Subcritical Fluid Extraction

Saturated liquid with unique characteristic is produced, which has a temperature between atmospheric and critical points. Such fluids are called subcritical fluids. The system should be maintained at high pressure, and this should be strictly followed to water. Other properties like diffusion rate, viscosity, as well as surface tension, should be maintained to extract the required compounds (Yizhak Marcus 2018; Xiao-Fei et al. 2019).

9.4.7 Ohmic-Assisted Extraction

The electric treatment should be carried out by using this method. Ohmic heating increases the temperature and moderate electric field decreases the temperature. Apart from these, the electro-portion and enhanced cell permeability can be used to extract bioethanol. These techniques are used as energy-saving methods (Gavahian et al. 2015). These also have many advantages, such as taking less time, and are more efficiency. The emergence of various ecofriendly technologies and distillation processes in the synthesis of bioethanol leads to the selection of different sources that can produce value-added applications. These are some of the promising strategies to develop the required biofuel (Mohsen et al. 2016a,b). The electrical treatment of material is specified with terms like ohmic heating and moderate electric field. Ohmic heating thoroughly increases the temperature of materials, but a moderate electric field is generally carried out at relatively lower temperatures (Gavahian et al. 2018). Mechanisms, like electroporation and enhanced cell permeability, are found to expedite the separation of valuable compounds and are further extreme in a moderate electric field. The separations of bioactive compounds using these techniques are successful as green energy-saving methods.

9.5 Advancement in Bioethanol Production from Microalgae

9.5.1 Surface Methods

The yield of this method depends on yeast volume; time is taken as well as the amount of biomass, while the microalgae are an essential source for synthesis. This is a promising source (Wolske and Stern 2018). Various parameters can be optimized in order to improve the yield. Statistical optimization of features is done to obtain maximum yield and to find the model that can predict the response and the output with the value of initial parameters (El-Mekkawi et al. 2019). Thus, the semi-pilot scale is used to analyze the result, and continuous-fermentation techniques are carried out. It is more effective to use an integrated system that is high-rate algal pound (HRAP), which is having a dimensional area of about 5.4 m^3 (Markou et al. 2013). Usually, the fed is physically pretreated before fed into the reactor. Algal biomass grew in HRAP (El-Naggar et al. 2018).

The incubation was carried out in three containers simultaneously. The required product is separated from water by using a rotary evaporator, and the bioethanol obtained is collected under the ice to avoid evaporation. The purity has to be tested with the help of calorimeter. The procedure is potassium dichromate added to 10 ml of the sample and mixed. It is then kept in a water bath at 60 °C for 20 minutes, and then incubated at room temperature (de Farias Silva and Bertucco 2016). The absorbances were measured at 600 nm with a spectrophotometer, and with the values, a curve is drawn and compared with standard. Bioethanol production at anaerobic condition depends on parameters like algal biomass; time is taken to carry out fermentation as well as yeast volume.

9.5.2 Ligno Celluloic Bio Ethanol Production

9.5.2.1 Membrane Technology

This has several advantages over other methods, such as less space is required, less labor cost, and the process more flexible. The reactor studies, as well as simulations, are carried out in which the membrane couples process can contribute to the production of bioethanol in more ways. These include filtration, optimization, self-cleaning, and so on. Advanced techniques are mainly based on nanofiltration-based methods with high-density recycling of cells. This system of integrated systems can make the process more efficient and can save the economy (Dey et al. 2020).

Various parameters are considered along with these flow characteristics in which membrane bioreactor (MBR) and mechanical manufacturing processes are used instead of other systems (Mutamim et al. 2013). In the case of lignocelluloic bioethanol, the most valuable one is the purification as well as accumulation of prehydrolyzates, saccharification, regulation, and regeneration process combinations, and integration between filtrations like micro (MF), ultra (UF), and nano (NF) processes have been developed by various researchers to make the complete process continuous, compact, and efficient. These materials are used to synthesize pure bioethanol by providing enrichment substances like sucrose, lactose, glucose, and so on. Another problem for the production is environmental insecurity and energy crisis (Mutamim et al. 2013). The usually used method for the production by using this lignocellulosic biomass, which can be carried out in a lab in the most feasible way. But in considering large-scale production, this technique cannot be used. So for commercial applications, new technology has to be used. The membrane process has the unique capacity to produce this biofuel with the most flexibility as well as the least energy requirements. The integration of membrane techniques can be much useful to solve present challenges in developing bioethanol for commercial applications from lignocellulosic biomass.

If we consider the two stages of the SSFF reactor integrated with UF, which can increase the yield, this process can also help separating as well as recycling and catalyzing. The enzyme is provided in the immobilized stage. To improve the process, the NF membrane is concentrated with ethanol before the fermentation is carried out. This can give more energy, all while cost will be lowered (Mutamim et al. 2013). This kind of novel method can enhance production by overcoming the present challenges. The major disadvantage of simultaneous saccharification and fermentation (SSF) was that the problem associated with recirculation (Dey et al. 2020). The SSF process is more beneficial comparatively. For proteinaceous enzymes, ultrafiltration (UF) is used for carrying out the separation (Brodeur et al. 2011). Substrates like straw microfiltration (MF) are used. In order to obtain the most purified form, the membrane-integrated process, distillation, or pervaporation is used, or it is integrated with the current system. The working principle of the MF-NF-DCMD system integrated membrane fermentation equipment for the production of fermentation and purification of bioethanol is shown in Figure 9.5. (Kumar et al. 2017).

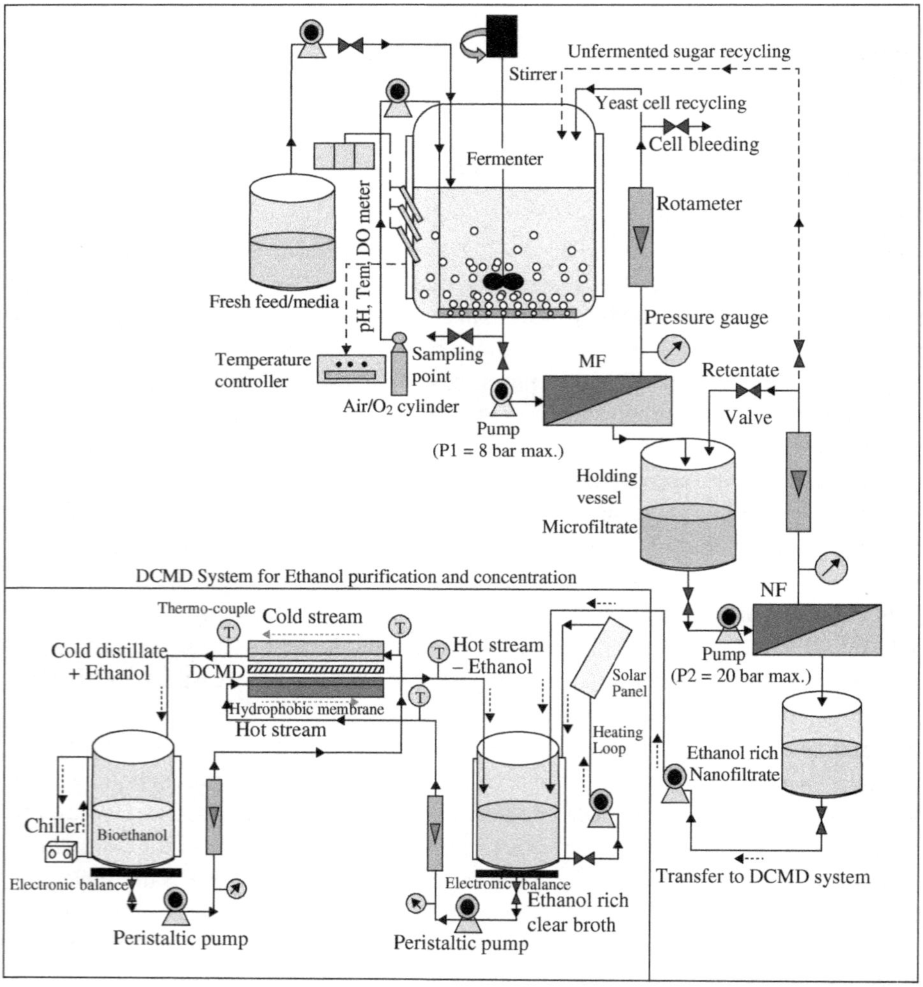

Figure 9.5 Diagram exhibited the working principle of the MF-NF-DCMD system integrated membrane fermentation equipment. Source: From Kumar et al. (2017). © 2017, Elsevier.

9.5.2.2 Microbial Technique

The use of microorganism is a critical technique, which is a fourth-generation biomass. This organism helps convert compounds into ethanol. The robust microbe is used for the efficient conversion of substances (Doma et al. 2018). Bioethanol production from lignocellulosic biomass conversion was mainly done by *S. passalidarum*. The ethanol production under aerobic conditions was lower and is a promising candidate strain for further metabolic engineering to develop robust industrial strains for the lignocellulosic ethanol (Yu et al. 2017).

9.5.2.3 Brown Algae

Nowadays, the economy is growing much faster than expected. Thus, the demand for energy also increases. The main reason is the depletion of fossil fuels and the pollution caused by it (Yu et al. 2017). The art of using these microalgae can produce more efficient biofuels, which is more sustainable and uses less freshwater. This third-generation biofuel is more effective and useful as compared to first and second generation fuels (Ravanal et al. 2019). So much of the research is going to develop more techniques based on these sustainable methods.

The saccharification and fermentation steps can be carried out via different configurations: separate hydrolysis and fermentation (SHF), SSF, simultaneous saccharification and co-fermentation (SSCF) of hexoses and pentoses, and consolidated bioprocessing (CBP). Initially, one SHF is in the hydrolysis reaction of polysaccharides from this algae, as this fermentation is carried out. In this process, both hydrolysis, as well as fermentation, occurs at optimum conditions. This also has a chance for contamination easily, and an inhibitory effect can occur (Ramachandra and Deepthi 2020).

Here, the amount of sugar generated is calculated as it inhibits the cellulase activity. The untreated biomasses of brown algae were fed to cellular enzymatic hydrolysis. From research, the yield of glucan was 29.8, 82.5, and 72.7 g kg 21 for the untreated, hot water pretended. Biomass acid, respectively, by using *S. cerevisiae* under anaerobic conditions, produced bioethanol of 34.6 kg per g of biomass fed into the reactor. Another example is by suing *L. digitata* for conversion in which yield of more than 70% is obtained theoretically, but when enzymatic hydrolysis is carried out, then yield obtained was more than 80% (Xiaoru et al. 2020).

9.5.2.4 Integrated Processes

Many hypotheses have been reported that the cellulases, as well as hemicellulase enzymes, were used for producing monomeric sugars. These microbes are used to convert the sugars into bioethanol with the help of the fermentation processes. This is known as separate hydrolysis and fermentation (SHF), in which the fermentation and hydrolysis occur simultaneously. SHF produces the sugar initially, and then the fermentation begins. From the fermentation product, the ethanol is separated initially, and then immobilized xylose is boiled with the ethanol to replenish it (Dalena et al. 2019). The main advantage of this method is all the processes are carried out under optimum conditions, but the only problem is associated with the accumulation of sugar. They are much costlier as compared to other processes. So in order to solve the issue, novel techniques have been introduced. This is the same saccharization and fermentation (SSF), where both processes take place simultaneously (Dalena et al. 2019). This method combines fermentation and hydrolysis, and occurs in one step. The sugar is formed in an underlying nutrient, and this can be easily converted into ethanol without getting accumulated. This requires maintaining optimum temperature, and microorganisms must be added. Enzymes like cellulase work at 50 °C for hydrolysis, but in case of fermentation, this works well in between temperatures like 28 and 37 °C. It is very difficult to reduce the temperature in this case, so better thermophilic bacteria are used for conversion. From these, it is clear that the physical properties of biomass can affect the yield as first-generation biomass can cause lots of pollution with a high carbon footprint (Rastogi and Shrivastava 2017).

9.5.2.5 Advances in Bioethanol Production from Agroindustrial Waste

Many methods are found to produce bioethanol from agroindustrial waste (AIW). Still, the main problem is associated with operating costs as uncertain substrate composition limits their implementation on a commercial scale. AIW is an abundant source that harms our environment, as it is difficult to dispose of. Thus, by using this waste, the pollution level across the globe can be controlled to an extent. Bioethanol synthesis from the cellulosic material of AIW includes various steps: milling, pretreatment, enzymatic hydrolysis, fermentation, distillation, as well as product recovery, and the final rectification. Each level requires an enzyme, which makes the process less cost-effective: pre-treatment and enzymatic hydrolysis, which all reduces the process efficiency. The leading producer of this bioethanol from AIW is Mexico, which had more than 800 agrofood products (Danay Carrillo-Nievesa et al. 2019). Considering many technologies, it is difficult to make this process efficient, thus integrated techniques can be used to make the process effective.

In many countries, investments are made in order to make bioethanol in the most effective way. This allows dedicating agricultural land to grow required food crops so that the feedstock needed to produce the biofuel can be increased. As the plant capacity increases, the investment required will be less as more outcomes can be obtained (Lynd et al. 2017). This biomass does not produce any harmful substances as a by-product. The integrated process can overcome the challenges for carrying out all the steps mentioned above. There are many factors that must be optimized, which include feedstock price, capital cost, as well as plant operating costs. The oil price is around USD 1.43/gal, while lignocellulosic ethanol could be around USD 0.84–0.91 per liter of gasoline equivalent (Lee et al. 2010).

The advanced design was made to produce biofuels like bioethanol, acetone, and hydrogen, etc., by using wheat straw, which is known as CBP principles (Mattam et al. 2016). CBP can be used to produce bioethanol, especially from lignocellulosic biomass and in this, the steps are integrated together, and a reaction occurs much faster. The major challenge is associated with the selection of the most appropriate microorganism. The yield mainly depends upon the organism that we are using. This method is identified for future production of bioethanol and their commercialization. The basic advantage of this AIW for bioethanol production is shown in Figure 9.6.

9.6 Fermentation Technique Advances

Nowadays, the demand for bioethanol is increasing, as it is one of the best fuels that can be used in the transport sector and along with this, the fossil fuel is becoming extinct.

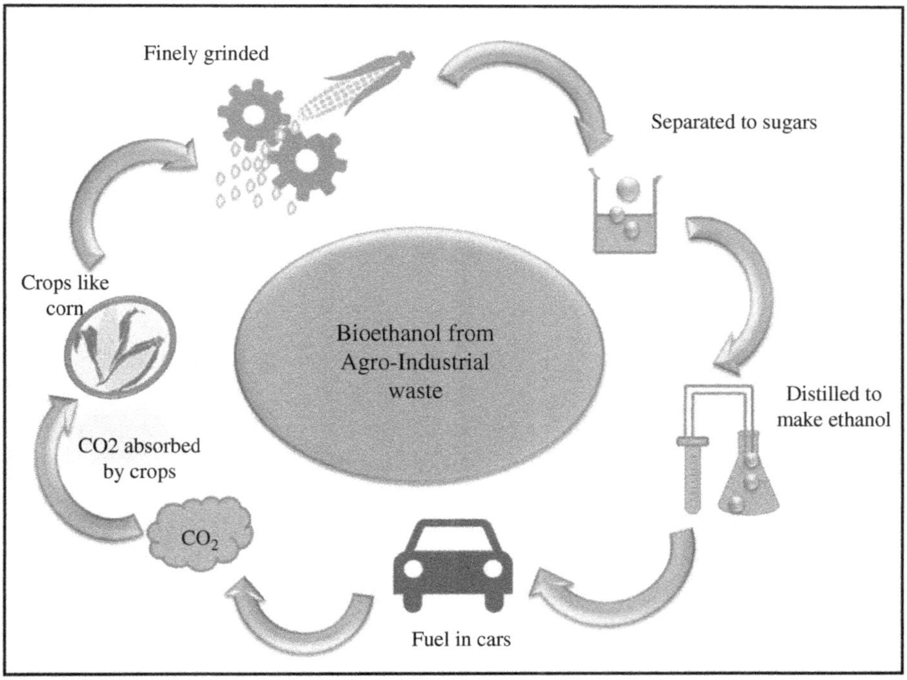

Figure 9.6 Sustainable production of bio ethanol from agroindustrial waste.

The economic importance of this fuel also increases (Pang et al. 2020). Many countries are taking measures to use this bioethanol in almost all the fields, and they have been investing a large amount of capital cost in improving the product in the most sustainable way. Thailand is one of the leading countries to use this bioethanol in Southeast Asia. Production of bioethanol, especially from fruit and vegetable peel, is also possible (Sharma and Aggarwal 2020). Considering an example of this biofuel production, from pineapple, the skin produces a yield of about 5.98 ± 1.01 g/after 48 hours of fermentation. All this can be increased, and challenges can be resolved by various emerging techniques (Casabar et al. 2019). In this case, the alkali pretreatment was carried out as pretreatment technique, but when the organism *T. harzianum* inoculum was used, the yield gradually increased as the reducing sugar is inversely proportional to the ethanol production increases. Similarly, a large number of microorganisms can be used to improve the production, and it is mostly advised to use this microbe after the hydrolysis technique (Kongkeitkajorn et al. 2020).

9.6.1 Synthesis from Municipal Wastes

9.6.1.1 Waste Paper

This is an attractive feedstock as compared to others for bioethanol production, as it is readily available (Nair et al. 2017). Paper sludge mainly contains cellulosic fibers, while paper mill uses different feedstocks. To carry out the process, two important factors should be considered that are an effective method to increase the content of fermentable sugars, as well as reducing the inhibitor concentration. The next one includes required performance and should be carried out to co-ferment mixed monosugars in the hydrolysate of ethanol. Various research is conducted to solve the issues associated with this process. The power of nonionic surfactants (Collins et al. 2020), such as NP-20, Tween 20, and Tween 80, to heighten sugar release from waste or recycled newspaper, has been reported. A large number of pretreatment techniques, along with this, are also reported. While this includes CO_2 explosion, chemical method, biological with the help of bacterial culture, steam explosion, and ozonolysis as well as by using liquids with high temperature, especially water. Apart from these, acids like phosphoric acid have also found to increase the production yield.

9.6.1.2 Coffee Residue

Coffee is found to be one of the most consumed beverages across the world. A large number of wastes that is coffee residue waste (CRW) were generated, and thrown into the dustbin directly. This can be a source to produce bioethanol to an amount. The coffee consumption in Kg per capita is very high when it was reported by the International Coffee Organization and this came to be noticed by many. This residue is classified into mucilage, pulp, and husk. Mucilage is used to produce pectin protein (Kulkarni, 2020). When analyzing the biomolecules present in this residue, the carbohydrate involves galactose, and glucose, as well as mannose. Pulp portion contains a large number of fermentable sugars, which can be converted to bioethanol. Research was carried out in order to use SSF for enzymatic hydrolysis of these sugars for better conversion (Kongkeitkajorn et al. 2020). This produced a yield of more than 85%. Similarly, spent coffee grounds have potential as raw material for integrated biorefineries. CRW also contains a large amount of hemicellulose and lignin, which can be a source for bioethanol production.

9.6.1.3 Food Waste

This is an essential source of bioethanol production, as it has a large number of organic compounds (Saha et al. 2015). The feasibility of ethanol is a major problem as it is limited to laboratory use. A model was predicted to obtain the maximum yield (Nair et al. 2017). Critical variables that affect the process were identified and they are optimized in order to improve the yield. The bacteria that can be used for the bioconversion are identifiable, and those with maximum capacity are used for producing the biofuel. When all the conditions are optimized, a yield of about $165\,g\,l^{-1}$ is obtained. The complex substances present in the food waste make the reaction more difficult. So a pretreatment technique has to be carried out, which has to hydrolyze the sugar. Enzymatic hydrolysis of sugar using carbohydrate enzyme with microorganism *S. cerevisiae* in batch mode produces 0.63 g of glucose per gram of feed (Ohia et al. 2020). This same study was carried out using SSF in which the open, as well as closed, fermentation was done. In the case of open fermentation, a maximum was obtained of about 33 g/L concentrated bioethanol was produced. The recent developments of bioethanol production from different sources are presented in the Table 9.4. Raveendran et al. studied the potential of food and cooking waste as an effective feedstock for bioethanol production. Food and cooking waste was also assessed in this study for bioethanol production (0.316 g) through saccharification and fermentation (Raveendran et al. 2020). For bioethanol production, Leonardo et al. used unripe plantain fruits (Musa paradisiaca L.). The biochemical component known as peel starch (39.4%) and

Table 9.4 Bioethanol production from various food wastes.

Food waste	Method	Production of bioethanol	References
Food and kitchen waste	Saccharification and fermentation	0.316 g	Raveendran et al. (2020)
Fruits	Saccharification and fermentation	$20.41\,l\,t^{-1}$	Leonardo et al. (2020)
Hamburger	Enzymatic hydrolysis and fermentation	$27.4\,g\,l^{-1}$	Wei et al. (2020)
Sweet potato	Enzymatic hydrolysis and fermentation	$10.95\,l\,t^{-1}$	Caroline et al. (2020)
Sweet potato	Enzymatic hydrolysis and fermentation	$161.4\,l\,t^{-1}$	Joab et al. (2018)
Potato peels	Saccharification and enzymatic hydrolysis	$0.49\,g\,g^{-1}$ consumed sugars	Imen et al. (2019)
Food waste	Enzymatic hydrolysis	$16.3\,g\,l^{-1}$	George et al. (2020)
Malaysian food waste	Enzymatic hydrolysis	$0.42\,g\,g^{-1}$	Halimatun et al. (2017)
Black tea waste	Saccharification and enzymatic hydrolysis	0.51 g	Dash et al. (2018)
Wild date palm	Enzymatic hydrolysis	$0.278\,g\,g^{-1}$	Swaraz et al. (2019)
Pomegranate peels	Saccharification and Enzymatic hydrolysis	80 g ethanol/kg	Sachin et al. (2018)
Sago pith waste	Enzymatic hydrolysis	$31.77\,kJ\,g^{-1}$ ethanol	Saravana et al. (2019)

pulp (84.2%) in unripe plantain fruits enabled this raw material to be produce flour and bioethanol (20.41 L/t) by saccharification and then fermentation (Leonardo et al. 2020).

Wei et al. studied waste hamburger (WH) for the bioethanol production via enzymatic hydrolysis (α-amylase) and fermentation. The entire ethanol production could be completed within 50 hours since the soluble nutrient pretreated with the enzyme amylase was straightforward to use with the yeast. The highest output of $27.4\,g\,l^{-1}$ of ethanol and 0.271 g of ethanol/g of WH (0.34 g of ethanol/g of TS) was achieved with $0.14\,ml\,l^{-1}$ of amylase (Wei et al. 2020).

9.6.1.4 Solid Waste

In some places, there will not be any agricultural biomass, MSW, etc., as a potential source. Two alternative feedstocks for bioethanol production, viz., refuse-derived fuel (RDF), and biodegradable municipal waste. An integrated system was used to produce bioethanol, which can carry out combined biocatalytic as well as gasification reactions (Abdeshahian et al. 2020). RDF can save up to 196 kg of CO_2 equivalents per ton MSW. Each of the solid wastes will be having different compounds present, and this requires different microorganisms. MSW fractions such as carrot and potato peels (typical kitchen waste), (Koponen et al. 2013) grass (typical garden waste), and newspaper and scrap paper

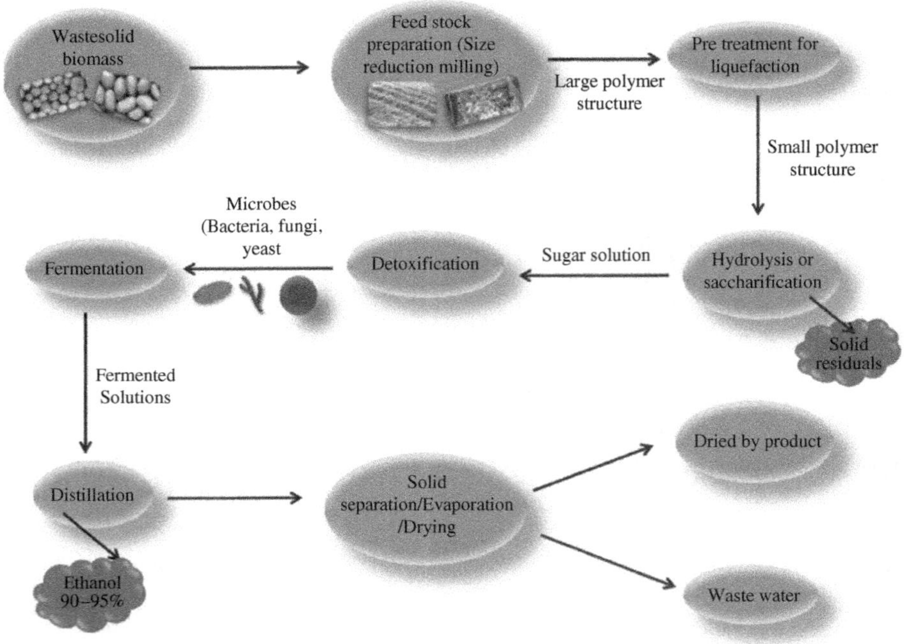

Figure 9.7 Flow chart showing bioethanol production from solid waste.

(typical paper/card fractions) were subjected to 15 different hydrolysis treatments in which 1% H_2SO_4 yields were the highest, followed by steam treatment at 121 °C and enzymatic hydrolysis with *Trichoderm aviride* at 60 FPU/g substrate (Park et al. 2010). Similarly, many studies on solid waste were carried out, and the maximum yield of ethanol was obtained. A flow chart has been demonstrated below, which shows the bioethanol production from solid waste in the most advanced way (Figure 9.7). The recent development of bioethanol production from various solid wastes, and the yield of the bioethanol is presented in Table 9.5.

9.7 Conclusion

In this chapter, we have been going through various advancements made in the present technologies to enhance the production of bioethanol. If the processes are not economical, then they will not happen. The major feedstock used to produce this bioethanol is coarse grain currently. Second-generation ethanol contributes to about 7% of total production. Each of the process parameters must be optimized so that the maximum yield can be obtained. Various research and developments are the key factors to improve the techniques for bioethanol production.

Various research and findings for the advancement of bioethanol production are the souls of the development of the nation. This screening technique allows optimizing the process to select better raw materials and instruments, which can make the process more

Table 9.5 Bioethanol production from various solid wastes.

Solid waste	Pretreatment	Features	Ethanol yield	References
Wheat straw	Prewashing and liquid hot water	Improved enzymatic ethanol yield	$0.41\,g\,g^{-1}$-cellulose	Chen et al. (2017)
Straw pulping	Sulfite	Improved enzymatic ethanol yield	76.8/20% solid	Chen et al. (2017)
Municipal solid waste	Hydrothermally at 100–160 °C for 0–60 min	Improved enzymatic ethanol yield	191.10 g ethanol/kg	Mahmoodi et al. (2018)
Municipal solid waste	Enzyme and alkali	Improved enzymatic ethanol yield	$41.41 \pm 0.06\,g\,l^{-1}$	Avanthi and Mohan (2019)
Municipal solid waste	Autoclaving	Improved enzymatic ethanol yield	1.5 kg/tonne	Fanran et al. (2019)
Carrageenan	Alkali	Improved enzymatic ethanol yield	$13.8\,g\,l^{-1}$	Kyriakou et al. (2019)
Rice hull and orange peel	Acid treatment	Bacterium improved ethanol yield	22.77 g	Jahanbakhshi et al. (2019)
Municipal solid waste	Hydrothermal	Improved enzymatic ethanol yield	$0.29\,l\,kg^{-1}$	Miezah et al. (2016)
Waste activated sludge	Alkaline	Improved enzymatic ethanol yield	1400–2200 mg COD/l	Zhu et al. (2019)
Starch	Bacterial amylase, Thermo-acidic	Improved enzymatic ethanol yield	$131.87\,g\,l^{-1}$	Kumar et al. (2016)
Sugarcane	Alkaline	Improved enzymatic ethanol yield	$82.83\,g\,l^{-1}$	Ye et al. (2018)
Bran	Autoclaved	Improved enzymatic ethanol yield	$135 \pm 10.8\,g\,kg^{-1}$	Canabarro et al. (2017)

efficient. Bioethanol produced is a liquid biofuel that can be easily synthesized from various biomass sources by using many technologies. This fuel is renewable and attractive, as well as oxygenated, which provides more environmental-friendly output. This can be combined with gasoline due to its high octane value and low cetane number. Blends of bioethanol, one of them known as E85, which has 85% of bioethanol and remaining gasoline, can be used as fuel in lightweight vehicles. In the case of Brazil, the bioethanol produced from cane, which is blended with gasoline, is called gasohol, which is extensively used in the transport sector. This bioethanol is blended with gasoline as it becomes more advantageous.

Bioethanol, which is synthesized in a sustainable manner, has to lead to the growth of the nation as well as solving the issues associated with the availability of oil and its fluctuating prices. On the basis of feedstock, the production quantity varies. Better methods are carried out to make the synthesis more efficient, cheap, and ecofriendly. First generation and second generation are not used as they become competitive for food sources, water, and soil quality, etc. It is important to make advancement in the present strategies. Research gave much importance to the study of various aspects and sustainable techniques for bioethanol production. Different merging techniques are discussed in this chapter that could demolish the existing challenges.

A significant challenge for the third generation of biomass from small algae is that biomass composition varies continuously. Various scientific, as well as technological, knowledge, has been used to solve the challenges. Some of the emerging advances were harvesting, drying, and other operations carried out to pretreat biomass, which can improve the efficiency, and implementing these methods becomes much more comfortable than conventional ones. Along with this, the government provided various subsidies and investments to develop bioethanol using third- or fourth-generation biomass.

From the above study, we can conclude that to meet the rapid industrial growth and global population growth, the need for the production of bioethanol as an alternative to green biofuel is gaining prominence. The bioethanol production process is still considered to be one of the largest energy-consuming and challenging methods. Therefore, there was a need to improve the bioethanol production process through reduced energy efficiency. The benefits of reduced power requirements, low labor costs, low space requirements, a wide range of operational flexibility, and building-based technologies are receiving more attention and may soon replace conventional high-power techniques.

References

Abdeshahian, P., Kadier, A., Rail, P.K., and da Silva, S.S. (2020). Lignocellulose as a renewable carbon source for microbial synthesis of different enzymes. In: *Lignocellulosic Biorefining Technologies*, 185–202. https://doi.org/10.1002/9781119568858.ch9.

Al-Hilphy, A.R., Al-Shatty, S.M., Al-Mtury, A.A.A., and Gavahian, M. (2020). Infrared-assisted oil extraction for valorization of carp viscera: Effects of process parameters, mathematical modeling, and process optimization. LWT-Food Science and Technology. 129:109541. doi: https://doi.org/10.1016/j.lwt.2020.109541

Almohammed, F., Mhemdi, H., and Vorobiev, E. (2016). Pulsed electric field treatment of sugar beet tails as a sustainable feedstock for bioethanol production. *Appl. Energy* 162: 49–57. https://doi.org/10.1016/j.apenergy.2015.10.050.

Almohammed, F., Koubaa, M., Khelfa, A. et al. (2017). Pectin recovery from sugar beet pulp enhanced by high-voltage electrical discharges. *Food Bioprod. Process.* 103: 95–103. https://doi.org/10.1016/j.fbp.2017.03.005.

Ao, Y., Shirui, S., Tao, S. et al. (2019). Energy-efficient extractive pressure-swing distillation for separating binary minimum azeotropic mixture dimethyl carbonate and ethanol. *Sep. Purif. Technol.* 229: 115817. https://doi.org/10.1016/j.seppur.2019.115817.

Archana, M. and Sanjoy, G. (2020). Saccharification of kans grass biomass by a novel fractional hydrolysis method followed by co-culture fermentation for bioethanol production. *Renew. Energy.* 146: 750–759. https://doi.org/10.1016/j.renene.2019.07.016.

Arup, J.B., Mayank, A., Arun, G., and Vijayanand, S.M. (2019). Physical insights of ultrasound-assisted ethanol production from composite feedstock of invasive weeds. *Ultra. Sonics. Chem.* 51: 378–385. https://doi.org/10.1016/j.ultsonch.2018.07.046.

Avanthi Althuri, A. and Mohan, S.V. (2019). Single pot bioprocessing for ethanol production from biogenic municipal solid waste. *Biores. Tech.* 283: 159–167. https://doi.org/10.1016/j.biortech.2019.03.055.

Barba, F.J., Grimi, N., and Vorobiev, E. (2015). Evaluating the potential of cell disruption technologies for green selective extraction of antioxidant compounds from *Stevia rebaudiana Bertoni* leaves. *J. Food Eng.* 149: 222–228. https://doi.org/10.1016/j.jfoodeng.2014.10.028.

Bobinaitė, R., Pataro, G., Lamanauskas, N. et al. (2015). Application of pulsed electric field in the production of juice and extraction of bioactive compounds from blueberry fruits and their by-products. *J. Food Sci. Technol.* 52 (9): 5898–5905. https://doi.org/10.1007/s13197-014-1668-0.

Bot, F., Verkerk, R., Mastwijk, H. et al. (2018). The effect of pulsed electric fields on carotenoids bioaccessibility: the role of tomato matrix. *Food Chem.* 240: 415–421. https://doi.org/10.1016/j.foodchem.2017.07.102.

Bouras, M., Grimi, N., Bals, O., and Vorobiev, E. (2016). Impact of pulsed electric fields on polyphenols extraction from Norway spruce bark. *Ind. Crop. Prod.* 80: 50–58. https://doi.org/10.1016/j.indcrop.2015.10.051.

Boussetta, N. and Vorobiev, E. (2014). Extraction of valuable biocompounds assisted by high voltage electrical discharges: a review. *C. R. Chim* 17 (3): 197–203. https://doi.org/10.1016/j.crci.2013.11.011.

Brodeur, G., Yau, E., Badal, K. et al. (2011, 2011). Chemical and physicochemical pretreatment of lignocellulosic biomass: a review. *Enzyme. Res.*: 787532. https://doi.org/10.4061/2011/787532.

Canabarro, N.I., Alessio, C., Foletto, E.L. et al. (2017). Ethanol production by solid-state saccharification and fermentation in a packed-bed bioreactor. *Renewable Energy.* 102: 9–14. https://doi.org/10.1016/j.renene.2016.10.026.

Caroline, T.W., Luciane, F.T., and Jorge, O.T. (2020). Food waste biorefinery advocating circular economy: bioethanol and distilled beverage from sweet potato. *J. Clean. Prod.* 268: 121788. https://doi.org/10.1016/j.jclepro.2020.121788.

Carrillo-Nieves, D., Alanís, M.J.R., de la Cruz, Q.R. et al. (2019). Current status and future trends of bioethanol production from agro-industrial wastes in Mexico. *Renew. Sust. Energ. Rev.* 102: 63–74. https://doi.org/10.1016/j.rser.2018.11.031.

Casabar, J.T., Unpaprom, Y., and Ramaraj, R. (2019). Fermentation of pineapple fruit peel wastes for bioethanol production. *Biomass. Convers. Bior.* 9: 761–765. https://doi.org/10.1007/s13399-019-00436-y.

Catarina, I.S., Constança, C.S., Solange, I.M. et al. (2018). Integrated 1st and 2nd generation sugarcane bio-refinery for jet fuel production in Brazil: techno-economic and greenhouse gas emissions assessment. *Renewable Energy* 129: 733–747. https://doi.org/10.1016/j.renene.2017.05.011.

Chen, J., Yuan, Z., Zanuso, E., and Trajano, H.L. (2017). *Response of biomass species to hydrothermal pretreatment*. In Hydrothermal Processing in Biorefineries (pp. 95–140). Cham: Springer https://doi.org/10.1007/978-3-319-56457-9_4.

Collins, F., Rozhkovskaya, A., Outram, J.G., and Millar, G.J. (2020). A critical review of waste resources, synthesis, and applications for Zeolite LTA. *Microporous Mesoporous Mater.* 291: 109667. https://doi.org/10.1016/j.micromeso.2019.109667.

Cristina, M.R.R., Zlatina, G., Pedro, F. et al. (2018). Electric field-based technologies for valorization of bioresources. *Biores. Tech.* 254: 325–339. https://doi.org/10.1016/j.biortech.2018.01.068.

Dalena, F., Senatore, A., Basile, M. et al. (2019). From sugars to ethanol—from agricultural wastes to algal sources: An overview. In: *Second and Third Generation of Feed stocks*, 3–34. https://doi.org/10.1016/B978-0-12-815162-4.00001-X.

Dash, I., Bhaskar, D., Harshita, B. et al. (2018). Investigation on the production of bioethanol from black tea waste biomass in the seawater-based system. *Bio. Resour. Tech. Rep.* 4: 209–213. https://doi.org/10.1016/j.biteb.2018.11.003.

Deng, Y., Ju, T., and Xi, J. (2018). Circulating polyphenols extraction system with high-voltage electrical discharge: design and performance evaluation. *ACS Sustain. Chem. Eng.* 6 (11): 15402–15410. https://doi.org/10.1021/acssuschemeng.8b03827.

Dey, P., Pal, P., Kevin, J.D., and Das, D.B. (2020). Lignocellulosic bioethanol production: prospects of emerging membrane technologies to improve the process–a critical review. *Rev. Chem. Eng.* 36: 333–367. https://doi.org/10.1515/revce-2018-0014.

Dias, M.O., Modesto, M., Ensinas, A.V. et al. (2011). Improving bioethanol production from sugarcane: evaluation of distillation, thermal integration and cogeneration systems. *Energy.* 36: 3691–3703. https://doi.org/10.1016/j.energy.2010.09.024.

Dias, M.O., Junqueira, T.L., Cavalett, O. et al. (2012). Integrated versus stand-alone second generation ethanol production from sugarcane bagasse and trash. *Bio. Resour. Tech.* 103: 152–161. https://doi.org/10.1016/j.biortech.2011.09.120.

Díaz, V.H.G. and Tost, G.O. (2016). Ethanol and isobutanol dehydration by heat-integrated distillation. *Chem. Eng. Process.* 108: 117–124. https://doi.org/10.1016/j.cep.2016.07.005.

Doma, H.S., Abdo, S.M., Hemdan, B.A., and Ali, G.H. (2018). Enhancing biomass, energy and value added compounds yield from pilot scale pond system. *J. Environ. Sci. Technol.* 11: 199–208. https://doi.org/10.3923/jest.2018.199.208.

El-Mekkawi, S.A., Abdo, S.M., Samhan, F.A., and Ali, G.H. (2019). Optimization of some fermentation conditions for bioethanol production from microalgae using response surface method. *Bull. Nat. Res. Cen.* 43: 164. https://doi.org/10.1186/s42269-019-0205-8.

El-Naggar, M.E., Samhan, F.A., Salama, A.A. et al. (2018). Cationic starch: safe and economic harvesting flocculant for microalgal biomass and inhibiting *E. coli* growth. *Int. J. Biol. Macromol.* 116: 1296–1303. https://doi.org/10.1016/j.ijbiomac.2018.05.105.

Fanran, M., Ibbett, R., de Vrije, T. et al. (2019). Process simulation and life cycle assessment of converting autoclaved municipal solid waste into butanol and ethanol as transport fuels. *Waste Management.* 89: 177–189. https://doi.org/10.1016/j.wasman.2019.04.003.

de Farias Silva, C.E. and Bertucco, A. (2016). Bioethanol from microalgae and cyanobacteria: a review and technological outlook. *Process Biochem.* 51: 1833–1842. https://doi.org/10.1016/j.procbio.2016.02.016.

Fincan, M. (2015). Extractability of phenolics from spearmint treated with pulsed electric field. *J. Food Eng.* 162: 31–37. https://doi.org/10.1016/j.jfoodeng.2015.04.004.

Gavahian, M., Farhoosh, R., Javidnia, K. et al. (2015). Effect of applied voltage and frequency on extraction parameters and extracted essential oils from Menthapiperita by ohmic assisted hydro distillation. *Innov. Food Sci. Emerg. Technol.* 29: 161–169. https://doi.org/10.1016/j.ifset.2015.02.003.

Gavahian, M., Farahnaky, A., and Sastry, S. (2016). Multiple effect concentration of ethanol by Ohmic-assisted Hydrodistillation. *Food Bioprod. Process.* 100: 85–91. https://doi.org/10.1016/j.fbp.2016.06.002.

Gavahian, M., Chu, Y.H., and Sastry, S. (2018). Extraction from food and natural products by moderate electric field: mechanisms, benefits, and potential industrial applications. *Compr. Rev. Food Sci. Food Saf.* 17: 1040–1052. https://doi.org/10.1111/1541-4337.12362.

Gavahian, M., Munekata, P.E., Eş, I. et al. (2019). Emerging techniques in bioethanol production: from distillation to waste valorization. *Green Chem.* 21: 1171–1185. https://doi.org/10.1039/C8GC02698J.

Gavahian, M., Sastry, S., Farhoosh, R., and Farahnaky, A. (2020). Ohmic heating as a promising technique for extraction of herbal essential oils: understanding mechanisms, recent findings, and associated challenges. *Adv. Food Nutr. Res.* 91: 227–273. https://doi.org/10.1016/bs.afnr.2019.09.001.

George, P., Aggelos, G., Aikaterini, K. et al. (2020). Bioethanol production from food waste applying the multienzyme system produced on-site by Fusarium oxysporum F3 and mixed microbial cultures. *Ferment.* 6: 39. https://doi.org/10.3390/fermentation6020039.

Gray, K.A., Zhao, L., and Emptage, M. (2006). *Bioethanol. Curr. Opin. Chem. boil.* 10: 141–146. https://doi.org/10.1016/j.cbpa.2006.02.035.

Gupta, V.K., Potumarthi, R., O'Donovan, A. et al. (2014). Bioenergy research: an overview on technological developments and bioresources. In: *Bioenergy Research: Advances and Applications* (eds. V.K. Gupta, M.G. Tuohy, C.P. Kubiceck, et al.), 23–47. Elsevier https://doi.org/10.1016/B978-0-444-59561-4.00002-4.

Halder., P., Azad, K., Shah, S., and Sarker, E. (2019). Prospects and technological advancement of cellulosic bioethanol ecofuel production. *Adv.Eco-Fuel. Sustain. Environ.*: 211–236. https://doi.org/10.1016/B978-0-08-102728-8.00008-5.

Halimatun, S.H., A.R., N.'.A., Mohd, N.M. et al. (2017). Over production of fermentable sugar for bioethanol production from carbohydrate-rich Malaysian food waste via sequential acid-enzymatic hydrolysis pretreatment. *Waste Manag.* 67: 95–105. https://doi.org/10.1016/j.wasman.2017.05.017.

Harish, B.B. and Senthil, K.R. (2020). Kinetics and optimization of microwave-assisted lignin fractionation with Protic low transition temperature mixture of *Sesamum indicum* straw for enhanced bioethanol production. *J. Molecular Liquids.* 303: 112660. https://doi.org/10.1016/j.molliq.2020.112660.

Herrero, M., Cifuentes, A., and Ibañez, E. (2006). Sub-and supercritical fluid extraction of functional ingredients from different natural sources: plants, food-by-products, algae and microalgae: a review. *Food Chem.* 98 (1): 136–148. https://doi.org/10.1016/j.foodchem.2005.05.058.

Hossain, M.B., Aguiló-Aguayo, I., Lyng, J.G. et al. (2015). Effect of pulsed electric field and pulsed light pre-treatment on the extraction of steroidal alkaloids from potato peels. *Innovative Food Sci. Emerg. Technol.* 29: 9–14. https://doi.org/10.1016/j.ifset.2014.10.014.

Imen Ben, A., Georgia, A., Ioanna, N. et al. (2019). On the evaluation of different saccharification schemes for enhanced bioethanol production from potato peels waste via a newly isolated yeast strain of *Wickerhamomyces anomalus. Bioresour. Technol.* 289: 121614. https://doi.org/10.1016/j.biortech.2019.121614.

Izaskun, D., Beatriz, G., Jalel, L., and Patricia, G. (2019). Multiproduct biorefinery from vine shoots: bio-ethanol and lignin production. *Renew. Energy.* 142: 612–623. https://doi.org/10.1016/j.ree ne.2019.04.131.

Jahanbakhshi, A. and Salehi, R. (2019). Processing watermelon waste using Saccharomycescerevisiae yeast and the fermentation method for bioethanol production. *J Food Proc. Eng.* 42 (7): e13283. https://doi.org/10.1111/jfpe.13283.

Jan, B., Qian, K., Lise, A. et al. (2015). Challenges and opportunities in improving the production of bio-ethanol. *Prog. Energy Combust. Sci.* 47: 60–88. https://doi.org/10.1016/j.pecs.2014.10.003.

Jana, A.K. (2010). Heat integrated distillation operation. *Appl. Energy* 87: 1477–1494. https://doi.org/10.1016/j.apenergy.2009.10.014.

Jasreen, K.S., Kurt, A.R., Stephanie, J., and Tong, W. (2018). Effect of co-products of enzyme-assisted aqueous extraction of soybeans, enzymes, and surfactant on oil recovery from integrated corn-soy fermentation. *Ind. Crop. Prod.* 121: 441–451. https://doi.org/10.1016/j.indcrop.2018.05.033.

Joab Oliveira, V.S., Manuel, F.A., da C.A., M., and Joana, M.D. (2018). Integrated production of biodiesel and bioethanol from sweet potato. *Renew. Energy.* 124: 114–120. https://doi.org/10.1016/j.renene.2017.07.052.

Kaur, M., Kumar, M., Sachdeva, S., and Puri, S.K. (2020). An efficient multiphase bioprocess for enhancing the renewable energy production from almond shells. *Energ. Convers. Manage.* 203: 112235. https://doi.org/10.1016/j.enconman.2019.112235.

Kirupa, S.M., Ravikumar, R., Naresh, K.M., and Sivakumar, U. (2018). Development of co-immobilized tri-enzyme biocatalytic system for one-pot pretreatment of four different perennial lignocellulosic biomass and evaluation of their bioethanol production potential. *Bioresou. Tech.* 269: 227–236. https://doi.org/10.1016/j.biortech.2018.08.091.

Koel, S., Uma, M.R., Jaya, S. et al. (2017). Membranes as a tool to support biorefineries: applications in enzymatic hydrolysis, fermentation and dehydration for bioethanol production. *Renew. Sust. Energ. Rev.* 74: 873–890. https://doi.org/10.1016/j.rser.2017.03.015.

Kongkeitkajorn, M.B., Sae-Kuay, C., and Reungsang, A. (2020). Evaluation of Napier grass for bioethanol production through a fermentation. *Process.* 8: 567. https://doi.org/10.3390/pr8050567.

Koponen, K., Soimakallio, S., Tsupari, E. et al. (2013). GHG emission performance of various liquid transportation biofuels in Finland in accordance with the EU sustainability criteria. *Appl. Energy.* 102: 440–448. https://doi.org/10.1016/j.apenergy.2012.07.023.

Koubaa, M., Barba, F.J., Grimi, N. et al. (2016). Recovery of colorants from red prickly pear peels and pulps enhanced by pulsed electric field and ultrasound. *Innovative Food Sci. Emerg. Technol.* 37: 336–344. https://doi.org/10.1016/j.ifset.2016.04.015.

Kulkarni, S.J. (2020). Studies, efforts and investigations on various aspects of solid waste management with emphasis on developing countries. In: *Sustainability Concept in Developing Countries*, 1–13. https://doi.org/10.5772/intechopen.91942.

Kumar, V., Nanda, M., and Singh, A. (2016). Effect of bacterial amylase pretreatment on bioethanol production from starch-based solid waste (SBSW). *Energy Sources, Part A: Recovery, Utilization, and Environmental Effects.* 38 (17): 2604–2609. https://doi.org/10.1080/15567036.2015.1098745.

Kumar, R., Ghosh, A.K., and Pal, P. (2017). Fermentative energy conversion: renewable carbon source to biofuels (ethanol) using *Saccharomyces cerevisiae* and downstream purification through solar driven membrane distillation and nanofiltration. *Energy Convers. Manag.* 150: 545–557. https://doi.org/10.1016/j.enconman.2017.08.054.

Kyriakou, M., Chatziiona, V.K., Costa, C.N. et al. (2019). Biowaste-based biochar: a new strategy for fermentative bioethanol overproduction via whole-cell immobilization. *Applied Energy.* 242: 480–491. https://doi.org/10.1016/j.apenergy.2019.03.024.

Lee, D.Y., Ebie, Y., Xu, K.Q. et al. (2010). Continuous H_2 and CH_4 production from high-solid food waste in the two-stage thermophilic fermentation process with the recirculation of digester sludge. *Bioresour. Technol.* 101: S42–S47. https://doi.org/10.1016/j.biortech.2009.03.037.

Leonardo, A.A., Juan, C.S., Luis, A.B., and Carlos, A.C. (2020). Performance evaluation and economic analysis ofthe bioethanol and flour production using rejectedun ripe plantain fruits (*Musa paradisiaca* L.) as rawmaterial. *Food Bioprod. Process.* 121: 29–42. https://doi.org/10.1016/j.fbp.2020.01.005.

Li, G. and Bai, P. (2012). New operation strategy for separation of ethanol–water by extractive distillation. *Ind. Eng. Chem. Res.* 51: 2723–2729. https://doi.org/10.1021/ie2026579.

Li, X., Luo, X., Jin, Y. et al. (2018). Heterogeneous sulfur-free hydrodeoxygenation catalysts for selectively upgrading the renewable bio-oils to second generation biofuels. *Renew. Sust. Energ. Rev.* 82: 3762–3797. https://doi.org/10.1016/j.rser.2017.10.091.

Li, P., Sakuragi, K., and Makino, H. (2019). Extraction techniques in sustainable biofuel production: A concise review. *Fuel Process. Technol.* 193: 295–303. https://doi.org/10.1016/j.fuproc.2019.05.009.

Liu, D., Vorobiev, E., Savoire, R., and Lanoisellé, J.L. (2011). Intensification of polyphenols extraction from grape seeds by high voltage electrical discharges and extract concentration by dead-end ultrafiltration. *Sep. Purif. Technol.* 81 (2): 134–140. https://doi.org/10.1016/j.seppur.2011.07.012.

Liu, J., Chen, J., Zhan, X. et al. (2015). Preparation and characterization of ZSM-5/PDMS hybrid pervaporation membranes: laboratory results and pilot-scale performance. *Sep. Purif. Technol.* 150: 257–267. https://doi.org/10.1016/j.seppur.2015.06.036.

Lü, J., Sheahan, C., and Fu, P. (2011). Metabolic engineering of algae for fourth generation biofuels production. *Energy Environ. Sci.* 4 (7): 2451–2466. https://doi.org/10.1039/C0EE00593B.

Luengo, E., Martínez, J.M., Álvarez, I., and Raso, J. (2016). Effects of millisecond and microsecond pulsed electric fields on red beet cell disintegration and extraction of betanines. *Ind. Crop. Prod.* 84: 28–33. https://doi.org/10.1016/j.indcrop.2016.01.016.

Lynd, L.R., Liang, X., Biddy, M.J. et al. (2017). Cellulosic ethanol: status and innovation. *Curr. Opin. Biotechnol.* 45: 202–211. https://doi.org/10.1016/j.copbio.2017.03.008.

Ma, X., Gao, M., Yin, Z. et al. (2020). Lactic acid and animal feeds production from Sophoraflavescens residues by Rhizopusoryzae fermentation. *Process Biochem.* 92: 401–408. https://doi.org/10.1016/j.procbio.2020.01.030.

Mahmoodi, P., Karimi, K., and Taherzadeh, M.J. (2018). Hydrothermal processing as pretreatment for efficient production of ethanol and biogas from municipal solid waste. *Biores. Tech.* 261: 166–175. https://doi.org/10.1016/j.biortech.2018.03.115.

Markou, G., Angelidaki, I., Nerantzis, E., and Georgakakis, D. (2013). Bioethanol production by carbohydrate-enriched biomass of Arthrospira (Spirulina) platensis. *Energies* 6: 3937–3950. https://doi.org/10.3390/en6083937.

Martínez, J.M., Delso, C., Álvarez, I., and Raso, J. (2019). Pulsed electric field permeabilization and extraction of phycoerythrin from *Porphyridium cruentum*. *Algal Res.* 37: 51–56. https://doi.org/10.1016/j.algal.2018.11.005.

Martínez, J.M., Delso, C., Álvarez, I., and Raso, J. (2020). Pulsed electric field-assisted extraction of valuable compounds from microorganisms. *Compr. Rev. Food Sci. Food Saf.* https://doi.org/10.1111/1541-4337.12512.

Mattam, A.J., Kuila, A., Suralikerimath, N. et al. (2016). Cellulolytic enzyme expression and simultaneous conversion of lignocellulosic sugars into ethanol and xylitol by a new Candida tropicalis strain. *Biotechnol. Biofuels* 9: 157. https://doi.org/10.1186/s13068-016-0575-1.

Meng, K., Obiri-Danso, K., Kádár, Z. et al. (2017). Municipal solid waste management in a low income economy through biogas and bioethanol production. *Waste and Biomass Valorization.* 8 (1): 115–127. https://doi.org/10.1007/s12649-016-9566-5.

Mohsen, G. and Asgar, F. (2018). Ohmic-assisted hydrodistillation technology: a review. *Trends Food Sci. Technol.* 72: 153–161. https://doi.org/10.1016/j.tifs.2017.12.014.

Mohsen, G., Asgar, F., Mohammad, S., and Sudhir, S. (2016a). Ethanol concentration of fermented broth by ohmic-assisted hydrodistillation. *Innov. Food Sci. Emerg. Technol.* 98: 44–49. https://doi.org/10.1016/j.ifset.2016.04.001.

Mohsen, G., Asgar, F., and Sudhir, S. (2016b). Ohmic-assisted hydrodistillation: a novel method for ethanol distillation. *Food Bioprod. Process.* 98: 44–49. https://doi.org/10.1016/j.fbp.2015.11.003.

Mutamim, N.S.A., Noor, Z.Z., Hassan, M.A.A. et al. (2013). Membrane bioreactor: applications and limitations in treating high strength industrial wastewater. *Chem. Eng. J.* 225: 109–119. https://doi.org/10.1016/j.cej.2013.02.131.

Nagappan, S. and Nakkeeran, E. (2020). Protein engineering approaches for lignocellulosic ethanol biorefinery. In: *Biovalorisation of Wastes to Renewable Chemicals and Biofuels* (eds.

N.K. Rathinam and R.K. Sani), 243–260. Amsterdam: Elsevier https://doi.org/10.1016/B978-0-12-817951-2.00012-2.

Nair, R.B., Lennartsson, P.R., and Taherzadeh, M.J. (2017). Bioethanol production from agricultural and municipal wastes. In: *Current developments in biotechnology and bioengineering* (eds. J.W.-C. Wong, R.D. Tyagi and A. Pandey), 157–190. Amsterdam: Elsevier https://doi.org/10.1016/B978-0-444-63664-5.00008-3.

Ohia, G.N., Ohia, N.P., Ekwueme, S.T., and Nwankwo, I.V. (2020). Hydrolysis of cellulose wastes: feasibility of fuel ethanol as alternative to gasoline from petroleum as a usable energy source in Nigeria. *Pet. Sci. Eng.* 4: 16–22. https://doi.org/10.11648/j.pse.20200401.12.

Pandiyan, K., Arjun, S., Surender, S. et al. (2018). Technological interventions for utilization of crop residues and weedy biomass for second generation bio-ethanol production. *Renew. Energy.* 132: 723–741. https://doi.org/10.1016/j.renene.2018.08.049.

Pang, B., Cao, X.F., Sun, S.N. et al. (2020). The direct transformation of bioethanol fermentation residues for production of high-quality resins. *Green Chem.* 22: 439–447. https://doi.org/10.1039/C9GC03568K.

Park, I., Kim, I., Kang, K. et al. (2010). Cellulose ethanol production from waste newsprint by simultaneous saccharification and fermentation using Saccharomyces cerevisiae KNU5377. *Process Biochem.* 45: 487–492. https://doi.org/10.1016/j.procbio.2009.11.006.

Parniakov, O., Roselló-Soto, E., Barba, F.J. et al. (2015). New approaches for the effective valorization of papaya seeds: extraction of proteins, phenolic compounds, carbohydrates, and isothiocyanates assisted by pulsed electric energy. *Food Res. Int.* 77: 711–717. https://doi.org/10.1016/j.foodres.2015.03.031.

Pei-Yao, Z., Wen-Hai, Z., L., C. et al. (2019). Efficient bio-ethanol recovery by non-contact vapor permeation process using membranes with tailored pore size and hydrophobicity. *Chem. Eng. Sci.* 207: 448–455. https://doi.org/10.1016/j.ces.2019.06.047.

Postma, P.R., Fernandes, D.A., Timmermans, R.A.H. et al. (2017). Pulsed electric field for protein release of the microalgae *Chlorella vulgaris* and *Neochloris oleoabundans*. *Algal Res.* 24: 181–187. https://doi.org/10.1016/j.algal.2017.03.024.

Prasoulas, G., Gentikis, A., Konti, A. et al. (2020). Bioethanol production from food waste applying the multienzyme system produced on-site by Fusariumoxysporum F3 and mixed microbial cultures. *Ferment.* 6: 39. https://doi.org/10.3390/fermentation6020039.

Puértolas, E. and Barba, F.J. (2016). Electrotechnologies applied to valorization of by-products from food industry: Main findings, energy and economic cost of their industrialization. *Food Bioprod. Process.* 100: 172–184. https://doi.org/10.1016/j.fbp.2016.06.020.

Rajiv, C.R. and Rintu, B. (2020). An innovative approach of mixed enzymatic venture for 2G ethanol production from lignocellulosic feedstock. *Energy Convers. Manag.* 207: 112504. https://doi.org/10.1016/j.enconman.2020.112504.

Ramachandra, T.V. and Deepthi., H. (2020). Bioethanol from macroalgae: prospects and challenges. *Renew. Sust. Energ. Rev.* 117: 109479. https://doi.org/10.1016/j.rser.2019.109479.

Rastogi, M. and Shrivastava, S. (2017). Recent advances in second generation bioethanol production: an insight to pretreatment, saccharification and fermentation processes. *Renew. Sust. Energ. Rev.* 80: 330–340. https://doi.org/10.1016/j.rser.2017.05.225.

Ravanal, M.C., Camus, C., Buschmann, A.H. et al. (2019). Production of bioethanol from Brown algae. In: *Advances in Feedstock Conversion Technologies for Alternative Fuels and*

Bioproducts (ed. M. Hosseini), 69–88. Amsterdam: Elsevier https://doi.org/10.1016/B978-0-12-817937-6.00004-7.

Raveendran, S., Athulya, M., Pooja, M. et al. (2020). Valorization of food and kitchen waste: an integrated strategy adopted for the production of poly-3-hydroxybutyrate, bioethanol, pectinase and 2, 3-butanediol. *Bioresour. Technol.* 310: 123–515. https://doi.org/10.1016/j.biortech.2020.123515.

Ruan, R., Zhang, Y., Chen, P. et al. (2019). Biofuels: introduction: biofuels: alternative feed stocks and conversion processes. *Biofuel.*: 3–43. https://doi.org/10.1016/B978-0-12-816856-1.00001-4.

Sachin, T., Antonio, F.P., Vijayraghavan, R., and Amit, A. (2018). An integrated green biorefinery approach towards simultaneous recovery of pectin and polyphenols coupled with bioethanol production from waste pomegranate peels. *Bioresour. Technol.* 266: 322–334. https://doi.org/10.1016/j.biortech.2018.06.072.

Sadeghinezhad, E., Kazi, S.N., Sadeghinejad, F. et al. (2014). A comprehensive literature review of bio-fuel performance in internal combustion engine and relevant costs involvement. *Renew. Sust. Energ. Rev.* 30: 29–44. https://doi.org/10.1016/j.rser.2013.09.022.

Saha, B.C., Nichols, N.N., Qureshi, N. et al. (2015). Pilot scale conversion of wheat straw to ethanol via simultaneous saccharification and fermentation. *Bioresour. Technol.* 175: 17–22. https://doi.org/10.1016/j. biortech.2014.10.060.

Saldaña, G., Cebrián, G., Abenoza, M. et al. (2017). Assessing the efficacy of PEF treatments for improving polyphenol extraction during red wine vinifications. *Innovative Food Sci. Emerg. Technol.* 39: 179–187. https://doi.org/10.1016/j.ifset.2016.12.008.

Sarkis, J.R., Boussetta, N., Tessaro, I.C. et al. (2015). Application of pulsed electric fields and high voltage electrical discharges for oil extraction from sesame seeds. *J. Food Eng.* 153: 20–27. https://doi.org/10.1016/j.jfoodeng.2014.12.003.

Sawatdeenarunat, C., Wangnai, C., Songkasiri, W. et al. (2019). Biogas production from industrial effluents. Biofuels: alternative feed stocks and conversion processes for the production of liquid and gaseous. *Biofuel.*: 779–816. https://doi.org/10.1016/B978-0-12-816856-1.00032-4.

Scherer, D., Krust, D., Frey, W. et al. (2019). Pulsed electric field (PEF)-assisted protein recovery from *Chlorella vulgaris* is mediated by an enzymatic process after cell death. *Algal Res.* 41: 101536. https://doi.org/10.1016/j.algal.2019.101536.

Sharma, A. and Aggarwal, N.K. (2020). Bioethanol production from water hyacinth. Water hyacinth: a potential. In: *Lignocellulosic Biomass for Bioethanol*, 91–106. https://doi.org/10.1007/978%3-030-35632-3_7.

Soave, G.J.a. and Feliu, A. (2002). Saving energy in distillation towers by feed splitting. *Appl. Therm. Eng.* 22: 889–896. https://doi.org/10.1016/S1359-4311(02)00006-6.

Song, Y., Oh, C., and Bae, H.J. (2017). Simultaneous production of bioethanol and value-added D-psicose from Jerusalem artichoke (Helianthus tuberosus L.) tubers. *Bioresour. Technol.* 24 (4): 1068–1072. https://doi.org/10.1016/j.biortech.2017.08.079.

Swaraza, A.M., Mohammed, A.S., Md Mahfuzur, R. et al. (2019). Bioethanol production potential in Bangladesh from wild date palm (*Phoenix sylvestris* Roxb.): an experimental proof. *Ind. Crop. Prod.* 139: 111507. https://doi.org/10.1016/j.ind crop.2019.111507.

Taghizadeh-Alisaraei, A., Hosseini, S.H., Ghobadian, B., and Motevali, A. (2017). Biofuel production from citrus wastes: a feasibility study in Iran. *Renew. Sust. Energ. Rev.* 69: 1100–1112. https://doi.org/10.1016/j.rser.2016.09.102.

Tavan, Y. and Shahhosseini, S. (2016). Feed-splitting as energy-saving technique in the heterogeneous distillation of ethanol–water Azeotropes. *Energ. Technol.* 4: 424–428. https://doi.org/10.1002/ ente.201500287.

Ueno, K., Negishi, H., Okuno, T. et al. (2017). High-performance silicalite-1 membranes on porous tubular silica supports for separation of ethanol/water mixt ures. *Sep. Purif. Technol.* 187: 343–354. https://doi.org/10.1016/j.seppur.2017.06.071.

Vaida, K., Darius, P., Vaida, K. et al. (2017). Fractionation of sea buckthorn pomace and seeds into valuable components by using high pressure and enzyme-assisted extraction methods. 85: 534–538. https://doi.org/10.1016/j.lwt.2017.02.041.

Vane, L.M., Alvarez, F.R., Rosenblum, L., and Govindaswamy, S. (2013). Efficient ethanol recovery from yeast fermentation broth with integrated distillation-membrane process. *Ind. Eng. Chem. Res.* 52: 1033–1041. https://doi.org/10.1021/ie2024917.

Wang, C., Dong, D., Wang, H. et al. (2016). Metagenomic analysis of microbial consortia enriched from compost: new insights into the role of Actinobacteriain lignocellulose decomposition. *Biotechnol. Biofuels* 9: 22. https://doi.org/10.1186/s13068-016-0440-2.

Wei, P., Cheng, L.H., Zhang, L. et al. (2014). *Renew. Sust. Energ. Rev.* 30: 388–400. https://doi .org/10.1016/j.rser.2013.10.017.

Wei, H., Yanxu, L., Xiaobin, X. et al. (2020). Bioethanol production from waste hamburger by enzymatic hydrolysis and fermentation. *J. Clean. Prod.* 264: 121658. https://doi.org/10.1016/j .jclepro.2020.121658.

Wolske, K.S. and Stern, P.C. (2018). Contributions of psychology to limiting climate change: opportunities through consumer behavior. In: *Psychology and Climate Change Human Perceptions, Impacts, and Responses* (eds. S. Clayton and C. Manning), 127–160. Amsterdam: Elsevier https://doi.org/10.1016/B978-0-12-813130-5.00007-2.

Xiao-Fei, W., Shuang-Shuang, Y., Qian, Z. et al. (2019). Subcritical liquefaction of lignocellulose for the production of bio-oils in ethanol/water system. *Renewable Energy* 136: 865–872. https://doi.org/10.1016/j.renene.2019.01.041.

Xiaoru, H., Jonas, H.H., and Anne-Belinda, B. (2020). Integrated bioethanol and protein production frombrown seaweed. Laminaria *digitata*. 197: 310–317. https://doi.org/10.1016/j .biortech.2015.08.091.

Ye, G., Zeng, D., Zhang, S. et al. (2018). Ethanol production from mixtures of sugarcane bagasse and Dioscorea composita extracted residue with high solid loading. *Biores. Tech.* 257: 23–29. https://doi.org/10.1016/ j.biortech.2018.02.008.

Yizhak, M. (2018). Extraction by subcritical and supercritical water, methanol, ethanol and their mixtures. *Separations.* 5: 4. https://doi.org/10.3390/separations5010004.

Yongteng, Z., Kang, M., Wenting, B. et al. (2018). Energy-saving thermally coupled ternary extractive distillation process by combining with mixed entrainer for separating ternary mixture containing bioethanol. *Energy.* 148: 296–308. https://doi.org/10.1016/j.energy.2018 .01.161.

Yu, X., Bals, O., Grimi, N., and Vorobiev, E. (2015). A new way for the oil plant biomass valorization: polyphenols and proteins extraction from rapeseed stems and leaves assisted by pulsed electric fields. *Ind. Crop. Prod.* 74: 309–318. https://doi.org/10.1016/j.indcrop.2015.03.045.

Yu, H., Guo, J., Chen, Y. et al. (2017). Efficient utilization of hemicellulose and cellulose in alkali liquor-pretreated corncob for bioethanol production at high solid loading by *Spathasporapassalidarum* U1-58. *Bioresour. Technol.* 232: 168–175. https://doi.org/10.1016/j.biortech.2017.01.077.

Zhu, Y., Zhao, Z., and Zhang, Y. (2019). Using straw as a bio-ethanol source to promote anaerobic digestion of waste activated sludge. *Biores. Tech.* 286: 121388. https://doi.org/10.1016/j.biortech.2019.121388.

10

Algal Biofuels – Types and Production Technologies

Sreedevi Sarsan and K. Vindhya Vasini Roy

Department of Microbiology, St. Pious X Degree and PG College, Hyderabad, India

10.1 Introduction

The combustion of petroleum, coal, natural gas, and other fossil fuels generates the energy needed for various purposes in the entire world. Burning of these fuels emits carbon dioxide and other gases, causing heavy pollution and global warming. Release of these pollutants can be decreased either by reducing the use of fossil fuels or replacing them with alternative renewable fuels. The most important renewable energy sources include solar, wind, and biomass. The fuels produced from renewable resources are called renewable fuels. Renewable resources, such as vegetables and biomasses, are replenished naturally and recurrently in less time and thus seem to be a potential solution. In addition, they reduce the emission of greenhouse gases like carbon dioxide, methane, nitrous oxide, etc., into the air, thus decreasing pollution.

Biofuels are the most popular sources of renewable energy made from biological raw materials, such as organic matter or biomass from agricultural crops or wastes. Biofuels are gaining increased interest and attention by many governments, the general public and scientific researchers, owing to increasing oil prices, the necessity for enhanced energy security, and concerns over increasing pollution due to greenhouse gas emissions from fossil fuels. Biofuels are broadly of two types: those produced from agricultural crops are called conventional biofuels and those produced from waste, inedible crops, or forestry products by using new technologies and processes are known as advanced biofuels. Some of these advanced biofuels may be blended with conventional fuels to become more compatible with current vehicles. Advanced biofuels are sustainable and are the primary form of biofuels to be employed in the future. Biofuels are classified into four generation types:

- The conventional biofuels prepared from agricultural food crops grown specifically for fuel production are considered as first-generation biofuels. The sugar, starch, or oils are extracted from these food crops and converted into bioethanol or biodiesel using fermentation or transesterification. These are also called agrofuels, which use exclusively cultivated plants and adversely impacts food and water sectors to a great extent (Saad et al. 2019).

Bioenergy Research: Evaluating Strategies for Commercialization and Sustainability, First Edition.
Edited by Neha Srivastava and Manish Srivastava.

- The biofuels produced from different plant- and animal-derived biomasses are included under second-generation biofuels (Service 2007). They include lignocellulosic biomass from woody plants, grasses, oil/oil seed crops, and other agriculture residues. These fuels are advantageous, as their production does not exclusively require cultivable land but the limitation lies in the difficulty of extraction of the fuel.
- Biofuels produced from algal contents or algal biomasses are known as third-generation biofuels. These biofuels came into existence with an intention to reduce the use of land and water. The algal cultivation requires neither farmland nor freshwater and thus do not lead to a decrease in food production.
- The new emerging biofuels that are produced from genetically modified algae or biomass using noncultivable lands are considered as fourth-generation biofuels. These metabolically engineered biofuels are produced with minimum cost and maximum yields (Saad et al. 2019). These fuels produce zero-carbon emissions and include fuels, such as electrofuels and photobiological solar fuels (Aro 2016).

10.2 Algal Biofuels

Algae are eukaryotic, photosynthetic organisms that can grow prominently in aquatic (freshwater/marine) and terrestrial environments. Light, water, carbon dioxide, nitrogen, phosphorus, potassium, and some other nutrients are essential requirements for the algal growth. They produce biomass very rapidly, almost 10 or even 100 times more productive than traditional bioenergy crops(Sheehan et al. 1998; Chisti 2007; Huesemann et al. 2009). Algae possess large amounts of carbohydrates and oil contents that can be processed into various biofuels (Brennan and Owende 2010; Nigam and Singh 2011; Behera et al. 2015). Algae are broadly categorized into two types: macroalgae and microalgae. Macroalgae or seaweeds are multicellular algae and generally grow in ponds. They are large, measuring about few inches to few feet in length. Phaeophyceae (brown), Rhodophyceae (red), and Chlorophyceae (green) seaweeds are the important groups of macroalgae. Microalgae are tiny, unicellular photosynthetic microorganisms that are less than 0.4 mm in diameter and normally grow in saline or fresh water environments. They are unicellular, existing singly or in colonies. Most importantly, microalgae belong to the classes: Bacillariophyceae (diatoms), Chlorophyceae (green algae), and Chrysophyceae (golden algae).

Algal biofuel is an alternative renewable fuel produced from algal biomass and its contents. The algal biomass contains lipids, carbohydrates, and proteins, which are processed and converted to biofuels. Macroalgae and microalgae are employed for the production of biofuels. Macroalgaeor seaweeds are highly available and the biofuels produced from them are known as seaweed fuels or seaweed oils. Macroalgae have several advantages over terrestrial plants, which makes them an important alternative for biofuel production such as growth using minimum nutrients, presence of easily fermentable carbohydrate substrates, easy harvesting techniques, and increased production per unit area. Some of the potential genera of macroalgae used for biofuel production are *Macrocystis*, *Gracilaria*, *Sargassum*, *Laminaria*, and *Ulva* (Chen et al. 2015; Suutari et al. 2015). Owing to its quick growth and several harvestings made per year, *Macrocystis pyrifera* may be exploited in biofuel production studies in future.

Microalgae are more preferred for production of biofuels because of their simple structural organization, rapid growth rates, and high amounts of oil. Some of the most potential microalgae suitable for mass production of biofuels are *Botryococcus braunii*, *Dunaliella tertiolecta*, *Chlamydomonas*, *Scenedesmus* sp., and *Chlorella* sps. There are numerous unique properties of microalgae that tend to recognize them as potentially good sources of biofuels over other terrestrial plants. Microalgae cultivation does not require either fertile land or irrigation and can be grown using noncultivable lands and wastewater or saline water. They grow very quickly and can be grown throughout the year as their growth is not dependent on seasonal variations. Microalgae species are easily amenable to gene manipulation techniques, which can be used to increase yields. Large-scale cultivation of microalgae may reduce the carbon emissions from the combustion exhaust of power plants and thus help in the remediation process (Rosenberg et al. 2008).

Based upon the species and culture conditions employed for growing, the composition microalgal biomass varies and usually comprises 7–23% of lipids, 6–71% of proteins, and 5–64% of carbohydrates (Brown 1991; Becker 2007; Mata et al. 2010; Singh and Dhar 2011). Different microalgae species produce various lipids, carbohydrates, or natural oils that can be harnessed and processed into fuels for vehicles (Banerjee et al. 2002; Metzger and Largeau 2005; Guschina and Harwood 2006; Chisti 2007; Rodolfi et al. 2009; Michael et al. 2010). The type and the relative composition of fatty acids differ among various microalgal species (Gouveia and Oliveira 2009). Most of the natural oil from microalgae is extracted as triacylglycerol, which is transformed into biodiesel, biohydrogen, biogas, and other biofuels. Glycerol, an important by-product of biodiesel production has enormous industrial applications. Also, the residual biomass, left after oil extraction from microalgae, can be used as a feed for animals. Thus the microalgae offer many avenues and has garnered enormous interest to use it as a potential alternative biofuel.

10.3 Production of Algal Biofuels

Algae have many important characteristics that make them advantageous over other biomass sources for production of biofuels. Some of them are that they have high reproducibility, high biomass yield, high lipid or carbohydrate content, do not require farmland, and can grow using any water resources – wastewater or saline (Mutanda et al. 2011; Pittman et al. 2011; Lam et al. 2019; Goh et al. 2019). The usual process of biofuel production involves cultivation of algae, separation of biomass by harvesting, drying, and cell disruption (optional), and then conversion into biofuel. Both macroalgae and microalgae are used for biofuels production and they differ in their cultivation systems and processing techniques. Owing to their small size, cultivation of microalgae is much easier and more controllable but harvesting is more complicated. Macroalgae, on the other hand, are less versatile and only few species are used for biofuels and anaerobic digestion to produce biogas is the main viable technology available.

10.3.1 Algae Cultivation Systems

Macroalgae and microalgae differ in their methods of cultivation. Macroalgae or seaweeds are mostly obtained naturally or cultivated in the open sea directly, whereas microalgae can be cultivated easily in different systems either on land or on water owing to their small size. The cultivation of macroalgae was first started in 1690 in Japan and is considered as the primary producer of cultured seaweed even today (Buck and Buchholz 2004). The cultivation of microalgae for use as a protein source was started in the 1950s, but its use as a renewable energy fuel was considered and investigated only in the 1970s. Light, water, carbon dioxide, nutrients, and optimum temperature are some of the minimum requirements for growing algae. Many such different systems were designed and developed for algae cultivation but scaling their production to industrial level is difficult. They include simple land-based or sea-based open systems to expensive yet highly controllable and optimized closed systems. New and innovative systems and technologies need to be developed to grow algae commercially on a large scale for profitable algae-based fuel generation.

10.3.1.1 Cultivation of Macroalgae

Macroalgae or seaweeds can be either obtained from natural populations or cultivated. In Europe, seaweed is mainly obtained by harvesting natural stocks while in Asian countries seaweed is cultivated. Harvesting natural populations/beach cast wild seaweed is not suitable for large-scale use and also the ecosystem gets disturbed. Therefore, it is advisable to culture macroalgae in dedicated systems and many macroalgal cultivation methods have come into practice (Raikova et al. 2019). Based on the source, seaweeds are of different types: natural seaweeds (wild and drift) and cultured seaweeds (land-based or aqua-cultured).

10.3.1.1.1 Natural Seaweed Sea water is an important source of naturally growing seaweeds. There are two types of natural seaweeds – wild seaweeds and drift seaweeds, which are harvested, collected, and processed into biofuels.

Wild Seaweeds Wild seaweeds are collected from natural biomass washed ashore from the sea. Brown seaweeds are predominant in cold waters while red seaweeds are dominant in warmer waters. The algae harvested are mainly *Laminaria hyperborea* and *Ascophyllum nodosum*. Wild seaweed harvesting is usually done by cutting monospecific strands of seaweeds like rockweeds and kelps. Another method involves assemblage of storm-cast fronds using nets, and this technique usually results in a mixture of species (Micheal Mac et al. 2017).

Drift Seaweeds These are another common source of natural seaweed, which drifts to seashores. The seaweed are found growing attached but later they get detached and grow further and are left behind by the receding waves. The accumulation of seaweed mass is quite large in certain places like Massachusetts, Bermuda, Ireland, and the Venice lagoon. Their availability is dependent on season and location, and therefore cannot be depended on for large-scale commercial use.

10.3.1.1.2 Seaweed Cultivation A number of methods are available to cultivate seaweeds depending upon the kind of algal species cultivated, form of life cycle, and various biotic and abiotic factors present. Seaweeds can be cultured in land-based or aquatic-based systems either in open or closed systems. Many types of seaweed can be propagated vegetatively in a one-step process, e.g., *Eucheuma*, *Kappaphycus*, and *Gracilaria*. Vegetative cultivation is initiated from small pieces of seaweed and growing them in a suitable environment. While harvesting, a small part is left from which new growth takes place. Some other seaweed cannot be propagated by vegetative means and need multistep processes as they exhibit an alternation of generations, e.g., *Porphyra*, *Ulva*, *Laminaria*, and *Undaria* (Pereira and Yarish 2008). They are propagated from spores and their life cycle alternates between large sporophyte and microscopic gametophyte. A new sporophyte is only formed after the fusion of haploid gametes and that is harvested. The mature sporophyte releases spores (conchospores from conchocelis of *Porphyra* and Zoopores in case of *Undaria*) that germinate and grow into microscopic gametophytes.

Land-Based Cultivation Systems Some macroalgae species like *Ulva rotundata*, *Ulva rigida*, *Gracilaria*, *Porphyra*, *Laminaria digitata*, *Gracilariopsis longissima*, and *Chondrus crispus* are grown using land-based cultivation systems. They are cultivated either in tanks or ponds. Seaweeds are grown in tanks which are provided with a continuous supply of seawater and aeration. Strong aeration provides adequate mixing of the algal thalli and also permits uptake of dissolved carbon dioxide. The productivity in tanks is more efficient than other systems. However, the efficiency is dependent on factors like aeration, temperature, pH and salinity, carbon dioxide, nutrients, and supply of seawater that can be controlled by monitoring the seawater pumping. Tank-type cultivation systems can also be employed for growing macroalgae using polluted waters and is thus helpful in remediation. In cultivation of seaweeds in ponds, the water exchange is done by use of tide gates resulting in low yields because of the lack of proper movement of water but their operating costs are less. Open raceway ponds are commonly used for cultivating seaweeds since the 1950s, e.g., *Ulva* sp. Open raceway ponds are shallow artificial ponds and have a closed loop for recirculation and are made up of several oval channels. Each channel has a paddlewheel for mixing and recirculation and baffles to guide the flow at bends (Brockmann et al. 2015).

Maricuture (Sea-Based Cultivation Systems) Many macroalgal species like *Porphyra*, *Eucheuma*, *Macrocystis*, etc., are cultivated in sea-based systems. Four basic methods are employed in mariculture: they are floating raft method, semi-floating raft method, off-bottom (fixed-bottom) method, and bottom planting method (Figure 10.1) (Bast 2013). In floating raft method or long line method, rafts are kept afloat with the help of buoys,

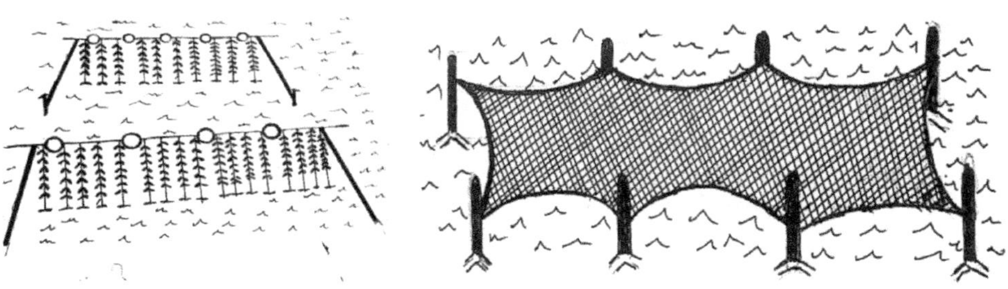

Long-Line Cultivation
(a)

Net style farming
(b)

Figure 10.1 Types of mariculture methods used in seaweed cultivation. a. Long-line cultivation. b. Net style farming.

which may be fixed near shore or at offshore sites. Seaweed is cultivated in nets made of nylon or polyethylene or coir, interwoven into a frame of wooden raft and held in place by deadweight mooring. The sporelings of algae are first produced in cold water in greenhouses and then planted out in the oceans. This type of method has been very successfully employed for cultivating *Laminaria* by the Chinese from as early as the 1950s. Net style farming is used for the cultivation of *Porphyra*, *Monostroma*, and *Ulva*. In this method the spores of algae are seeded onto nets that are attached to the top ends of the poles, which are either fixed to support systems as in the off-bottom method or attached to buoys as in the semi-floating method. In this method the nets are exposed to low tides allowing periodic exposure to sunlight and water improves the growth of the seaweed. The bottom planting method is usually seen in areas with a low water level during low tide and is useful for benthic genera like *Gracillaria*, *Sarcodiotheca*, and *Caulerpa*. In this method, macroalgae are grown on a substratum placed directly on the shoal and they are found always immersed in water. The offshore or deep-water method is used to cultivate the seaweed *M. pyrifera*. The growing seaweeds get their nutrition from the cold deep waters of the sea. The main advantage of this method over cultivation in shallow waters is that the growth of the algae is not restricted by nutrient limitation.

10.3.1.2 Cultivation of Microalgae

Microalgae have great potential to produce various kinds of biofuels. Microalgae can grow easily, as they are photoautotrophic and need simple nutritional requirements like water, salts, and light for their growth. Selection of a suitable species and strain of algae is an imperative factor determining the success of a biofuel venture. A number of parameters influence choosing the optimal strain and the kinds of biofuel intended to be produced is an important consideration. Microalgae containing high lipids or natural oils are the most appropriate strains for the production of biodiesel, while those with high polysaccharide content are best for bioethanol production. Other parameters such as high growth rate, ease of harvesting, and amenability to strain improvement are also important (Griffiths and Harrison 2009; Abu-Ghosh et al. 2016).

Microalgae develop using different growth modes such as phototrophic, heterotrophic, and mixotrophic (Table 10.1). Although phototrophic mode of growth is the preferred method for large-scale production of algal biomass, each method has its own advantages. Most algae use photoautotrophic mode while some species also exhibit heterotrophic growth and depend on organic compounds for their carbon and energy source. The heterotrophic mode of algal growth has several advantages when compared to phototrophic mode. Heterotrophic algae are grown in fermenters and the huge knowledge of fermentation technology is already available and consistent production is reproducible

Table 10.1 Microalgal growth types.

Growth type	Source of energy	Source of carbon	Light requirement
Photoautotrophic	Light	CO2/inorganic C	Obligatory
Heterotrophic	Organic	Organic C	Not required
Photoheterotrophic	Light	Organic	Obligatory
Mixotrophic	Light and organic	Inorganic and organic	Not obligatory

because of the high degree of process controls. The additional benefits of heterotrophic mode of algal growth come from exclusion of light requirements. In addition, they are unaffected by environmental conditions and also the costs of harvesting are less (Barclay et al. 1994; Behrens and Kyle 1996; Ceron Garcia et al. 2000; Chen and Chen 2006). Moreover, the total content of lipids are found to be more in algae grown heterotrophically than in phototrophic algae and thus results in higher yields (Miao and Wu 2004, 2006; Li et al. 2007; Yu et al. 2009). But the limitation of heterotrophic mode of cultivation of algae is that it requires adequate amount of oxygen for the catabolism of organic substrates (Clark et al. 1995). Several algae can also switch between these two modes of growth – photoautotrophy and heterotrophy – and are called as mixotrophic, i.e., they can simultaneously utilize both organic and inorganic carbon sources and light as their energy source. Some studies have shown that the lipid content and biomass produced are higher in mixotrophic mode of cultivation of algae and thus show increased productivity (Wen and Chen 2003; Fang et al. 2004; Li et al. 2014; Gilmour 2019).

The efficiency and production cost of microalgal biofuels is dependent upon the method of cultivation system employed (Lee 2001; Pulz 2001; Carvalho et al. 2006). The cultivation of microalgae varies among different species and its success is primarily based on the method employed and existing environmental conditions. A number of factors influence the production of algal biomass such as light, temperature, CO_2, and O_2 concentration, macro and micronutrients, vitamins, salinities, and mixing/aeration conditions. The three commonly used cultivation systems employed for growth of microalgae are open pond systems, closed systems, and hybrid systems (Figure 10.2). Each method has its own advantages and limitations (Table 10.2).

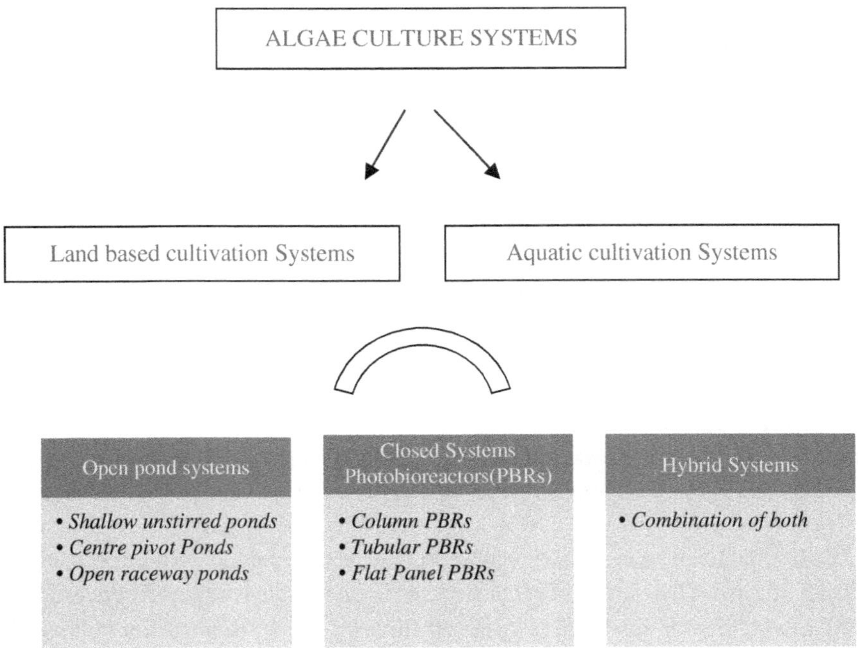

Figure 10.2 Types of microalgal culture systems.

Table 10.2 Types of algae cultivation systems.

Algae cultivation system	Advantages	Disadvantages
Open systems		
Unstirred ponds	Cheap, low maintenance, and low energy consumption	• Controlling of factors like pH, temperature difficult. • No mixing device leads to accumulation of dead algae and limits nutrient and light transfer • Prone to contamination with other algae. • Large area needed • Restricted microalgal strains can be cultivated
Circular ponds	Cheap, low maintenance and low energy consumption. provides better mixing	Small size, which causes improper mixing especially due to too long rotating arm
Raceway ponds	Low construction cost and energy consumption, simple to operate	• Evaporation losses high • Prone to contamination with other algae • Large area needed, • Restricted microalgal strains can be cultivated
Closed systems		
Column/vertical (bubble or airlift photobioreactor)	Compact, controlled growth conditions, better mixing and mass transfer, relatively simple to operate	Surface area less for illumination, decrease in surface area for illumination on scale up, scale up difficult
Tubular photobioreactor	Can be used in open areas, large surface area for illumination, higher biomass yield, culture conditions can be controlled	Large open area required, oxygen build up, mixing difficult leading to pH CO_2 gradients
Flat plate bioreactor	Larger surface area available for light transfer, can be used in open areas, good light transfer.	Scale up difficult and expensive, temperature control difficult, loss of water due to higher evaporation

10.3.1.2.1 Open Pond Systems Open ponds are the simplest, oldest, and cheapest systems used in outdoor cultivation of microalgae on a large scale since the 1950s (Shen et al. 2009). They are used for culturing many microalgae species like *Chlorella*, *Spirulina*, and *Dunaliella* (Borowitzka 2005; Chisti 2007; Singh and Dhar 2011; Farrokh et al. 2019). Ponds can be excavated and built up with walls and may be either lined with impermeable materials or left unlined. They are usually constructed in nonarable lands, and hence, don't compete with food crops for agricultural lands. In these open ponds, algae are cultivated as similar to as external environment conditions. Maintenance and cleaning are thus

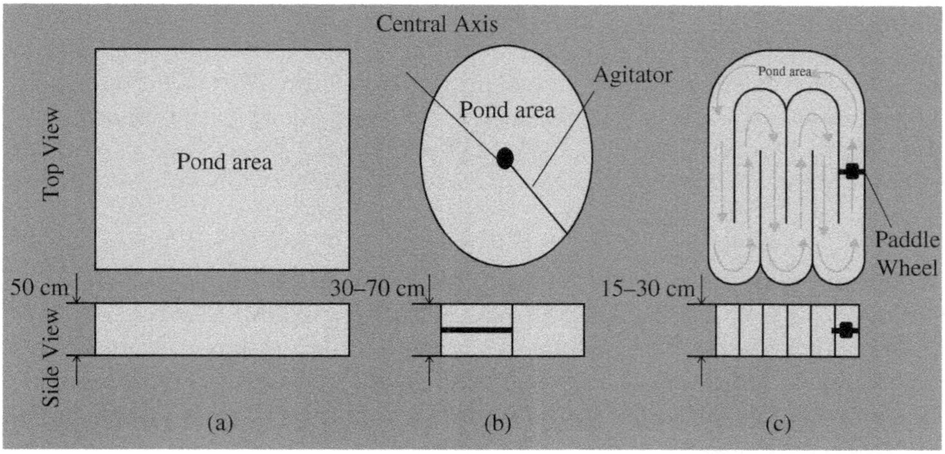

Figure 10.3 Types of open pond systems for microalgae cultivation. a. Unstirred Ponds b. Circular Ponds c. Raceway Ponds. Source: Adapted from Armin (2015).

comparatively easy and energy consumption is low. These open ponds have numerous advantages: low costs of construction and maintenance, easy scale up, and the possible remediation of wastewaters used in ponds. However, factors like temperature, pH, and light intensity are difficult to control in open systems. Another major drawback of open pond systems is contamination with other algal species, thus severely reducing the yields. This contamination problem may be prevented by growing extremophiles, i.e., which can tolerate and grow in high or low pH or high salinity environments thus leading to sustainable and reliable cultivation of the desired algal species (Medipally et al. 2015; Singh and Dhar 2011). There are many different designs of open ponds available for microalgal cultivation but unstirred ponds, circular ponds, and raceway ponds are the major types employed (Figure 10.3).

Shallow Unstirred Ponds These are the simplest form of open pond systems ranging from a few m^2 to 250 Ha in size and are usually less than 50 cm depth. Simple natural lakes are considered as unstirred open ponds or natural water ponds with uncovered beds can be used to construct them. They are the least expensive and technically easy among all the cultivation methods available. Unstirred open ponds are used for cultivation of microalgae, which can grow even with limited facilities and/or also in presence of other microbial contaminants. They have been used for culturing microalgal species *Dunaliellasalina* but yields are very low usually below 1 g d^{-1} (Borowitzka and Borowitzka 1990; Borowitzka 2005; Singh and Dhar 2011). The growth of algae in these types of ponds is limited due to low dissolved CO_2 levels. Also, there is no mixing in these ponds, which causes an accumulation of dead biomass on the surface limiting the penetration of sunlight and slower diffusion of nutrients ultimately resulting in less growth. To overcome these limitations, open ponds with some sort of agitation were developed.

Circular Ponds These are circular ponds constructed of concrete and are lined with plastic sheets or other inert membranes. Their size usually ranges about 45 m diameter and

30–70 cm depth and is facilitated with a pivoted agitator in the center. Hence these are also called as centre pivot ponds (Moheimani 2005). The mechanical arm brings about stirring in a circular motion. Circular ponds produce higher yields ranging from 8.5 to 21grams/day but one of the limitations is their small size, which causes improper mixing especially due to a too long rotating arm (Shen et al. 2009; Singh and Dhar 2011). These types of ponds are commercially employed in the cultivation of microalgae species of *Chlorella* and *Chlamydomonas.*

Open Raceway Ponds Raceway ponds are the most commonly used open pond systems for cultivation of algae on a commercial scale. Raceway ponds are constructed either in single channels or individual raceways are joined together to build a groups of channels. They are usually 15–30 cm in depth and are constructed either of concrete or compacted earth or lined with plastics. In open pond systems, stirring or mixing is one of the primary factors considered in design and operation, in order to provide proper sunlight and CO_2 to the algal cells. This is facilitated by paddlewheels or pumps or airlifts, which often provides proper water circulation and agitation thus keeping the microalgal cells along with nutrients and water in continuous motion around the raceway (Moheimani and Borowitzka 2006; Schenk et al. 2008; Shen et al. 2009). Sparging with carbon dioxide enhances CO_2 consumption and also mixing. These ponds are run continuously and harvesting of microalgal cells is done behind the paddle wheel after completion of every circulation loop around the raceway. They are commercially used for the cultivation of algae such as *Spirulina*, *Chlorella, Haematococcus*, and *Dunaliella* (Singh and Dhar 2011). They are most widely used owing to their relatively low costs of construction and maintenance but seasonal light and temperature changes make it difficult to achieve high productivities (Borowitzka 2005; Shen et al. 2009).

10.3.1.2.2 Closed Systems In order to overcome many limitations of open pond systems such as large area requirement, uncontrolled environments, evaporation, contamination, suitability to only limited species and low productivities, the closed systems also known as photobioreactors (PBRs) were developed and encouraged. PBRs are made using transparent materials like glass, plastic, polyvinyl chloride (PVC), or polycarbonate to allow light penetration. They can be located indoors or outdoors and are enclosed within a transparent tube so as to avoid direct exposure to the atmosphere. PBRs have many advantages when compared to open pond systems such as high biomass production and cell density, controlled growth parameters, and reduced contamination (Shen et al. 2009). PBRs have high surface area: volume ratio resulting in better diffusion of sunlight, which is a major factor in photoautotrophic cultivation of algae. Another advantage of closed PBRs is that mixotrophic and heterotrophic growth of algae is feasible (Armin 2015). However, the cost of construction, maintenance, and operation are high. The higher productivities obtained can to some extent offset these drawbacks. Diverse designs of PBRs are developed such as tubular, column, helical, and flat plate PBRs (Figure 10.4) (Tredici and Zittelli 1998; Sanchez et al. 1999; Ugwu et al. 2005; Singh and Dhar 2011).

Annular/Column PBRs These are vertical PBRs and include bubble columns, airlift, or stirred tank reactors types. The air is supplied from below and the rising air bubbles

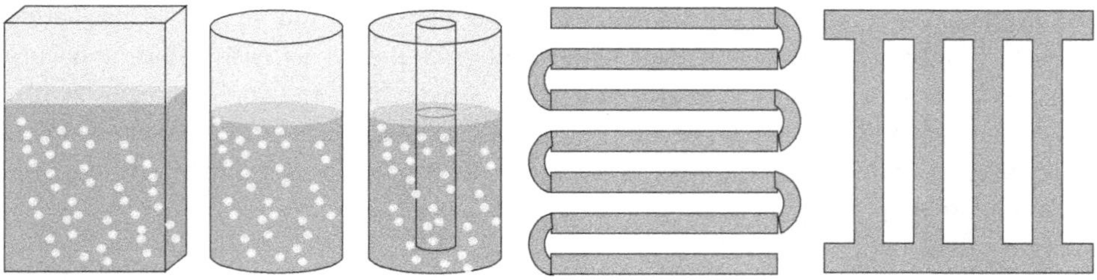

Figure 10.4 Types of photobioreactors. a. Flat Plate reactor b. Bubble Column Reactor c. Bubble Column Annular Reactor d. Serpentine Tubular Reactor e. Manifold Tubular Reactor. Source: Adapted from Nappa et al. (2016).

allow mixing as well as aeration. In airlift reactors, the presence of an inner chamber or draft tube provides for circular flow. The columns are made of transparent material and light diffuses through the walls. In some designs, internal lighting is also provided. The advantages of column PBRs include controlled growth conditions, better mixing, and mass transfer (Medipally et al. 2015).

Tubular PBRs Tubular PBRs are commercially preferred for cultivation of algae due to ease of construction, high surface area : volume ratio, better gas transfer control, and greater biomass yield (Moser 1991; Pulz 2001; Ugwu et al. 2008; Shen et al. 2009). Tubular PBRs are employed in the cultivation of algae such as *Chlorella sorokiniana*, *Phaeodactylum tricornutum*, and *Haematococcus*. Tubular PBRs are made up of transparent tubes arranged in different configurations – straight, vertical, horizontal, inclined, and helical. The culture flows through these tubes, which are exposed to sunlight and aeration and mixing brought about by mechanical pumps or airlift systems. The culture is allowed to flow through a tank or bubble column to remove oxygen. The tank or bubble column is sparged with air and also CO_2 for carbonation. Heat exchangers may also be included in the design to control temperature (Acien Fernandez et al. 2013). There is variant geometry of tubular PBRs such as helical flow reactors, helical tubular reactors, floating horizontal reactors, serpentine PBRs, and manifold PBRs (Armin 2015). Serpentine PBRs are made of straight transparent tubes arranged in a horizontal or vertical position and connected by means of U-bends to form a flat loop. Manifold PBRs, on the other hand, consist of parallel tubes connected to two manifolds used for distribution and collection of the algal culture (Acien et al. 2017).

Flat Panel Photobioreactors (FPP) These are also called as plate PBRs and are usually inclined or vertically aligned. They are made up of transparent rectangular container having many flat panels, each of which is made from joining two transparent sheets. The panels are usually about less than 1.5 m in height and 0.10 m in width. The culture is grown in these panels, which can be illuminated from both sides and aeration is provided by mixing. In some designs, inbuilt baffles are present. A 6000 l flat-plate photobioreactor is commercially available (Pulz 2001). Some flat plate photobioreactors consists of flat vessels prepared of glass sheets and joined by means of silicon rubber (Hu et al. 1996). Some other type of vertical flat-panel photobioreactors(FPP) are made of plastic bags supported on metal frames. These reactors are constructed at a reduced cost and also have the added advantage of replacing the plastic bags when contamination or excessive fouling occurs (Acien Fernandez et al. 2013; Schenk et al. 2008). Researchers and some companies like Solix, Accordion, and Algenol are developing and testing many new and innovative closed culture systems with an aim to lower the cost of construction and maintenance without compromising the productivity.

10.3.1.2.3 Hybrid Systems Hybrid PBRs are designed by combining open and closed systems in order to give better productivity. In these systems, axenic cultures are grown in closed PBRs and the algal culture from these reactors is used to inoculate open ponds. However, this method is not suitable for growing algae for biofuel as it is more costly and is operated as a batch process (Medipally et al. 2015).

10.3.2 Harvesting of Algae

The method of harvesting of microalgae is important in biofuel production, as the separation process accounts for 20–30% of production cost. Therefore, an inexpensive and energy-efficient method of the harvesting must be chosen for efficient biofuel production (Medipally et al. 2015).

10.3.2.1 Harvesting of Macroalgae

Harvesting of the macroalgae can be done either manually or by mechanical means. There are various methods employed for harvesting based on the type of algal species. Macroalgae may be free floating or grow by attaching to a substratum. Free-floating algae are usually harvested by installing a net in the pond or tank. Attached forms are either removed completely from the base, or for some species, they are cut and detached from the substratum either above the holdfast as in the case of *Laminaria* or the stipe as in *Himanthalia*. Specialized cutting and collecting equipment, like sickles, hand rakes, drag rakes, winchers, and suction harvesters have been employed by different seaweed companies for harvesting.

10.3.2.2 Harvesting of Microalgae

Microalgae remain in suspension during culture due to their small size and hence harvesting of microalgae is comparatively difficult (Lam and Lee 2012; Sualiand Sarbatly 2012; Lam et al. 2019; Gilmour 2019). Microalgae can be harvested by separating it from the liquid and then drying. There are many methods employed for the separation of microalgae like sedimentation, flocculation, filtration, and centrifugation (Figure 10.5). The method chosen depends upon the algal strain used for the biofuel production.

10.3.2.2.1 Flocculation Microalgae flocculate to form large clumps, which help to readily separate them from the culture medium. These flocs are formed due to the assemblage of biomass caused by neutralization of the negative cell surface charge. This method is considered as the best method as it is a less expensive and has a high yielding process (Augustine et al. 2019). Some species of algae tend to flocculate naturally when left still in cultivation ponds without disturbing. This may be attributed to limited nitrogen, levels of dissolved oxygen or pH, and many other environmental factors. In many cases, flocculation of algal

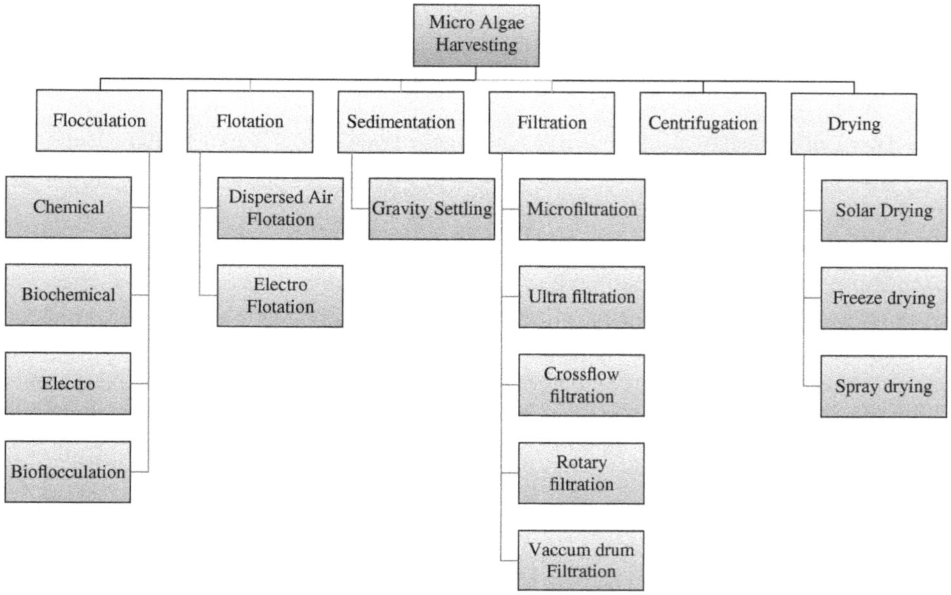

Figure 10.5 Overview of microalgae harvesting methods. Source: Adapted from Kalpesh et al. (2013).

species can be brought about by various ways, such as chemical flocculation, bioflocculation, and electroflocculation. The chemical flocculation uses some of the inorganic and organic chemicals, such as alum, ferric sulfate, ferric chloride, or lime and polyelectrolytes (Singh and Dhar 2011). The method of flocculation using chemicals is less preferred as it has certain limitations, such as the presence of enormous metal salts in the biomass harvested, disposal problems, and also is a too-expensive method. Bioflocculation is a process where two different species of microbes are co-cultured together, such that the sedimentation of the nonflocculating microalgal species is promoted in the presence of the other organism. For example, the flocculation of microalgae *Chlorella vulgaris* is enhanced due to its associated bacteria (Lee et al. 2013). *Nannochloropsis* also tends to flocculate in the presence of a naturally flocculating algal species *Skeletonema*. The electroflocculation method is mainly due to the movement of electrically charged particles in an electric field. In this method, the oxidation of a metal anode such as aluminum or iron or magnesium produces the active coagulant algal species resulting in flocculation (Poelman et al. 1997).

10.3.2.2.2 Flotation In this process, air is bubbled through the algal cultures in cultivation ponds. These air bubbles (>10 μm) adhere to the biomass and increase their buoyancy, which makes them rise rapidly to the surface and float there, and finally the top layer loaded with algal aggregates is skimmed off (Levin et al. 1961; Alkarawi et al. 2018; Singh and Dhar 2011; Sati et al. 2019; Lam et al. 2019).

10.3.2.2.3 Sedimentation Sedimentation, though inexpensive as a method, is a slow process and results in low biomass. The cells have to be dense and nonmotile (Armin 2015). However, sedimentation following flocculation yields better results. Sonication is another method that uses ultrasound waves ranging from 20 to 100 MHz, which help in the destruction of algal cells and facilitating their sedimentation. In this method, the sound waves pass through the algal cells and creates a cavitation effect without inducing any shear stress on the algal biomass (Lam et al. 2019).

10.3.2.2.4 Filtration Filtration is a well-known method employed for separation of solid-liquid mixtures. It involves the passage of particles through filtration screens of certain pore size under pressure or vacuum. Based on the particle size, they are either passed through the screen or are retained on it. The filtration process is a cost-effective process of harvesting, but clogging of screens and fouling of membranes makes it less suitable to larger species of microalgae. Clogging of pores is counteracted in the cross-flow filtration method, which employs membranes, and the flow is tangential to the membrane and the pile up of material is also avoided in this method. Cross-flow filtrations were shown to achieve dewatering of algal slurry up to 24% solids. Novel ceramic coated membrane sheets and nickel alloy metal sheet membranes have been developed for algal harvesting. A more advanced membrane-filtration system with improved harvest efficiency has been developed, which combines harvest/dewatering filtration without the use of flocculants (Sati et al. 2019).

10.3.2.2.5 Centrifugation Centrifugation is another solid-liquid separation method used for harvesting microalgae, which separates particles based on the gravitational force. The recovery and production costs of this method are dependent upon the sizes and morphology of the cells being harvested. For example, single smaller algal cells settle more readily and faster than filamentous cells and large colonial cells. Although industrial scale centrifuges are generally used for efficient separation of algal biomass, the capital and operating energy costs are too high. However, better results will be achieved when centrifugation is used in the second or final stage of filtration in biofuel production process (Armin 2015; Singh and Dhar 2011).

10.3.3 Drying

Drying is primarily done after algal separation in order to obtain high biomass concentrations and also ensures proper storage and ease of transportation. Different methods are available like solar drying, spray drying, conveyor belt dryers, and freeze drying. Each of the methods has inherent advantages and limitations and hence the selection of the method depends upon the cost, the algal species, and the end-product desired.

10.3.4 Cell Disruption

A wide variety of polymers are seen in different groups of algal cells. The cell walls of algae are made of carbohydrates and glycoproteins, and cell disruption is required in certain processes of biofuel production to release their cell contents. A variety of physical, chemical, and enzymatic methods are available for cell disruption. Various mechanical methods include homogenization, bead milling, ultra-sonication, microfluidization, and autoclaving. Some of the chemical methods include treatment with detergents, acids, alkalis, antibiotics, solvents, etc. Enzymes like glycosidases, glucanases, peptidases, and lipases can also be used for bringing about cell lysis, but may be costly.

10.3.5 Conversion into Biofuel

Algae with high oil and carbohydrate contents and fast growth rate can be easily transformed into different biofuels (Abomohra and Elshobary 2019). The algal biomass or the cellular contents obtained from algae species are utilized to generate different kinds of biofuels such as biodiesel, bioethanol, biogas, biomethanol, biobutanol, biohydrogen, and biosyngas.

- Biodiesel by transesterification of lipid fraction
- Bioethanol by fermentation of the carbohydrate components
- Biogas by anaerobic digestion
- Biomethanol production by gasification and anaerobic digestion
- Biohydrogen by photolysis or photo-fermentation
- Biosyngas by direct combustion

10.4 Types of Algal Biofuels

The algal biofuels are diverse in nature and different kinds are produced based on the component of the algal cells used and the technique employed in the production (Table 10.3). There are three main components of algal biomass that can be processed into biofuels: lipids/natural oils, carbohydrates, and proteins. The carbohydrate component of algal biomass is used in the production of bioethanol via fermentation process, whereas oil extracted from algae is subjected to the transesterification process to produce biodiesel. The residual biomass is processed further to produce biomethane, biogas, and other biofuels through hydrolysis, fermentation, and gasification. The residual biomass can even be utilized as animal feed, fertilizers, biocontrol agents, nutraceuticals, therapeutics, and other value-added products (Pulz and Gross 2004; Harun et al. 2010; Archana and Thomas 2017; Tiwari and Kiran 2019).

Table 10.3 Types of algal biofuels.

Source	Conversion process	Biofuel
Algal biomass	Anaerobic digestion	Biomethanol, biogas
Algae rich incarbohydrates	Fermentation	Bioethanol, biohydrogen
Algae with high lipid content	Transesterification	Biodiesel
Algal biomass	Thermo conversion	Biogasoline, Biochar, jet fuel
Algae	Biophotolysis	Biohydrogen

10.4.1 Biodiesel

Biodiesel is a mixture of monoalkyl esters of long chain fatty acids, such as fatty acid methyl esters (FAMEs) obtained from vegetable oils, fats, and grease by means of transesterification. They are the commonly used biofuels in Europe and are used as a substitute for petroleum-based fuels. It is used as a fuel for vehicles either in its pure form or as diesel additive upon blended in small quantities with diesel. Biodiesel is a nontoxic, biodegradable alternative fuel and advantageous because of high combustion efficiency and less CO_2 and other particulate matter emissions (Hossain et al. 2008). However, it has limitations such as higher nitrogen oxide emissions, higher viscosity, reduced power output, greater engine wear, and higher prices. The major feedstock sources employed in biodiesel production are soybean, corn, canola, mustard, sunflower, palm, rapeseed, jatropha, hemp, *Pongamia pinnata*, vegetable oils, animal fats, and algae. Microalgae are the most excellent sources for conversion into biodiesel due to their high content of oils. *Chlorella pyrenoidosa*, *D. tertiolecta*, *Nannochloropsis oculata*, *Spirulina* sp. *Schizochytrium limacinum*, *Dictyochloropsis splendid*, *Desmodesmus quadricaudatus*, *B. braunii*, *Scenedesmus* sps., *Ankistrodesmus* sps., *Chlamydocapsa bacillus*, and *Kirchneriella lunaris* are some of the algae that accumulate large quantities of lipids (Nascimento et al. 2013; Behera et al. 2015; Saad et al. 2019; Tiwarl and Kiran 2018). The lipids from these algae are extracted and further processed to biofuel and hence are preferred for biodiesel production. The process of biodiesel production from algae involves harvesting of biomass, drying, extraction of oil, and its transesterification.

There are many methods available for extraction of oil from algae: presses, supercritical fluid extraction, solvents, and/enzymatic extraction, acid/base hydrolysis, ultrasonic- or microwave-assisted extraction, osmotic shock, and pulsed electric field. Some of these methods like microwave and acid-base hydrolysis are mainly used as biomass pretreatment methods. Supercritical carbon dioxide fluid extraction is highly efficient but not a commercially viable method due to high operation costs. This method involves the use of CO_2, which gets liquefied under high pressure and heat and thus possess both liquid and gas properties. The lipid extraction is usually done using a filter press or an organic solvent like methanol, isopropanol, and petroleum ether, which gives 75% and 90% of oil yield respectively (Nagle and Lemke 1990). Solvent extraction with hexane is also the most common method used for extraction. The residual biomass that contains starch and proteins is either processed by anaerobic fermentation to yield ethanol or used as animal feed or feedstock.

The biodiesel is produced from algae either by directly subjecting the algal biomass to transesterification or via a multistep process wherein the algae are first concentrated, oil extracted, and later on transesterified. The process of biodiesel production by direct transesterification is fast and less expensive (Lewis et al. 2000; Miao and Wu 2006; Johnson and Wen 2009; Mulbry et al. 2009; Kumar et al. 2017). Transesterification can be carried out with or without a catalyst and involves conversion of highly viscous oils into lower viscosity alkyl esters. During the manufacturing process, triglycerides react with a monohydric alcohol (CH_3OH or C_2H_5OH) in the presence of KOH catalyst to yield biodiesel comprising mainly FAMEs and glycerol (Figure 10.6) (Johnson and Wen 2009; Oh et al. 2018). The products formed are let into a separator to separate methyl esters and glycerol. The by-product glycerol is then removed. Evaporation is done to recover the excess methanol. The final product biodiesel is rinsed with water, neutralized, and then dried (Xu et al. 2006).

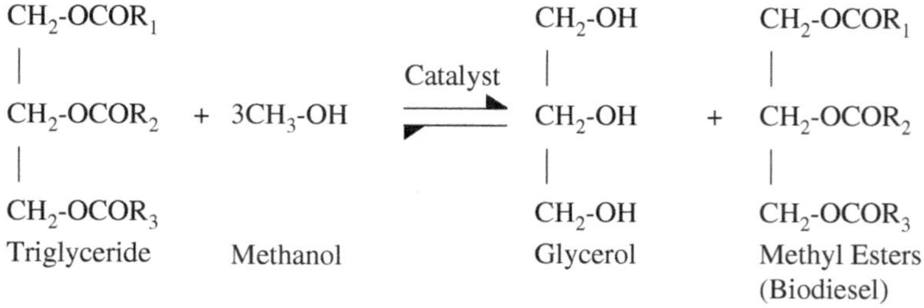

Figure 10.6 Transesterification of triglyceride to biodiesel.

Enzyme-mediated transesterification is an attractive option for producing biodiesel as it is more environmentally friendly and can be carried out in mild conditions. One of the enzymes used as a catalyst in transesterification is lipase. The enzymes cause simultaneous esterification and transesterifications eliminating by-product formation. The main drawback of using enzymes for this conversion is their high cost, which can be overcome via immobilized technology where enzymes are reused many times.

Some macroalgae such as *Chaetomorpha*, *Ulva*, and *Enteromorpha* are used for production biodiesel through transesterification of their oils (Milledge et al. 2014). Microalgae with high content of oil and rapid growth are efficient to produce an excellent alternative biodiesel. The oil content of microalgae strains varies widely and typically ranges between 20–50% but some strains may contain high amounts exceeding up to 80% (Table 10.4) (Metting 1996; Kojima and Zhang 1999; Spolaore et al. 2006; Demirbas and Demirbas 2011; Singh and Dhar 2011; Cuellar-Bermudez et al. 2014). Many factors influence the production of microalgal biomass and oil accumulation, such as inherent potential of algal species, temperature, pH, salinity, light intensity, mineral, and nitrogen sources. The type of fatty esters present and their properties will influence and determine the properties of the biodiesel fuel. The lipid productivity (mg/l/d) and the % of lipid content/unit dry weight of algal biomass is highest in green algae than other red or brown microalgal species (Ota et al. 2009). Green algae (Chlorophyceae) are the most promising species of algae and considered as the excellent sources for production of biodiesel. These are aquatic, unicellular photosynthetic eukaryotes that can increase and double their biomass within a day. They possess exceptionally high growth rates and large population densities. They also have high lipid contents (>50%), which makes them ideal and high yielding species of biodiesel production (Demirbas and Demirbas 2011).

Chlorella, *Dunaliella*, and *Scenedesmus* are considered as potential algal species among the Chlorophyceae members for production of biodiesel. *Chlorella* is a unicellular, non-motile, spherical, approximately 2–10 μm diameter in size. Chlorella is a green alga possessing chl-a and chl-b pigments in its chloroplast to carry out photosynthesis. They reproduce and multiply rapidly using carbon dioxide, water, light, and a few minerals. Because of its photosynthetic efficiency and around 32% of oil yield, chlorella is a prospective food and energy source and highly efficient for production of biofuels. *D. tertiolecta* is a unicellular marine green alga, rod- to oval-shaped, size ranging from 10 to 12 μm in dm. They do not form clumps or chains and can be cultivated easily. It is a fast-growing strain having a high carbon dioxide rate and has about 37% of oil yield. *Scenedesmus dimorphus* is a unicellular alga and possess 45% of oil contents. It is considered as the best species for production of biodiesel, but has a drawback that it is too heavy and forms thick sediments during cultivation due to improper agitation.

Table 10.4 Oil content of microalgae species.

S. No:	Microalgae	Lipid content(% of dry weight)
1.	*Botryococcus braunii*	25–75
2	*Chlorella* sp.	28–32
3	*Crypthecodinium cohnii*	20
4	*Cylindrotheca* sp.	16–37
5	*Nitzschia* sp.	45–47
6	*Phaeodactylum tricornutum*	20–30
7	*Schizochytrium* sp.	50–77
8	*Tetraselmis suecia*	15–23
9	*Dunaliella primolecta*	23
10	*Isochrysis*sp.	25–33
11	*Nannochloris* sp.	20–35
12	*Nanochloropsis* sp.	31–68
13	*Neochloris oleoabundans*	35–54
14	*Ankistrodesmus*	28–40
15	*Chlorella protothecoides*	15–55
16	*Cyclotella*	35–42
17	*Dunaliella tertiolecta*	36–42
18	*Hantzschia*	60–66
19	*Phaeodactylum tricornutum*	31
20	*Scenedesmusdimorphous*	6–45
21	*Scenedesmus obliquus*	11–55
22	*Schizochytrium*	50–77
23	*Stichococcus*	9–59
24	*Tetraselmis suecica*	15–32
25	*Thalassiosira pseudonana*	21–31
26	*Chlorella vulgaris*	56
27	*Cryptecodinium cohnii*	20
28	*Monalanthus salina*	20–70
29	*Parietochloris incise*	62
30	*Monodus subterraneus*	39

Sources: Based on Chisti (2007); Singh and Dhar (2011).

10.4.2 Bioethanol

Bioethanol is a type of biofuel produced from sugars/polysaccharides through fermentation process. The technology of bioethanol production may include fermentation of pretreated biomass or fermentation of carbohydrate reserves under dark conditions or photo-fermentation using light energy (Lakatos et al. 2019). Sugars are obtained from various sources like corn, wheat, sugarcane, sweet sorghum, or molasses or cellulose-rich biomass from trees, grasses, as well as algae. The algae depend up on light, carbon dioxide, and nutrients to synthesize abundant quantities of starch, cellulose, and other sugars. Algae, both macro-and microalgae, have profiles of high sugar/carbohydrates and are excellent substrates for fermentation. The amount of carbohydrates per unit dry weight varies considerably among algal species and usually ranges from 30–70% in red algae, 25–40% in green algae, and 30–50% in brown algae. Many of the sugars present in algae exist as polymers and need to undergo pretreatment for conversion to fermentable sugars. However, they do not have lignin and hemicelluloses and hence the pretreatment methods would be simpler when compared to other lignocellulosic biomass substrates. The algal biomass is subjected to physical pretreatment where size reduction is done either by milling, crushing, steam explosion, or mechanical shear. This is followed by hot water wash, liquefaction, and enzymatic or chemical hydrolysis leading to saccharification of fermentable sugars (Khan et al. 2017). The sugars released act as substrates and are fermented by *Saccharomyces cerevisiae* or other microorganisms. However, pretreatment of algal biomass usually yields different types of monosaccharides. Hence their conversion rates will be different as the fermenting microorganisms cannot utilize the different types of sugars to the same extent. For example, *S. cerevisiae* mainly ferments hexose sugar-glucose but cannot ferment the pentose sugars like xylose. Hence using a combination of microorganisms may be preferable and effective, as they can ferment different types of sugars. The bioethanol obtained is further purified by distillation (Khan et al. 2017). The bioethanol produced is used in vehicles as a fuel either in its pure form or blended with petrol (up to 10% or more) as a gasoline additive.

Different macroalgae and microalgae species belonging to genera *Chlorococcum*, *Ulva*, *Gracilaria*, *Laminaria*, *Sargassum*, *Prymnesium*, *Gelidium*, *Chlorella*, and *Spirogyra* may be exploited and utilized for bioethanol production (Eshaq et al. 2011; Markou and Nerantzis 2013; Rajkumar et al. 2014; Chen et al. 2015; Armin 2015; Singh and Dhar 2019; Abdulla et al. 2020). Microalgae are highly potential sources of bioethanol as they have high productivity rate, high content of carbohydrates, and lack lignin, unlike many other crop plant sources (Singh and Dhar 2011; Odjadjare et al. 2015; Jambo et al. 2016). Biomass can be utilized readily as they have simple cellular composition in microalgae. Microalgae can be genetically engineered to obtain high yields of bioethanol (Deng and Coleman 1999; Hamelinck et al. 2005). After extraction of oil, the carbohydrate part of the microalgal dry biomass, which mainly consists of glucose, starch, cellulose, and hemicellulose, are employed for the production of bioethanol via fermentation process. In this process of fermentation, the various kinds of sugars are converted to ethanol and carbon dioxide by yeast or amylase enzymes. The fermented mash containing ethanol along with other nonfermentable solids is then distilled off to yield high-strength ethanol (Dismukes et al. 2008). Another method of bioethanol production using microalgae involves the anaerobic fermentation of starch contained within their cells (Hirano et al. 1997). Harun et al. (2009) reported the simultaneous production of biodiesel and bioethanol in which the algal oil is extracted and utilized for production of biodiesel while the residual biomass is subjected to fermentation by yeast to give bioethanol.

Different microalgae species have varying carbohydrate content and are used for bioethanol production (Table 10.5). *Porphyridium cruentum*, a red alga, is found to accumulate large amount of carbohydrates and thus considered as an ideal species for ethanol

Table 10.5 Carbohydrate content of microalgae species.

S. No:	Microalgae	Carbohydrate content (% of dry weight)
1.	*Chlorella pyrenoidosa*	26%
2.	*Chlorella vulgaris*	15%
3.	*Dunaliella salina*	32%
4.	*Scenedesmus obliquus*	15%
5.	*Porphyridium cruentum*	50%
6.	*Scenedesmus dimorphus*	80%

Source: Adapted from Tiwari and Kiran (2017).

production (Ali 2013). *S. dimorphus* consists of 80% fermentable sugars, which resulted in high yield of bioethanol (Cheng et al. 2017; Sivaramakrishnan and Incharoensakdi 2018). A red alga, *Gelidium amansii,* was shown to produce ethanol with a conversion yield of 74.7% in a sequential process of hydrolysis followed by fermentation, whereas the yield was 84.9% when both saccharification and fermentation occurred simultaneously in the process (Kim et al. 2015). Another red alga, *Gracilaria* sp., produced $4.72\,g\,l^{-1}$ of ethanol with 94% conversion efficiency in a study that employed acid hydrolysis followed by enzymatic hydrolysis and fermentation using immobilized *S. cerevisiae* (Wu et al. 2014).

10.4.3 Biogas/Biomethane

Biogas is a fuel majorly comprising 55–75% of methane and 25–45% carbon dioxide and may also include some other gases like hydrogen, hydrogen sulfide, and nitrogen. Biomethane is obtained from algal biomass by way of anaerobic digestion or gasification or pyrolysis. Pyrolysis and gasification processes need dry biomass, which needs a high input of energy and hence is not a feasible method for production of biogas. The process of biogas/biomethane production using anaerobic digestion is an attractive process as it eliminates many steps like drying and extraction, which accounts for high production costs of algal biofuels. In this process, the organic matter of the algal biomass is degraded under anaerobic conditions to produce certain gases like methane, carbon dioxide, and hydrogen sulfide. There are four stages in the process of anaerobic digestion: hydrolysis, fermentation, acetogenesis, and methanogenesis (Figure 10.7). In the first step, the insoluble organic material (lipids or carbohydrates or proteins) is hydrolyzed in the presence of microbial enzymes of some obligate anaerobic bacteria, such as *Clostridia* and *Streptococci* and get converted into soluble organic materials. In next step, the soluble organic compounds formed are enzymatically converted into volatile fatty acids (VFAs) and alcohols by acidogenic bacteria. The third step is called as acetogenesis, in which the VFAs and alcohols are further transformed by means of acetogenic bacteria to form acetic acid and hydrogen. These are further metabolized by the methanogens and finally form methane and carbon dioxide via methanogenesis (Cantrell et al. 2008; Vergara-Fernandez et al. 2008; Brennan and Owende 2010; Romagnoli et al. 2011; Oncel 2013; Behera et al. 2015).

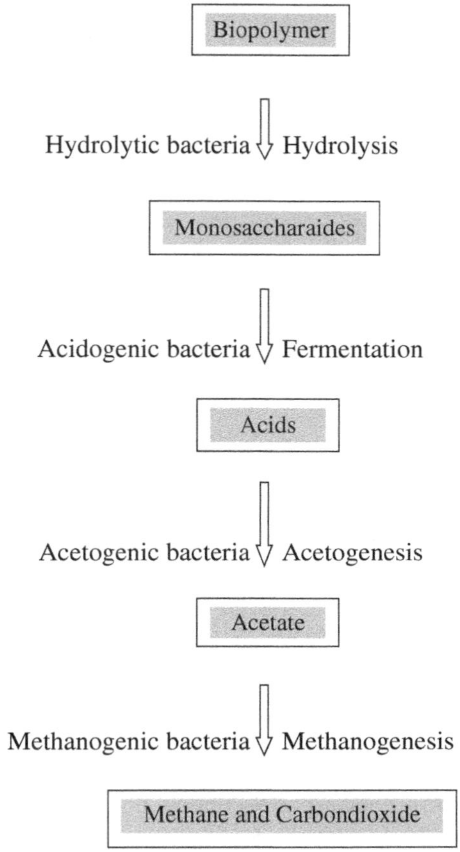

Figure 10.7 Stages of anaerobic digestion in the process of biogas production.

The potential yield of methane is more in algal cells with high energetic lipid contents than those with high amounts of carbohydrates and proteins. Hence microalgae are considered as attractive and the best sources for biogas production via anaerobic digestion (Li et al. 2002; Cirne et al. 2007). The yield may be less as the process of anaerobic digestion in algae is challenging, owing to their recalcitrant cell-wall constituents, low C : N ratio, toxicity due to ammonia, excess salinity, and some metal ions (Ward et al. 2014). Seaweeds/macroalgae with biomass having high moisture content were found to be feasible and suitable for production of biogas through anaerobic digestion (Sutherland and Varela 2014; Milledge et al. 2019). Currently, seaweeds are preferred for industrial production of biofuels via anaerobic digestion process. *Gracilaria* and *Ulva,* which possess high carbohydrate content for anaerobic digestion to methane gas, are found to be the most successful and suitable species. The production of biomethane by anaerobic digestion process from different microalgae is also reported by several researchers. Sangeetha et al. (2011) reported biogas production from algae *Chaetomorpha litorea* as high as 80.5 $l\,kg^{-1}$ of dry algal biomass via anaerobic digestion process. Vergara-Fernandez et al. (2008) reported a two-phase anaerobic digestion system in the marine algae *M. pyrifera* and *Durvillaea antarctica* and the yield of biogas was approximately 180.4 $ml\,g^{-1}$ dry algae/day. Similarly in *Chlamydomonas reinhardtii* and *Scenedesmus obliquus*, the amount of biogas produced was reported to be 587 and 287 $ml\,g^{-1}$

dry algae, respectively (Mussgnug et al. 2010). In *Scenedesmus* sp. it has been reported that the residual biomass, which is free from lipids and amino acids yields more biogas when compared to raw biomass. Therefore, the process of anaerobic digestion to produce biomethane from waste biomass is proposed to improve the microalgae cultivation practices and their potential yield (Ramos-Suárez and Carreras 2014; Singh and Dhar 2011).

10.4.4 Biomethanol

Methanol is chiefly manufactured from nonrenewable fossil fuels like natural gas. But algal biomass may also be used to produce it as biomethanol using anaerobic digestion, gasification, and pyrolysis methods. The production of biomethanol using biomass sources is less costly and also high yielding, and hence, is more preferred. This is considered as a promising alternative fuel with diverse applications, benefiting not only the economy but also the environment (Demirbas 2008b). In methods of gasification and pyrolysis employed for biomethanol production, high temperature and pressure are used for the extraction. In the anaerobic digestion method, algae are decomposed into simple components that are then transformed using microbes like acidogenic and methanogenic bacteria. Biomethanol is advantageous in that it has low viscosity and is easily combustible. But its limited use is because it has lower energy density than bioethanol and other bioalcohols. Moreover, biomethanol is much more volatile as well as very toxic, which makes it less valuable.

10.4.5 Biobutanol

Biobutanol can be produced from microalgae in a biorefinery using the residual biomass waste left after oil extraction. Butanol and other solvents can be produced from macroalgae through fermentation by *Clostridia* sp. (Potts et al. 2012). Butanol, apart from molecular similarity to gasoline and possessing high energy density, provides better performance and resistance to corrosion than ethanol blended with gasoline. Hence, biobuatanol is a better fuel than biomethanol or bioethanol and used in most of the gasoline engines. Butanol is not only used as a biofuel but also serves as an industrial solvent (Yeong et al. 2018). In spite of many significant uses, only limited works on the microalgal fermentation to produce butanol have been reported (Cheng et al. 2014; Gao et al. 2016; Wang et al. 2016). Microalgae possessing high concentrations of starch and sugars are considered as ideal candidates for biobutanol production. *C. vulgaris*, *Chlorella reinhardtii*, *Tetraselmis subcordiformis*, and *S. obliquus*are some of the prospective species that are employed in the production of biobutanol (Yeong et al. 2018; Singh and Dhar 2019).

10.4.6 Biohydrogen

Hydrogen can be a prospective and promising resource of biofuels, because water is the by-product during its manufacture. Moreover, greenhouse gases are not emitted leading to reduced pollution. Biohydrogen can be produced from various organic materials of microalgae through different means: (i) steam reformation, (ii) dark and photomicrobial fermentation, and (iii) biophotolysis (Kapdan and Kargi 2006; Ran et al. 2006; Wang et al. 2007; Shaishav et al. 2013; Khetkorn et al. 2017; Tiwari and Kiran 2019). Microalgae can directly produce hydrogen photosynthetically under anaerobic conditions with more than 80% of photon conversion efficiency (Melis and Happe 2001). There are many hydrogen producing cyanobacterial species belonging to genera such as *Gloeocapsa*, *Anabaena*, *Spirulina*, *Cyanothece*, *Nostoc*, etc. (Jeffries et al. 1978; Aoyama et al. 1997; Antal and Lindblad 2005; Bolatkhan et al. 2019). Among all, *Anabaena* species are found to generate significantly large amounts of hydrogen gas. These cyanobacterial species use nitrogenase and hydrogenase enzymes to generate hydrogen gas. Historically, prior incubation of the green algal cells under anaerobic conditions in the dark led to the evolution of hydrogen. But recently it has been shown that H_2 evolution also occurs with isolated hydrogenase enzyme upon incubation and catalyzed in presence of light. In studies carried out on *C. reinhardtii* and other green algae, it has been shown that sustained production of hydrogen occurs under conditions of sulfur deprivation. Photosystem II (PSII) is inhibited under limited levels of sulfur, causing depletion of oxygen resulting in higher respiration rate than photosynthesis and thus leading to more hydrogen production (Melis and Happe 2001; Armin 2015). *C. reinhardtii*, a unicellular green algae, is used as the model species to study biohydrogen production because of its high hydrogenase activity, ease of cultivation, sequenced genome, and low cost of production. There are certain limitations for large-scale production of hydrogen fuel. But different strategies such as isolating and screening for new organisms, optimizing their growth conditions, and biotechnological and genetic engineering techniques can be adopted for improving hydrogen production and thus unlock many avenues leading to sustainable biofuel in the future (Melis and Happe 2001; Pandey et al. 2007).

10.4.7 Biosyngas

Biosyngas is a gaseous mixture comprising carbon monoxide, hydrogen, methane, and other hydrocarbons, produced from algal biomass by gasification in the presence of oxygen, water vapor, or air. It is a low calorific gas that is used either in the gas turbines or as fuel directly. A high temperature of 800–1200 °C and dry biomass with less than 20% of moisture content are essential for the gasification process (Ghasemi et al. 2012). The biomass is dried and sometimes pyrolyzed and then subjected to partial combustion, i.e., combustion with insufficient amount of oxygen such that the biomass is not completely converted into carbon dioxide and water. The resulting gas mixture, syngas, contains more energy as fuel and is more efficient in comparison to direct combustion of the original biofuel.

10.4.8 Green Diesel

Green diesel is also well-known as renewable diesel and can be produced by algae through a hydrocracking refinery process of vegetable oils and animal fats or any other biological oil feedstocks. In the hydrotreating process, the larger molecules are catalytically broken down into shorter hydrocarbon chains at high temperatures and pressure (Knothe 2010). It is distributed similar to that of petroleum-based diesel without the need for new engines, pipelines, or infrastructure, but very costly to produce. As crude oil contains very low levels of oxygen, deoxygenation is not of much concern in petroleum refining but the hydrodeoxygenation of algae oil process is important and can be solved only with the advancement of research on effective catalysts, thus making it an economically feasible process (Zhou and Lawal 2015; Zhou 2016).

10.5 Advantages of Algal Biofuels

Algal biofuels considered as third-generation biofuels possess numerous advantages than first and second generations of biofuels. Algae promises to deliver many benefits in comparison to the existing biofuel production technologies. There are economic and environmental advantages to consider algal biofuel as an alternative fuel. Algal biofuels are potential transportation fuels as they are sustainable, nontoxic, biodegradable, and have low carbon footprint.

10.5.1 Ease of Growth

Microalgae inhabit diverse habitats and hence can be beneficially exploited for biofuel production. Currently, terrestrial plants are not considered as promising sources for biofuels because of their limited production capacity that is inadequate to meet the energy needs across the globe. Microalgae can be grown in many diverse habitats without competing for land or water resources with conventional plants and hence are considered as potential. The main advantage of employing microalgae as the substrate for production of biofuels is that they can be grown much more easily than the traditional crop plants. Thus the microalgae cultivation is regarded as a passive process due to the fact that they need very few resources for growth and do not require much attention (Mata et al. 2010). They have rapid growth rates, greater than most of the traditional crops used in biofuel production. Microalgae possess high photon conversion efficiency leading to better yields of biomass per hectare. They have a shorter harvesting cycle than other conventional crops resulting in mass production of algal biomass, which makes it advantageous and applicable in the biofuel arena (Schenk et al. 2008).

10.5.2 Impact on Food

Algaculture or algal farming is done using nonarable lands for producing biodiesel and other biofuels. Algae are grown on marginal lands, i.e., lands, which are unsuitable for agriculture and growing food crops. Hence, it would not be a competition to agriculture or food production as there will be no intrusion on cultivable land resources. The fact that large-scale cultivation of algae can be done on noncrop lands is advantageous and overcomes the concern of biofuel feedstock crops competing with food production. This may hopefully help in putting an end to the long-term food vs. fuel debate. Many traditional crop plants such as corn and palm used in production of biofuels usually serve as feed for livestock and are also an excellent source of human food. Hence when these feedstocks are used as sources for biofuels, the amount of animal feed and human food is reduced, as well as results in increased costs of food and fuel production. This problem can be alleviated by utilizing algae for biodiesel production. Since algae are not a primary food source of humans, they may be used exclusively for making biofuels with a little impact on the food industry (Vasudevan and Briggs 2008). Many of the by-products and extracts produced during the algal processing for biofuels find their use as animal feed. Algal biomass and waste by-products are much cheaper than conventional grain–based feeds and also are an effective means of minimizing waste (Demirbaş 2008a). Thus, the microalgae could be an answer to the energy crisis without compromising on the food and water resources or impacting the biodiversity conservation (Groom et al. 2008).

10.5.3 Environmental Impact

Algal biofuels are considered as an ecofriendly alternative to current biofuels as it provides numerous environmental benefits. The damage caused to ecosystem is minimal as algae can be grown in areas other than those used for agriculture and/or forests. One of the attractive characteristic features of algal fuels is that they can be effectively grown using saline and wastewater unfit for drinking or agriculture, thereby reducing the need for freshwater resources (Demirbas and Demirbas 2011; Philippa et al. 2014). Microalgae can thus be grown in wastewater leading to biomass production needed for biofuels in conjunction with remediation of wastewater. This opens up a new strategy in biofuels production leading to the pure water as by-product. Algae can grow while utilizing runoff water from agricultural lands polluted with fertilizers and other nutrients, thus preventing contamination of drinking water sources such as freshwater lakes and rivers (Demirbas and Demirbas 2011). Most of the industrial and domestic wastewater generated is usually discharged without treatment into surface water bodies, negatively impacting the environment and human health. Microalgae will absorb significant amounts of toxic chemicals and metals from industrial effluents, thus controlling pollution and harmful consequences. Pollutants like ammonia, nitrates, and phosphates, which are responsible for eutrophication problems and make water unsafe, may provide nutrients required for algal growth, thus minimizing the resources for cultivation of algae. The cultivation of algae does not require any pesticides or other amendments, thus avoiding the risk of pollution. Many algae species used in biodiesel production have the capability to sequester carbon dioxide from the air to use it as an energy source for their growth. Algal biofuels are nontoxic and biodegradable and hence relatively harmless to the environment than petroleum-based fuels (Schenk et al. 2008). In biofuel production, algal sources will lead to neutral/zero carbon emissions when compared to other sources of fossil fuels, thus reducing the incidence of global warming. The photosynthetic metabolism of microalgae allows for absorption of CO_2 and release of oxygen. Algae are very efficient in the uptake of CO_2 and algal CO_2 fixation accounts for approximately 40% of the global carbon fixation (Falkowski et al. 1998; Parker et al. 2008). If algal farms are built near power plants, the CO_2 emitted in the flue gases can be taken up as a source of carbon by algae, thus resulting in reduced carbon emissions. Algae have the ability to reduce up to 80% of CO_2 emissions from power plants (Acien Fernandez et al. 2012). The production and burning of algal biofuels will produce less or no harmful pollutants like sulfur oxides or nitrous oxides, carbon monoxide, and unburned hydrocarbons (Mata et al. 2010; Hemaiswarya et al. 2012). Microalgae thus possess high CO_2 sequestering ability concomitant with fuel production having neutral CO_2 and other greenhouse gas emissions.

10.5.4 Algal by Products

A number of by-products are produced by microalgae during processing of biofuels and have many diverse uses. The entire algal biomass may be used as a product, but more often many high value end-products of agricultural, pharmaceutical, nutraceutical, and industrial relevance are extracted from algae (Chen et al. 2019; Patel et al. 2019). These by-products find use as food, additives in food products, fish feed, natural dyes, antioxidants, and other purposes (Molina Grima et al. 2003; Pulz and Gross 2004; Singh et al. 2005; Khan et al. 2018). For example, the polysaccharides, dyes, and oils from algae have been used as thickeners and water-binding agents in cosmetics (Spolaore. et al. 2006). Antibiotics and antifungals and other natural products derived from microalgae have been playing a prominent role within the pharmaceutical industry. *Spirulina* contains omega 3 and omega 6 polyunsaturated fats, amino acids, vitamins, and pigments such as beta-carotene (Jaromir 2000; Tokusoglu and Uunal 2003). Some algal species are used as fertilizer for agricultural lands, thus reducing the crop-production costs (Stefan Kraan 2013).

10.5.5 Economic Benefits

The main advantage of algal biofuels is higher yields compared to other systems of biofuels production. Compared to land-based biofuel feedstock such as corn or soybeans, the volume of algal biomass produced is high and this results in higher oil productivity than other sources (Smith et al. 2010). Algal biofuels are versatile in that they can be cultivated on nonarable and marginal lands leading to enhanced production levels. Microalgae can be cultivated during the entire year and harvested batchwise, thus providing a reliable and incessant supply of fuel. Biodiesel derived from algae has more amounts of polyunsaturated fatty acids, which makes them suitable even for cold weather climates. The market competition for algae oil is very limited and this is another beneficial factor. Algae produce 10–100 times more fuel per unit area when compared to other sources of biofuels, but the cost per unit mass production is high because of the huge capital investment and operating costs. However, innovations in production technology and using waste resources will tend to make them more productive.

10.6 Limitations

Although algae are grown, harvested, and utilized as an efficient source for production of biofuels, there are still certain hindrances to implement the technology on a commercial scale to support the present fuel needs of the growing population. Algal biofuel production is a relatively new and expensive technology to be commercialized in the immediate future. The major factors limiting the production of algal biofuels are the difficulty in preserving the desired algal species, high harvesting costs, and low yields. The main problem lies in finding and developing an efficient algal species, which is fast-growing and easy to harvest. The algae should also possess high lipid content for extraction and processing into biofuels. Apart from the algal species possessing desirable characteristics, the type of cultivation system used for production should be suitable and cost-effective (Demirbas and Demirbas

2011). There are many challenges in growing algae, and constant temperature, optimum sunlight, and continuous removal of waste oxygen from the water, are necessary for proper algal growth. Moreover, open ponds are exposed to sunlight and rainfall that may cause evaporation or changes in salinity and pH levels, hindering the algal growth (Schenk et al. 2008). Hence, further research works should be taken up with a focus on novel materials and prototypes used in algal cultivation. Various techniques of screening for efficient strains, metabolic engineering to increase productivity, and recombinant techniques need to be employed to make cost-effective and efficient production of algae biofuels (Majidian et al. 2018; Khan and Fu 2020).

Algae produce biodiesel with many polyunsaturated fatty acids, which makes them unstable especially in hot climatic conditions (Vasudevan and Briggs 2008). Although algal biofuels are nontoxic, there are potential environmental hazards if ignited or spilled, because of its flammable nature like any other combustible fuel. Algal biofuels show reduced hazards when compared to fossil fuels as they are produced by a localized approach and are less toxic. Although there are fewer hazards, algal biofuels should be treated with proper safety measures when used for transportation or any other purposes. The production cost of algal biofuels is dependent upon various factors, such as biomass yield, percentage of lipid/oil content, the processing methods used for extraction of oil, and scale of production systems. The facilities bringing about conversion of algae to biofuels such as raceways, PBRs, and anaerobic fermenters are too expensive. Thus the capital investment, labor, and operational costs used in algal cultivation and processing are too high, making algal biofuels costlier and economically inaccessible than the conventional fuels (Radmer and Parker 1994; Kumar 2019).

10.7 Conclusion

Algae are considered as one of the chief renewable sources employed for biofuels, which can perhaps replace the conventional petroleum-based fuels. The algal biofuels could provide many potential environmental and economic advantages, as they require less land for cultivation and also confer high productivity. Apart from the tedious biomass harvesting processes, the high costs of production are the main disadvantages of algal biofuels. Production can be augmented by optimization of processes and environmental conditions required for growth, as well as genetic modifications of the producing algal species. Innovations and research in algae farming and harvesting techniques are required to reduce the costs of algae production so as to achieve economically viable algae biofuels. Exploring innumerable pathways for algae cultivation and processing may optimistically lead to breakthrough advances in the biofuels arena, leading to development of algal fuels as sustainable energy fuels.

References

Abdulla, R., King, T.K., Jambo, S.A., and Faik, A.A. (2020). Microalgae chlorella as a sustainable feedstock for bioethanol production. In: *Green Engineering for Campus Sustainability* (ed. A. Yaser), 81–103. Singapore: Springer.

Abomohra, A.E.F. and Elshobary, M. (2019). Biodiesel, bioethanol, and biobutanol production from microalgae. In: *Microalgae Biotechnology for Development of Biofuel and Wastewater Treatment* (eds. M. Alam and Z. Wang), 293–321. Singapore: Springer.

Abu-Ghosh, S., Fixler, D., Dubinsky, Z., and Iluz, D. (2016). Flashing light in microalgae biotechnology. *Bioresour. Technol.* 203: 357–363.

Acien Fernandez, F.G., Gonzalez-Lopez, C.V., Fernandez Sevilla, J.M., and Molina Grima, E. (2012). Conversion of CO_2 into biomass by microalgae: how realistic a contribution may it be to significant CO_2 removal? *Appl. Microbiol. Biotechnol.* 96 (3): 577–586.

Acien Fernandez, F.G., Fernandez Sevilla, J.M., and Molina Grima, E. (2013). Photobioreactors for the production of microalgae. *Rev. Environ. Sci. Biotechnol.* 12: 131–151.

Acien, F.G., Molina, E., Reis, A. et al. (2017). Photobioreactors for the production of microalgae. In: *Microalgae-Based Biofuels and Bioproducts: From Feedstock Cultivation to End-Products* (eds. R. Muñoz and C. Gonzalez-Fernandez), 1–44. Woodhead Publishing.

Ali, R. (2013). Effects of nitrogen on growth and carbohydrate formation in *Porphyridium cruentum*. *Open Life Sci.* 9 (2): 156–162.

Alkarawi, M.A.S., Caldwell, G.S., and Lee, J.G.M. (2018). Continuous harvesting of microalgae biomass using foam flotation. *Algal Res.* 36: 125–138.

Antal, T.K. and Lindblad, P. (2005). Production of H_2 by sulphur-deprived cells of the unicellular cyanobacteria *Gloeocapsa alpicola* and *Synechocystis* sp. PCC 6803 during dark incubation with methane or at various extracellular pH. *J. Appl. Microbiol.* 98 (1): 114–120.

Aoyama, K., Uemura, I., Miyake, J., and Asada, Y. (1997). Fermentative metabolism to produce hydrogen gas and organic compounds in cyanobacterium *Spirulina platensis*. *J. Ferment. Bioeng.* 83 (1): 17–20.

Armin, H. (2015). Algae biotechnology – green cell-factories on the rise. *Curr. Biotechnol.* 4: 389–415.

Aro, E.M. (2016). From first generation biofuels to advanced solar biofuels. *Ambio.* 45 (1): S24–S31.

Augustine, A., Tanwar, A., Tremblay, R., and Kumar, S. (2019). Flocculation processes optimization for reuse of culture medium without pH neutralization. *Algal Res.* 39: 101437.

Banerjee, A., Sharma, R., Chisti, Y., and Banerjee, U.C. (2002). *Botryococus braunii*: a renewable source of hydrocarbons and other chemicals. *Crit. Rev. Biotechnol.* 22 (3): 245–279.

Barclay, W.R., Meager, K.M., and Abril, J.R. (1994). Heterotrophic production of long chain omega-3 fatty acids utilizing algae and algae-like microorganisms. *J. Appl. Phycol.* 6 (2): 123–129.

Bast, F. (2013). Agronomy and cultivation methods for edible seaweeds. *Int. J. Agric. Food Sci. Technol.* 4 (7): 661–666.

Becker, E.W. (2007). Micro-algae as a source of protein. *Biotechnol. Adv.* 25 (2): 207–210.

Behera, S., Singh, R., Arora, R. et al. (2015). Scope of algae as third generation biofuels. *Front. Bioeng. Biotechnol.* 2 (90): 1–13.

Behrens, P.W. and Kyle, D.J. (1996). Microalgae as a source of fatty acids. *J. Food Lipids.* 3: 259–272.

Bolatkhan, K., Kossalbayev, B.D., Zayadan, B.K. et al. (2019). Hydrogen production from phototrophic microorganisms: reality and perspectives. *Int. J. Hydrogen Energy.* 44: 5799–5811.

Borowitzka, M.A. (2005). Culturing microalgae in outdoor ponds. In: *Algal Culturing Techniques* (ed. R.A. Andersen), 205–218. New York: Elsevier Academic.

Borowitzka, L.J. and Borowitzka, M.A. (1990). Commercial production of ß-carotene by *Dunaliella salina*in open ponds. *Bull. Mar. Sci.* 47 (1): 244–252.

Brennan, L. and Owende, P. (2010). Biofuels from microalgae-a review of technologies for production, processing, and extractions of biofuels and co-products. *Renewable Sustainable Energy Rev.* 14 (2): 557–577.

Brockmann, D., Pradinaud, C., Champenois, J. et al. (2015). Environmental assessment of bioethanol from onshore grown green seaweed. *Biofuels, Bioprod. Biorefin.* 9 (6): 696–708.

Brown, M.R. (1991). The amino-acid and sugar composition of 16 species of microalgae used in mariculture. *J. Exp. Mar. Biol. Ecol.* 145 (1): 79–99.

Buck, B.H. and Buchholz, C.M. (2004). The offshore-ring: a new system design for the open ocean aquaculture of macroalgae. *J. Appl. Phycol.* 16: 355–368.

Cantrell, K.B., Ducey, T., Ro, K.S., and Hunt, P.G. (2008). Livestock waste-to-bioenergy generation opportunities. *Bioresour. Technol.* 99 (17): 7941–7953.

Carvalho, A.P., Meireles, L.A., and Malcata, F.X. (2006). Microalgal reactors: a review of enclosed system designs and performances. *Biotechnol. Progr.* 22: 1490–1506.

Ceron Garcia, M.C., Fernandez Sevilla, J.M., Acien Fernandez, F.G. et al. (2000). Mixotrophic growth of *Phaeodactylum tricornutum* on glycerol: growth rate and fatty acid profile. *J. Appl. Phycol.* 12 (3): 239–248.

Chen, G.Q. and Chen, F. (2006). Growing phototrophic cells without light. *Biotehnol. Lett* 28 (9): 607–616.

Chen, J., Bai, J., Li, H. et al. (2015). Prospects for bioethanol production from macroalgae. *Trends Renewable Energy.* 1 (3): 185–197.

Chen, H., Li, T., and Wang, Q. (2019). Ten years of algal biofuel and bioproducts: gains and pains. *Planta.* 249: 195–219.

Cheng, H.H., Whang, L.M., Chan, K.C. et al. (2014). Biological butanol production from microalgae-based biodiesel residues by *Clostridium acetobutylicum*. *Bioresour. Technol.* 184: 379–385.

Cheng, L.M., Lee, K.T., and Chan, D.J. (2017). Evaluation on microalgae biomass for bioethanol production. *Mater. Sci. Eng., A.* 206: 12–18.

Chisti, Y. (2007). Biodiesel from microalgae. *Biotechnol. Adv.* 25 (3): 294–306.

Cirne, D.G., Paloumet, X., Bjornsson, L. et al. (2007). Anaerobic digestion of lipid-rich waste-effects of lipid concentration. *Renewable Energy.* 32 (6): 965–975.

Clark, G.J., Langley, D., and Bushell, M.E. (1995). Oxygen limitation can induce microbial secondary metabolite formation: investigations with miniature electrodes in shaker and bioreactor culture. *Microbiology.* 141 (3): 663–669.

Cuellar-Bermudez, S.P., Romero-Ogawa, M.A., Rittmann, B.E., and Parra-Saldivar, R. (2014). Algae biofuels production processes, carbon dioxide fixation and biorefinery concept. *J. Pet. Environ. Biotechnol.* 5 (4): 184.

Demirbaş, A. (2008a). Production of biodiesel from algae oils. *Energy Sources Part A.* 31 (2): 163–168.

Demirbaş, A. (2008b). Biomethanol production from organic waste materials. *Energy Sources Part A.* 30 (6): 565–572.

Demirbas, A. and Demirbas, M.F. (2011). Importance of algae oil as a source of biodiesel. *Energy Convers. Manage.* 52 (1): 163–170.

Deng, M.D. and Coleman, J.R. (1999). Ethanol synthesis by genetic engineering in cyanobacteria. *Appl. Environ. Microbiol.* 65 (2): 523–528.

Dismukes, G.C., Damian, C., Nicholas, B. et al. (2008). Aquatic phototrophs: efficient alternatives to land-based crops for biofuels. *Curr. Opin. Biotechnol.* 19 (3): 235–240.

Eshaq, F.S., Ali, M.N., and Mohd, M.K. (2011). Production of bioethanol from next generation feed-stock alga *Spirogyra* species. *Int. J. Eng. Sci. Technol.* 3: 1749–1755.

Falkowski, P.G., Barber, R.T., and Smetacek, V.V. (1998). Biogeochemical controls and feedbacks on ocean primary production. *Science.* 281 (5374): 200–207.

Fang, X., Wei, C., Zhao-Ling, C., and Fan, O. (2004). Effects of organic carbon sources on cell growth and eicosapentaenoic acid content of *Nannochloropsis* sp. *J. Appl. Phycol.* 16: 499–503.

Farrokh, P., Sheikhpour, M., Kasaeian, A. et al. (2019). Cyanobacteria as an eco-friendly resource for biofuel production: a critical review. *Biotechnol. Progr.* 35 (5): e2835.

Gao, K., Orr, V., and Rehmann, L. (2016). Butanol fermentation from microalgae-derived carbohydrates after ionic liquid extraction. *Bioresour. Technol.* 206: 77–85.

Ghasemi, Y., Rasoul-Amini, S., Naseri, A.T. et al. (2012). Microalgae biofuel potentials (review). *Appl. Biochem. Microbiol.* 48: 126–144.

Gilmour, D.J. (2019). Microalgae for biofuel production. In: *Advances in Applied Microbiology*, vol. 109 (eds. G.M. Gadd and S. Sariaslani), 1–30. Academic Press, Elsevier Inc.

Goh, B.H.H., Ong, H.C., Cheah, M.Y. et al. (2019). Sustainability of direct biodiesel synthesis from microalgae biomass: a critical review. *Renewable Sustainable Energy Rev.* 107: 59–74.

Gouveia, L. and Oliveira, A.C. (2009). Microalgae as a raw material for biofuels production. *J. Ind. Microbiol. Biotechnol.* 36 (2): 269–274.

Griffiths, M.J. and Harrison, S.T.L. (2009). Lipid productivity as a key characteristic for choosing algal species for biodiesel production. *J. Appl. Phycol.* 21: 493–507.

Groom, M.J., Gray, E.M., and Townsend, P.A. (2008). Biofuels and biodiversity: principles for creating better policies for biofuel production. *Conserv. Biol.* 22 (3): 602–609.

Guschina, I.A. and Harwood, J.L. (2006). Lipids and lipid metabolism in eukaryotic algae. *Prog. Lipid Res.* 45 (2): 160–186.

Hamelinck, C.N., Van Hooijdonk, G., and Faaij, A.P.C. (2005). Ethanol from lignocellulosic biomass: techno-economic performance in short, middle and long-term. *Biomass Bioenergy.* 28 (4): 384–410.

Harun, R., Danquah, M., and Forde, G.M. (2009). Microalgal biomass as a fermentation feedstock for bioethanol production. *J. Chem. Technol. Biotechnol.* 85 (2): 199–203.

Harun, R., Singh, M., Forde, G.M., and Danquah, M.K. (2010). Bioprocess engineering of microalgae to produce a variety of consumer products. *Renewable Sustainable Energy Rev.* 14 (3): 1037–1047.

Hemaiswarya, S., Raja, R., Carvalho, I.S. et al. (2012). An Indian scenario on renewable and sustainable energy sources with emphasis on algae. *Appl. Microbiol. Biotechnol.* 96 (5): 1125–1135.

Hirano, A., Ryohei, U., Shin, H., and Ogushi, Y. (1997). CO_2 fixation and ethanol production with microalgal photosynthesis and intracellular anaerobic fermentation. *Energy* 22 (2–3): 137–142.

Hossain, A., Salleh, A., Boyce, A. et al. (2008). Biodiesel fuel production from microalgae as renewable energy. *Am. J. Biochem. Biotechnol.* 4: 250–254.

Hu, Q., Guterman, H., and Richmond, A. (1996). A flat inclined modular photobioreactor for outdoor mass cultivation of photoautotrophs. *Biotechnol. Bioeng.* 51 (1): 51–60.

Huesemann, M.H., Hausmann, T.S., Bartha, R. et al. (2009). Biomass productivities in wild type and pigment mutant of *Cyclotella* sp. (diatom). *Appl. Biochem. Biotechnol.* 157 (3): 507–526.

Jambo, S.A., Abdulla, R., Azhar, S.H. et al. (2016). A review on third generation bioethanol feedstock. *Renewable Sustainable Energy Rev.* 65: 756–769.

Jaromir, L. (2000). *Spirulina platensis (Arthrospira): Physiol. Cell-Biol. Biotechnol.* 38 (4): 552–552.

Jeffries, T.W., Timourien, H., and Ward, R.L. (1978). Hydrogen production by *Anabaena cylindrica*: effect of varying ammonium and ferric ions, pH and light. *Appl. Environ. Microbiol.* 35 (4): 704–710.

Johnson, M. and Wen, Z. (2009). Production of biodiesel fuel from the microalga *Schizochytrium limacinum* by direct transesterification of algal biomass. *Energy Fuels.* 23 (10): 294–306.

Kalpesh, K.S., Garg, S., Li, Y. et al. (2013). Critical analysis of current microalgae dewatering techniques. *Biofuels* 4 (4): 397–407.

Kapdan, I.K. and Kargi, F. (2006). Bio-hydrogen production from waste materials. *Enzyme Microb. Technol.* 38 (5): 569–582.

Khan, S. and Fu, P. (2020). Biotechnological perspectives on algae: a viable option for next generation biofuels. *Curr. Opin. Biotechnol.* 62: 146–152.

Khan, M.I., Lee, M.G., Shin, J.H., and Kim, J.D. (2017). Pretreatment optimization of the biomass of *Microcystis aeruginosa* for efficient bioethanol production. *AMB Express* 7 (1): 19.

Khan, M.I., Shin, J.H., and Kim, J.D. (2018). The promising future of microalgae: current status, challenges, and optimization of a sustainable and renewable industry for biofuels, feed, and other products. *Microb. Cell Fact.* 17 (1): 36.

Khetkorn, W., Rastogi, R.P., Incharoensakdi, A. et al. (2017). Microalgal hydrogen production – a review. *Bioresour. Technol.* 243: 1194–1206.

Kim, H.M., Wi, S.G., Jung, S. et al. (2015). Efficient approach for bioethanol production from red seaweed *Gelidium amansii*. *Bioresour. Technol.* 175: 128–134.

Knothe, G. (2010). Biodiesel and renewable diesel: a comparison. *Prog. Energy Combust.* 36 (3): 364–373.

Kojima, E. and Zhang, K. (1999). Growth and hydrocarbon production of microalga *Botryococcus braunii* in bubble column photobioreactors. *J. Biosci. Bioeng.* 87 (6): 811–815.

Kraan, S. (2013). Mass-cultivation of carbohydrate rich macroalgae, a possible solution for sustainable biofuel production. *Mitig. Adapt. Strateg. Glob. Change* 18: 27–46.

Kumar, S. (2019). Algal biomass to bio-energy: recent advances. *J. Ecophysiol. Occup. Health* 19 (3 & 4): 78–85.

Kumar, G., Shobana, S., Chen, W.H. et al. (2017). A review of thermochemical conversion of microalgal biomass for biofuels: chemistry and processes. *Green Chem.* 19 (1): 44–67.

Lakatos, G.E., Ranglová, K., Manoel, J.C. et al. (2019). Bioethanol production from microalgae polysaccharides. *Folia Microbiol.* 64: 627–644.

Lam, M.K. and Lee, K.T. (2012). Microalgae biofuels: a critical review of issues, problems and the way forward. *Biotechnol. Adv.* 30 (3): 673–690.

Lam, M.K., Khoo, C.G., and Lee, K.T. (2019). Scale-up and commercialization of algal cultivation and biofuels production. In: *Biofuels from Algae-Biomass, Biofuels and Biochemicals* (eds. A. Pandey, J.-S. Chang, C.R. Soccol, et al.), 475–506. Elsevier Publishers.

Lee, Y.K. (2001). Microalgal mass culture systems and methods: their limitation and potential. *J. Appl. Phycol.* 13: 307–315.

Lee, J., Cho, D.H., Ramanan, R. et al. (2013). Microalgae-associated bacteria play a key role in the flocculation of *Chlorella vulgaris*. *Bioresour. Technol.* 131: 195–201.

Levin, G.V., Clendenning, J.R., Gibor, A., and Bogar, F.D. (1961). Harvesting of algae by froth flotation. *Appl. Environ. Microbiol.* 10 (2): 169–175.

Lewis, T., Nichols, P.D., and McMeekin, T.A. (2000). Evaluation of extraction methods for recovery of fatty acids from lipid-producing microheterotrophs. *J. Microbiol. Methods.* 43 (2): 107–116.

Li, Y.Y., Sasaki, H., Yamashita, K. et al. (2002). High-rate methane fermentation of lipid-rich food wastes by a high-solids co-digestion process. *Water Sci. Technol.* 45 (12): 143–150.

Li, X., Xu, H., and Wu, Q. (2007). Large-scale biodiesel production from microalga *Chlorella protothecoides* through heterotrophic cultivation in bioreactors. *Biotechnol. Bioeng.* 98 (4): 764–771.

Li, Y.-R., Tsai, W.-T., Hsu, Y.-C. et al. (2014). Comparison of autotrophic and mixotrophic cultivation of green microalgal for biodiesel production. *Energy Procedia.* 52: 371–376.

Majidian, P., Tabatabaei, M., Zeinolabedini, M. et al. (2018). Metabolic engineering of microorganisms for biofuel production. *Renewable Sustainable Energy Rev.* 82: 3863–3885.

Markou, G. and Nerantzis, E. (2013). Microalgae for high-value compounds and biofuels production: a review with focus on cultivation under stress conditions. *Biotechnol. Adv.* 31 (8): 1532–1542.

Mata, T.M., Martins, A.A., and Caetano, N.S. (2010). Microalgae for biodiesel production and other applications: a review. *Renewable Sustainable Energy Rev.* 14 (1): 217–232.

Medipally, S.R., Yusoff, F.M., Banerjee, S., and Shariff, M. (2015). Microalgae as sustainable renewable energy feedstock for biofuel production. *Biomed. Res. Int.* 2015: 1–13.

Melis, A. and Happe, T. (2001). Hydrogen production. Green microalgae as a source of energy. *Plant Physiol.* 127 (3): 740–748.

Metting, F.B. (1996). Biodiversity and application of microalgae. *J. Ind. Microbiol. Biotechnol.* 17 (5–6): 477–489.

Metzger, P. and Largeau, C. (2005). *Botryococcus braunii*: a rich source for hydrocarbons and related ether lipids. *Appl. Microbiol. Biotechnol.* 66 (5): 486–496.

Miao, X.L. and Wu, Q. (2004). High yield bio-oil production from fast pyrolysis by metabolic controlling of *Chlorella protothecoides*. *J. Biotechnol.* 110 (1): 85–93.

Miao, X. and Wu, Q. (2006). Biodiesel production from heterotrophic microalgal oil. *Bioresour. Technol.* 97 (6): 841–846.

Michael, H., Javier, G., Miller, T. et al. (2010). Biofuels from algae: challenges and potential. *Biofuels.* 1 (5): 763–784.

Micheal Mac, M., Lynn, C., Liam, M. et al. (2017). Sustainable harvesting of wild seaweed resources. *Eur. J.Phycol.* 52 (4): 371–390.

Milledge, J., Smith, B., Dyer, P., and Harvey, P. (2014). Macroalgae-derived biofuel: areview of methods of energy extraction from seaweed biomass. *Energies.* 7 (11): 7194–7222.

Milledge, J., Nielsen, B., Maneein, S., and Harvey, P. (2019). A brief review of anaerobic digestion of algae for bioenergy. *Energies* 12 (6): 1166.

Moheimani, N.R. (2005). *The Culture of Coccolithophorid Algae from Carbon Dioxide Bioremediation*. PhD diss. Perth, Australia: Murdoch University.

Moheimani, N.R. and Borowitzka, M.A. (2006). The long-term culture of the coccolithophore *Pleurochrysis carterae* (Haptophyta) in outdoor raceway ponds. *J. Appl. Phycol.* 18 (11): 703–712.

Molina Grima, E.M., Belarbi, E.H., Fernandez, F.G.A. et al. (2003). Recovery of microalgal biomass and metabolites: process options and economics. *Biotechnol. Adv.* 20 (7–8): 491–515.

Moser, A. (1991). Tubular bioreactors: case study of bioreactor performance for industrial production and scientific research. *Biotechnol. Bioeng.* 37 (11): 1054–1065.

Mulbry, W., Kondrad, S., Buyer, J., and Luthria, D. (2009). Optimization of an oil extraction process for microalgae from the treatment of manure effluent. *J. Am. Oil Chem. Soc.* 86: 909–915.

Mussgnug, J.H., Klassen, V., Schluter, A., and Kruse, O. (2010). Microalgae as substrates for fermentative biogas production in a combined biorefinery concept. *J. Biotechnol.* 150 (1): 51–56.

Mutanda, T., Ramesh, D., Karthikeyan, S. et al. (2011). Bioprospecting for hyper-lipid producing microalgal strains for sustainable biofuel production. *Bioresour. Technol.* 102 (1): 57–70.

Nagle, N. and Lemke, P. (1990). Production of methyl-ester fuel from microalgae. *Appl. Biochem. Biotechnol.* 24 (5): 355–361.

Nappa, M., Teir, S., Sorsamäki, L., and Karinen, P. (2016). Energy requirements of microalgae biomass production, CLIC Innovation Oy Carbon Capture and Storage Program (CCSP).

Nascimento, I.A., Marques, S.S.I., Cabanelas, I.T.D. et al. (2013). Screening microalgae strains for biodiesel production: lipid productivity and estimation of fuel quality based on fatty acids profiles as selective criteria. *Bioenergy Res.* 6: 1–13.

Nigam, P.S. and Singh, A. (2011). Production of liquid biofuels from renewable resources. *Prog. Energy Combust. Sci.* 37 (1): 52–68.

Odjadjare, E.C., Mutanda, T., and Olaniran, A.O. (2015). Potential biotechnological application of microalgae: a critical review. *Crit. Rev. Biotechnol.* 37 (1): 37–52.

Oh, Y.K., Hwang, K.R., Kim, C. et al. (2018). Recent developments and key barriers to advanced biofuels: a short review. *Bioresour. Technol.* 257: 320–333.

Oncel, S.S. (2013). Microalgae for a macroenergy world. *Renewable Sustainable Energy Rev.* 26: 241–264.

Ota, M., Kato, Y., Watanabe, H. et al. (2009). Fatty acid production from a highly CO_2 tolerant alga, *Chlorocuccum littorale*, in the presence of inorganic carbon and nitrate. *Bioresour. Technol.* 100 (21): 5237–5242.

Pandey, A., Tiwari, A., Srivastava, P., and Pandey, A. (2007). Using reverse micelle as microreactor for hydrogen production by coupled system of Nostoc/P4 and Anabaena/P4. *World J. Microbiol. Biotechnol.* 23: 269–274.

Parker, M.S., Mock, T., and Armbrust, E.V. (2008). Genomic insights into marine microalgae. *Annu. Rev. Genet.* 42: 619–645.

Patel, A., Matsakas, L., Rova, U., and Christakopoulos, P. (2019). A perspective on biotechnological applications of thermophilic microalgae and cyanobacteria. *Bioresour. Technol.* 278: 424–434.

Pereira, R. and Charles, Y. (2008). Mass production of marine macroalgae. In: *Encyclopedia of Ecology*, vol. 3 (ed. S.E. Jorgensen) and Fath Ecological Engineering (eds.), 2236–2247. Oxford: Elsevier.

Philippa, K., Andrew, B., Camargo-Valero, M.A. et al. (2014). An overview of the potential environmental impacts of large-scale microalgae cultivation. *Biofuels* 5 (3): 331–349.

Pittman, J.K., Dean, A.P., and Osundeko, O. (2011). The potential of sustainable algal biofuel production using wastewater resources. *Bioresour. Technol.* 102 (1): 17–25.

Poelman, E., De Pauw, N., and Jeurissen, B. (1997). Potential of electrolytic flocculation for recovery of micro-algae. *Resour. Conserv. Recycl.* 19 (1): 1–10.

Potts, T., Du, J., Paul, M. et al. (2012). The production of butanol from Jamaica Bay macro algae. *Environ. Prog. Sustain. Energy* 31 (1): 29–36.

Pulz, O. (2001). Photobioreactor: production systems for phototrophic microorganisms. *Appl. Microbiol. Biotechnol.* 57 (8): 287–293.

Pulz, O. and Gross, W. (2004). Valuable products from biotechnology of microalgae. *Appl. Microbiol. Biotechnol.* 65 (6): 635–648.

Radmer, R.J. and Parker, B.C. (1994). Commercial applications of algae: opportunities and constraints. *J. Appl. Phycol.* 6 (2): 93–98.

Raikova, S., Allen, M.J., and Chuck, C.J. (2019). Hydrothermal liquefaction of macroalgae for the production of renewable biofuels. *Biofuels, Bioprod. Biorefin.* 13 (6): 1483–1504.

Rajkumar, R., Yaakob, Z., and Takriff, M.S. (2014). Potential of the micro and macro algae for biofuel production: a brief review. *Bioresources* 9: 1606–1633.

Ramos-Suárez, J.L. and Carreras, N. (2014). Use of microalgae residues for biogas production. *J. Chem. Eng.* 242: 86–95.

Ran, C.Q., Chen, Z.A., Zhang, W. et al. (2006). Characterization of photo-biological hydrogen production by several marine green microalgae. *Wuhan Ligong Daxue Xuebao.* 28 (2): 258–263.

Rodolfi, L., Chini Zittelli, G., Bassi, N. et al. (2009). Microalgae for oil: strain selection, induction of lipid synthesis and outdoor mass cultivation in a low-cost photobioreactor. *Biotechnol. Bioeng.* 102 (1): 100–112.

Romagnoli, F., Blumberga, D., and Gigli, E. (2011). Biogas from marine macroalgae: a new environmental technology-life cycle inventory for a further LCA. *Environ. Clim. Technol.* 4: 97–108.

Rosenberg, J., Oyler, G., Wilkinson, L., and Betenbaugh, M. (2008). A green light for engineered algae: redirecting metabolism to fuel a biotechnology revolution. *Biotechnology* 19: 430–436.

Saad, M.G., Dosoky, N.S., Zoromba, M.S., and Shafik, H.M. (2019). Algal biofuels: current status and key challenges. *Energies* 12 (10): 1–22.

Sanchez, M.A., Contreras, G.A., Garcia, C.F. et al. (1999). Comparative evaluation of compact photobioreactors for large-scale monoculture of microalgae. *J. Biotechnol.* 70 (1–3): 249–270.

Sangeetha, P., Babu, S., and Rangasamy, R. (2011). Potential of green alga *Chaetomorpha litorea* (Harvey) for biogas production. *Int. J. Curr. Sci.* 1: 24–29.

Sati, H., Mitra, M., Mishra, S., and Baredar, P. (2019). Microalgal lipid extraction strategies for biodiesel production: a review. *Algal Res.* 38 (1): 101413.

Schenk, P., Thomas-Hall, S., Stephens, E. et al. (2008). Second generation biofuels: high-efficiency microalgae for biodiesel production. *Bioenergy Res.* 1: 20–43.

Service, R.F. (2007). Cellulosic ethanol. Biofuel researchers prepare to reap a new harvest. *Science* 315 (5818): 1488–1491.

Shaishav, S., Singh, R.N., and Satyendra, T. (2013). Biohydrogen from algae: fuel of the future. *Int. Res. J. Environ. Sci.* 2: 44–47.

Sheehan, J., Dunahay, T., Benemann, J., and Roessler, P.A. (1998). *Look Back at the U S Department of Energy's Aquatic Species Program – Biodiesel from Algae*, vol. 328. Denver, CO: National Renewable Energy Laboratory.

Shen, Y., Yuan, W., Pei, Z.J. et al. (2009). Microalgae mass production methods. *Trans. ASABE* 52 (4): 1275–1287.

Singh, N.K. and Dhar, D.W. (2011). Microalgae as second-generation biofuel. A review. *Agron. Sustainable Dev.* 31 (4): 605–629.

Singh, J. and Dhar, D.W. (2019). Overview of carbon capture technology: microalgal biorefinery concept and state-of-the-art. *Front. Mar. Sci.* 6: 29.

Singh, S., Kate, B.N., and Banerjee, U.C. (2005). Bioactive compounds from cyanobacteria and microalgae: an overview. *Crit. Rev. Biotechnol.* 25 (3): 73–95.

Sivaramakrishnan, R. and Incharoensakdi, A. (2018). Utilization of microalgae feedstock for concomitant production of bioethanol and biodiesel. *Fuel.* 217: 458–446.

Smith, V.H., Sturm, B.S.M., Denoyelles, F.J., and Billings, S.A. (2010). The ecology of algal biodiesel production. *Trends Ecol. Evol.* 25 (5): 301–309.

Spolaore, P., Joannis-Cassan, C., Duran, E., and Isambert, A. (2006). Commercial applications of microalgae. *J. Biosci. Bioeng.* 101 (2): 87–96.

Suali, E. and Sarbatly, R. (2012). Conversion of microalgae to biofuel. *Renewable Sustainable Energy Rev.* 16 (6): 4316–4342.

Sutherland, A.D. and Varela, J.C. (2014). Comparison of various microbial inocula for the efficient anaerobic digestion of *Laminaria hyperborea*. *BMC Biotech.* 14: 7.

Suutari, M., Leskinen, E., Fagerstedt, K. et al. (2015). Macroalgae in biofuel production. *Phycol. Res.* 63: 1–18.

Tiwari A and Kiran T (2017) Biofuels from Microalgae. In: *Advances in Biofuels and Bioenergy* (eds. S.K. Gupta and F. Bux), 239–249.

Tiwari, A., Kiran, T., and Pandey, A. (2019). Algal cultivation for biofuel production. In: *Second and Third Generation of Feedstocks*, 383–403. Elsevier.

Tokusoglu, O. and Uunal, M.K. (2003). Biomass nutrient profiles of three microalgae: *Spirulina platensis, Chlorella vulgaris,* and *Isochrisis galbana*. *J. Food Sci.* 68 (4): 1144–1148.

Tredici, M.R. and Zittelli, G.C. (1998). Efficiency of sunlight utilization: tubular versus flat photobioreactors. *Biotechnol. Bioeng.* 57 (2): 187–197.

Ugwu, C.U., Ogbonna, J.C., and Tanaka, H. (2005). Characterization of light utilization and biomass yields of *Chlorella sorokiniana* in inclined outdoor tubular photobioreactors equipped with static mixers. *Process Biochem.* 40 (11): 3406–3411.

Ugwu, C.U., Aoyagi, H., and Uchiyama, H. (2008). Photobioreactors for mass cultivation of algae. *Bioresour. Technol.* 99 (10): 4021–4028.

Vasudevan, P.T. and Briggs, M. (2008). Biodiesel production – current state of the art and challenges. *J. Ind. Microbiol. Biotechnol.* 35 (5): 421–430.

Vergara-Fernandez, A., Vargas, G., Alarcon, N., and Velasco, A. (2008). Evaluation of marine algae as a source of biogas in a two-stage anaerobic reactor system. *Biomass Bioenergy.* 32 (4): 338–344.

Wang, Z., Pan, Y., Dong, T. et al. (2007). Production of hydrogen from catalytic steam reforming of bio-oil using C12A7-O-based catalysts. *Appl. Catal., A* 320: 24–34.

Wang, Y., Guo, W., Cheng, C.L. et al. (2016). Enhancing bio-butanol production from biomass of *Chlorella vulgaris* JSC-6 with sequential alkali pretreatment and acid hydrolysis. *Bioresour. Technol.* 200: 557–564.

Ward, A.J., Lewis, D.M., and Green, F.B. (2014). Anaerobic digestion of algae biomass: a review. *Algal Res.* 5: 204–214.

Wen, W.G. and Chen, F. (2003). Heterotrophic production of eicosapentaenoic acid by microalgae. *Biotechnol. Adv.* 21: 273–294.

Wu, F.C., Wu, J.Y., Liao, Y.J. et al. (2014). Sequential acid and enzymatic hydrolysis in situ and bioethanol production from *Gracilaria* biomass. *Bioresour. Technol.* 156: 123–131.

Xu, H., Miao, X., and Wu, Q. (2006). High quality biodiesel production from a microalga *Chlorella protothecoides* by heterotrophic growth in fermenters. *J. Biotechnol.* 126 (4): 499–507.

Yeong, T.K., Jiao, K., Zeng, X. et al. (2018). Microalgae for biobutanol production–technology evaluation and value proposition. *Algal Res.* 31: 367–376.

Yu, H., Jia, S., and Dai, Y. (2009). Growth characteristics of the cyanobacterium *Nostoc flagelliforme* in photoautotrophic, mixotrophic and heterotrophic cultivation. *J. Appl. Phycol.* 21 (1): 127–133.

Zhou, L. (2016). Hydrodeoxygenation of microalgae oil to green diesel over Pt, Rh and presulfided NiMo catalysts. *Catal. Sci. Technol.* 6 (5): 1442–1454.

Zhou, L. and Lawal, A. (2015). Evaluation of presulfided NiMo/γ-Al2O3 for hydrodeoxygenation of microalgae oil to produce green diesel. *Energy Fuels.* 29 (1): 262–272.

11

Biomethane Production and Advancement

Rajeev Singh[3], *P K Mishra*[1], *Neha Srivastava*[1], *Akshay Shrivastav*[2] *and K R Srivastava*[1]

[1] *Department of Chemical Engineering and Technology, Indian Institute of Technology(BHU), Varanasi, India*
[2] *Department of Chemical Engineering, Madan Mohan Malaviya University of Technology, Gorakhpur*
[3] *Department of Environmental Studies, Satyawati College, University of Delhi, Delhi, India*

11.1 Introduction

Fossil fuels are the materials present in different state of matters, such as solid, gas, or liquid, which goes through the process of combustion in order to produce energy. These fossil fuels are generally produced from dead remains of the carboniferous era's decayed plants and animals (Salehabadi et al. 2020; Arutyunov et al. 2017). Different types of fossil fuels, like coal, oil, and natural gas, etc., are used as a source of energy in the transport and working of almost every industry. Serious health, as well as environmental problems, like asthma, air and water pollution, etc., occurs because of its high consumption (Basu et al. 2020; Nema et al. 2012; Meetham et al. 2016). Fossil fuels release different hazardous gases, such as carbon monoxide, sulfur dioxide, nitrogen dioxide, and carbon dioxide, etc., which results in many deadly consequences on the habitats (Nwankwoala et al. 2015). This nonrenewable energy produced from dead remains (fossils) is decreasing with higher rates as per the assumption and that results in a complete consumption of fossil fuels in future. The exponential growth in the consumption of energy source at a global level and the abundance of available source of energy indicates rejecting the nonrenewable energy source and finding cheaper and renewable source of energy, like solar energy, wind energy, hydroelectric, and bioenergy biofuels, etc. (Höök et al. 2013). Renewable energy sources, such as biofuels produced from biodegradable waste, has been in high demand at the global level in recent days (Banos et al. 2011).

Biofuel is one of the best sources of energy that provides renewable energy, along with reducing carbon energy, greenhouse emissions (carbon dioxide and methane), and local air pollution, such as particulates, sulfur, and lead (Dincer 2000). The production of biofuel is focused on not only fulfilling the requirements of energy production at the decentralized level but also for fulfilling the requirements of transport. This generates interest in regional groups as well as involving the lands of regional communities (Grau et al. 2010). This creates incentives for the local communities, especially if community lands are involved. There are two types of biofuels and the main biofuels used now are ethanol and biodiesel.

Bioenergy Research: Evaluating Strategies for Commercialization and Sustainability, First Edition.
Edited by Neha Srivastava and Manish Srivastava.

The basic difference between ethanol and biodiesel are that one is fuel and the other one is oil (Demirbas 2008). Ethanol is an alcohol created through fermentation and can be used as a substitute or along with gasoline, while biodiesel is generated by extracting naturally occurring oils from different plants and seeds by the process known as transesterification (Naik et al. 2010). Apart from these two biofuels, biogas is considered to be the one of the most renewable source of energy that is generally produced from most biological treatment plants. Biogas can be produced using both aerobic as well as anaerobic techniques, but anaerobic biotechnology has been found to be the best alternative against the strategies used for waste disposal (Choong et al. 2018; Brémond et al. 2018). The technique utilizes the bulk of biomass for energy production and the by-product can be used as compost in the soil of nutrient stability. The produced biogas is the main source for methane whose production are verified using ultimate biochemical oxygen demand (BOD) or chemical oxygen demand (COD) (Sigurnjak et al. 2020; Bhagwat et al. 2020). These BOD and COD data are calculated using a standard result that is 0.35 m^3 and methane gas will be liberated from 1 kg COD destroyed at STP. Apart from methane gas, libration of these anaerobic biotechnology are also responsible for the production of biohydrogen and bioethanol using organic biomass as a substrate (Pant et al. 2012; Ntaikou et al. 2010; Rao et al. 2020).

The quantity as well as the quality of the biogas generated throughout the anaerobic digestion totally depends upon the type of substrate and it also affects the same factor during methane production, too (Villa et al. 2020; Zheng et al. 2020). The process of methane production is classified into two different steps: first, production of biogas using a different substrate, whereas in the next step, purification of biogas in order to obtain biomethane can be used in transport sectors (Rotunno et al. 2017). In the first step, organic matters are degraded by the action of different microorganisms in the absence of air. Different groups of symbiotic microbes plays a different role at different steps of degradation of complex biomolecules (Olaniran et al. 2013; Quigg et al. 2016). As per the activities performed by the symbiotic microbes they are classified in to four different groups: (i) microbes responsible for hydrolysis, (ii) microbes responsible for fermentation and production of different organic acids, (iii) acidogenic microbes that degrade acids into hydrogen, acetate, and CO_2, and (iv) methanogenic microorganisms that generates biogas using CO_2, hydrogen, and acetic acids (Ali shah et al. 2014; Chojnacka et al. 2015; Li et al. 2013; Luo et al. 2018).

The anaerobic degradation of biodegradable waste liberates about 60–65% of methane gas, which can be used in different sectors across the industries for energy production. The produced methane gas has a high potential for replacing coal, natural gas, or electricity (Agrafiotis et al. 2014). In this technique of methane production, the by-product produced after the process is rich in different minerals, like nitrogen, phosphorus, and sulfur, which increases nutrition in the soil. Biomethane is a highly efficient and easily combustible type of fuel among different types of fuel that are present currently (Bharathiraja et al. 2018). Different types of substrate can be used for producing methane gas, and even wet biomasses can be used as a substrate in this case, which was found to be harmful in the case of other biofuel generation. Some common substrates that are generally used in producing methane gas using anaerobic biotechnology are manure, oil residue, livestock waste, harvest surplus, and much more (Chandra et al. 2012). Application of these types of substrate of producing biomethane have many more advantages, such as reduction of animal waste and its odors, and utilization of agricultural waste, which controls environmental pollution (Atelge et al. 2020). This anaerobic degradation also helps to eliminate the environmental hazards like extra production of sludgy manures. This is why production of methane was found to be one of the unique and best ways for the farmers to follow all the rules and regulation framed by the government in the field of animal waste. Apart from all these, the process also plays a significant role in controlling and destroying different harmful microorganisms that were generated in the animal waste (Bhatt et al. 2020; Logan et al. 2019). Nowadays, waste generated from the poultry farms plays an excellent role in biogas production, as well as the by-product in these processes, which have large amount sof nitrogen that can be used in farming (Freitas et al. 2019).

This review gives an overview of the process involved in production of biogas and its purification to obtain methane gas. It explains different advancements and developments going on across the world in the techniques of biomethane production. It also gives a brief explanation on benefits as well as the limitations in the process of biomethane production. Apart from all this, it also explains all the possible advantages as well as limitations of producing and applying methane gas.

11.1.1 Process Involved in Biomethane Production

Anaerobic biotechnology of biomass degradation is one of the biological activities in which microorganisms play an important role in degrading organic matter available around them in absence of oxygen. In the climatic-controlled anaerobic degradation of biodegradable waste, there occurs liberation of methane gas as one of the last functions (Londhe et al. 2019; Narwal et al. 2017). These anaerobic biotechnologies of degrading biomass can be classified into three different groups as per the activities played by the microbes. These three different groups are as follows: hydrolysis, acidogenesis, and methanogenesis (Zhang et al. 2017a,b; Rettenmaier et al. 2020). Degradation in the absence of oxygen results in the mass reduction of biomass and liberates energy along with liquid or solid biofertilizers. On the basis of climatic conditions maintained during the degradation of organic matter, the process is classified into three different groups: (i) thermophilic conditions in which the temperature is found to be 50–60 °C, (ii) mesophilic conditions in which the temperature was found to be in range of 35–37 °C, and (iii) psychrophilic conditions, in which the temperature was found to be in range of 12–16 °C (Devi et al. 2020; Mani et al. 2020). Some disadvantages that are generally found in thermophilic conditions are the reduction of process stability and the reduction in amount of water released, which will affect the further fermentation process (Ryue et al. 2020). Apart from these, it also requires a high amount of thermal energy as compared to the other two condition that is also responsible for the thermal destruction of degrading microorganisms at high temperatures. Hydrolysis, as well as fermentation in thermophilic conditions, was found to be in a higher rate as compared to others, but these rates cannot boost up the production of methane gas during the process (Liu et al. 2019; Hans and kumar 2019). Liberation of biomethane did not vary when anaerobic degradation occured in the temperature range of 30–60 °C (Alcaraz-lbarra et al. 2020).

In the first step, the process of biomethane production biomass gets hydrolyzed or liquefied by the action of active microorganisms. This hydrolyzed biomass further gets fermented by the action of help of fermentative microbes, which converts these complex biomolecules structures, such as cellulose and hemicelluloses into soluble molecules, such as fatty acids, amino acids, and sugar (Banu et al. 2020; Kamani et al. 2019; Ahmad et al. 2019). Complex biomolecules get degraded into simple monomers, such as protein to amino acids or peptides and cellulose to alcohol or sugar with the help of degrading microorganisms, such as enzymes like protease, lipase amylase, and cellulase, etc. (Liu and Kokare 2017). The hydrolysis step is found to be a rate-limiting step, which has significant importance in the high biodegradable waste content and is responsible for increasing the amount of methane gas produced in the anaerobic process (Barcelos et al. 2020). These limitations can be uplifted in different industries by using many chemical reagents that brighten the process of hydrolysis. The application of chemical reagents not only provides the benefit of reducing the hydrolysis time, but also it increases the yield of biomethane gas during the process (Ning et al. 2016).

In the next step of biomethane production, hydrogen, carbon dioxide, and different organic aids are liberated from the hydrolyzed biomasses by the action of acetogenic microorganisms. Some of the acid produced throughout this step are butric acid ($CH_3CH_2CH_2COOH$), propionic acids (CH_3CH_2COOH), acetic acid (CH_3COOH), and ethanol (C_2H_5OH) (Wainaina et al. 2019). These different products produced during the process of acetogenesis is because of the many types of microorganisms available in the reactions, such as *Sytrophomonas wolfei*, which is responsible for degrading butyrate, and *Syntrophobactor wolinii*, which is responsible for degrading propionate. Other acids are liberated because of microbes, such as *lactobacillus*, clostridium sp., *acinomyces*, and *peptococcus anerobus* (Scapini et al. 2019). The reaction that occurs during the process of acetogensis are shown below:

$$C_6H_{12}O_6 \rightarrow 2C_2H_5OH + 2CO_2$$

At last, the methane gas produced by the action of microorganisms in the process is known as methanogenesis. This process can occur in two different ways on the basis of action performed by the microbes: in the first way, the degradation of molecules of acetic acid, the liberation of methane and carbon dioxide, and in the other step, the reduction of carbon dioxide by the help of hydrogen for production of methane (Ali et al. 2020; Park et al. 2019). Production of methane gas is found to be higher in the case of a reduction process and occurs as a result of carbon dioxide, but is due to a limited amount of hydrogen in the reaction chamber and results in an acetate reaction as the initial step in the production of methane. This degradation is possible because of different methanogenic bacteria, such as methanococcus, methanobacillus, methanosarcina, and methanobacterium (Mostafazadeh et al. 2017; Jiang et al. 2018). On the basis of raw material used by microbes, these methanogens are classified into two different groups, which are H_2/CO_2 consumer and acetate consumer. Species such as methanothrix spp. and methanosarcina spp. are important in anaerobic degradation of both H_2/CO_2 consumer and acetate consumer.

The reactions that occur during the process of methanogenesis process are shown below:

$$2CH_3COOH \rightarrow CH_4 + CO_2$$

$$2C_2H_5OH + CO_2 \rightarrow CH_4 + 2CH_3COOH$$

$$CO_2 + 4H_2 \rightarrow CH_4 + 2H_2O$$

The produced biogas in the process of anaerobic digestion contains different gases, such as hydrogen, ammonia, hydrogen sulfide, siloxanes, and carbon dioxide, and the major section is methane. There are other substances that can cause corrosion in the pipes of industries and that also inhibit anaerobic digestion (Awe et al. 2017). Different research activities have developed many technologies for the purification of produced biogas and removal of unwanted products, like siloxane, ammonia, and hydrogen sulfide. After the purification and removal of unwanted gases from the main stream, biogas still contains some traces of sulfidric acid and ammonia along with hydrogen and carbon dioxide, which are removed from the main stream from production of biomethane (Li et al. 2019).

11.1.2 Purification of Biogas for Methane Production

Purification of produced biogas are generally performed by using technology of membrane separation, which separates carbon dioxide from the main stream of biogas and produces biomethane as a final product. Using this technique leads to production of methane gas of suitable quality that can be transported to different parts of the world through predefined distribution channels (Vrbova et al. 2017; Yousef et al. 2018). The application of polymeric membrane separation technique is found to be economical in order to remove unwanted parts, such as hydrogen sulfide and carbon dioxide, as compared to preexisting technologies in both the factor's operating cost as well as capital cost (Li et al. 2018). However, performance of this membrane at the commercial level are found to be susceptible as the main stream contains different gases, such as ammonia, which ruptures the membrane. In order to overcome this issue, requirements of extensive pretreatment of the main stream is done for protecting its polymeric membrane, which at the end increases the cost of operation (Clarke et al. 2018).

The process of separation includes a compressor system of biogas used to increase the pressure inside the anaerobic degradation technique process from atmospheric pressure to 31 bar. The compressed biogas is transferred to the tank of stabilizing pressure in the feed as well as to regulate its flow toward the polymeric membrane module (Sahoo et al. 2019; Shelford et al. 2019). These pressurized biogases are passed through the first model of polymeric membrane module for removal of gases like carbon dioxide and water vapors. It also helps in the removal of other unwanted substances like oxygen, hydrogen sulfide, ammonia, and hydrogen by using permeation, and which produces nonpermeate biogas that has methane as a primary component (Miltner et al. 2017). The output gas is dried and produced at 30 bar pressure and moved along to the next step targeting the purification of methane gas. In order to increase the yield of methane gas, the permeate gas derived from the first polymeric membrane module is moved to the next polymeric membrane module system (Saadabadi et al. 2019). The first polymeric membrane module system distributes the main stream into two different parts, into which the first one is the nonpermeate gas stream, having a concentration of methane about 95% by volume and is stored in a slightly high-pressure tank that has a pressure of 30 bar (Kapoor et al. 2019). The second one is the permeate stream having carbon dioxide and water vapor as a major portion along with some additional impurities and are collected down at 2 bar pressure (Angelidaki et al. 2019).

The permeate stream is recirculated in the stage of compression and finally transferred to the second stage of polymeric membrane module system for purification. Recovery of other components can be done using a second polymeric membrane module system having methane concentration more than 85% and the permeate produced from this stage can be used as a fuel, as it is has a low methane content (Kumar et al. 2020; Lee and Kim 2020). This stream contains all the unwanted substances that were removed from the main biogas stream for obtaining methane; these impurities are carbon dioxide, hydrogen, water vapor, sulfuric acids, and ammonia, as well as nitrogen.

Jusoh et al. (2020), developed an innovative membrane for purification biogas and obtained methane as a major product. In this experiment, coupling was done between fluorinated polyimide and Linde T to develop a new polymeric membrane that can upgrade the quality of methane gas produced. The fabrication was performed for two hours in which Linde T gets homogenously distributed on the surface of fluorinated polyimide. The developed membrane increases the quality of methane gas by 172% and feasible to be applied at the industrial level (Jusoh et al. 2020). Hidalgo et al. (2020) performed an experiment to understand the behavior of fiber membrane made up of hollow fibers, which is microporous in nature and used for enhancing the quality of biogas. Physical solution of sodium chloride and deionized water were used as a physical solvent to absorb carbon dioxide and in other setups, sodium hydroxide as used as a chemical absorbent. When contractors with a two membrane module system were used, both processes gave more than 99% pure methane. When the single membrane was used, sodium hydroxide was found to be only a absorbent, which increased the time of solvent recirculation cycles and was found to be a very economical process (Hidalgo et al. 2020). Pepper et al. (2019) developed a membrane to remove the unwanted gases and to purify the methane gas so that it could be used at the commercial level. The developed membrane removes a majority of carbon dioxide from the feed stream of biogas. Apart from all these, it also removes traces of materials like ketones, aldehydes, siloxanes, sulfur-rich molecules, and many halocarbons from the biogas. The final composition of the methane gas stream matches the requirement for pipeline injection (Pepper et al. 2019).

11.2 Advancement Undergoing in the Process of Methane Production

Purification of biogas to obtain methane can be obtained by different advanced technologies. These techniques include chemical and physical absorption, as well as adsorption cryogenic separation and many more (Yousef et al. 2019). Among these biological techniques, it is found to be more suitable for the upgrade of biogas to obtain methane, but its maturity level is still low as compared to other mentioned technologies (Adnan et al. 2019). These advanced techniques that are currently being used are water scrubbing, scrubbing by chemicals, pressure swing adsorption cryogenic separation, and membrane separation.

11.2.1 Adsorption by Pressure Swing

Adsorption using pressure swing is the one advanced technology that is based on selective adhesion of different compounds present in the biogas mixture, over the exposed surface of a solid, which is microporous in nature (Qian et al. 2020). The size of these micropores is designed in such a way that molecules of carbon dioxide can easily pass through it, whereas the filtering has methane molecules of larger size (Yang et al. 2019). Substance like activated carbon and zeolites can be used in preparation of adsorptive material that can enhance the quality of methane gas. Apart from its benefit, the pressure swing adsorption technique was found to be costlier as compared to other steps (Kailasa et al. 2020). In order to reduce the cost and to upgrade the quality of biogas produced in the process, the process occurs though two to three different columns. Canevesi et al. (2019) designed an experiment with different columns and the equalization tank in order to perform steps of equalization of asynchronous pressure and keep the process continuous. Carbon materials were used for preparing the sieve, which was further used as an absorbent. In this experiment, the feed was used with the composition of 60% methane and 40% carbon dioxide between the pressures ranging from 0.1 to 5 bar. The performance of these processes is verified on the basis of four different factors: purity of methane, production, recovery, and energy consumed. In this process, the purity of methane was found to be higher that 97.5% and its recovery was found to be more than 90%. Application of the third column increased the recovery by 4%, which represented the importance of equalization of pressure and reduction in the waste of methane gas (Canevesi et al. 2019).

11.3 Adsorption Methods

Adsorption of unwanted substances from the biogas for producing methane can be done using the nature of solubility of different components available in the biogas in specific liquids. For absorption of carbon monoxide from the biogas stream, generally by water or organic solvent, such as methanol, polyethylene glycol ether, or N methyl pyrrolidone in the physical process, whereas amine scrubbing can be done in case of the chemical absorption technique. The water adsorption technique is also beneficial in case of hydrogen sulfide adsorption (Farooq et al. 2017; Canevesi et al. 2019; Zhang et al. 2017a,2017b), whereas in the process of chemical scrubbing, reversible reaction occurs among absorbed material as well as solvent. A common solution, which is generally used in this process is piperazine, methyl diethanolamine, monoethanolamine, and diethanolamine (Kulkarni et al. 2019). These types of amine-scrubbing solvents contains a absorber tank in which carbon dioxide gets adsorbed at a suitable climatic condition, which is 20–65 °C and pressure in the range of 1–2 bar. After this adsorption process, the solvent moved on from stripping where adsorbed carbon dioxide gets released from the mixture when it gets heated (Manyuchi et al. 2016). Chemical scrubbing techniques, in which amines are used as a solvent, helps in obtaining the methane gas with more than 99% purity. There are some issues in this process, is it requires a pretreatment method in order to remove hydrogen sulfide from the biogas mixture. Chemical scrubbing techniques involve both high operation costs, as well as high investment cost (Collet et al. 2017; Florio et al. 2019).

11.4 Separation by Membrane

Polymeric membrane is a thin layer of polymers that acts as a semipermeable barrier, designed for targeting some specific molecules. The operation is incurred by some external force or factor such as temperature, pressure, concentration, and electric charges (Bayrakdar et al. 2017). In the current scenario, three different types of polymeric membrane re generally used, which are inorganic, polymeric, as well as mixed matrix. The application of inorganic membrane presents many other advantages as compared to polymeric membranes because of resistance applied by chemical and mechanical strength, as well as stability toward temperature (Solowski et al. 2018).

Mixed matrix membrane is mostly used in an industrial setting and as for different technology, pretreatment is mostly required from the removal of hydrogen sulfide whose presence negatively affected the quantity as well as quality of the yield (Nemestóthy et al. 2018). Currently, the multistage membrane technique is mostly preferred for the recovery of methane gas up to 99.5% pure. Mixed matrix membrane has the benefit of both inorganic as well as polymeric materials, and it is only responsible for the trade-off occurring among gas selectivity as well as permeability and this mixed matrix polymeric membrane would be hampered (Kim et al. 2016). Application of this mixed matrix membrane at the industrial level its needed for the investigation of effect caused by carbon dioxide plasticization and increments in the permeability nature of the membrane.

11.5 Cryogenic Separation

Cryogenic separation on the most known technique is used for separation of different gases, both at a lab scale as well as at an industrial scale. The main principle that is being used in the process is the difference in the required climatic condition for liquefaction (Yousef et al. 2018). The gases, like hydrogen sulfide and carbon dioxide, have different liquefying points at different temperature and pressure, and this plant of cryogenic separation can occur at a very low temperature of −170 °C and at very high pressure of 80 bar (Mehrpooya et al. 2017). Using these techniques lowers the losses occurring in the process of methane production, but the technique is found to be costlier as compared to others. Even with the high cost, there is high demand for this process in production of road fuel as well as in the production of liquefied natural gas (Spitoni et al. 2019).

11.6 Biological Technique for Purification of Biogas

One of the most advanced and interesting techniques that can be used as the alternative against the current ongoing technique of biogas production is commonly known as the biological technique of methane production (Li et al. 2020). This process includes separation of unwanted gases by using hydrogenotropic methanogenesis and consists of applying hydrogenotrophic methanogens that covert carbon dioxide and hydrogen to methane (Cano et al. 2018). Even if this process has a potential to be used in a large scale, different problems and challenges reduces its market deployment and industrial interest (Morgan et al. 2018).

11.6.1 Advantage and Limitation of Biomethane Production

Production of biomethane has lots of advantages as it reduces the demand of fossil fuel. Methane gas is somewhat similar to different natural gases, in that it can easily replace the position of natural gases, oil, and coal to a large extent (Schiavon et al. 2017). Combustion of methane for production of heat energy, light energy, or as a fuel releases carbon dioxide along with many harmful gases, but it did not contribute to an increase in the level of carbon dioxide in the atmosphere (Herbes et al. 2018). This is possible because similar amounts of carbon dioxide gas will be liberated as compared to what it is left for natural degradation. Production of methane causes less amounts of soil, air, as well as water pollution; it reduces the chances of contamination of biomasses and reduces the application of fossil fuel, but releases of a high amount of slurry produced in this process can cause water pollution and that results in serious damage to our environment (Barbot et al. 2016; Wu et al. 2020). Application of biomethane reduces the release of greenhouse gases, which are found to be one of the best environmental benefits on the earth's atmosphere. Unlike the traditional method for producing energies, the application of methane would not increase the concentration of greenhouse gases because carbon dioxide or other greenhouse gases will be released in the environment during the degradation of biodegradable waste (Kamusoko et al. 2019).

In other words, we can say that application of biomethane is found to be one of the best alternatives for satisfying the demands of the world without contributing any harm to living beings, including humans. Biomethane has minimal intervention effect on nature. Development of methane gas requires degrading reactors and tanks and some storage facilities (Horschig et al. 2016), but apart from these, as it visually affects the landscape, production of these gases do not cause any adverse effect on the environment and local ecosystems. Biogas production plants require a minimal intervention in nature, which are not valid for coal, oil, natural gases, and even some of the renewable sources of energy (Jingura et al. 2017). The main limiting factors of methane production is its improper utilization in transportation sectors. These limiting factors can be classified to different groups, such as issues related with technology and potential of production, economy, and the policy of government. Some technical parameters that are responsible for failure of methane production are climatic conditions (Koupaie et al. 2019). The required temperature for suitable production lies between 30–37 °C, but to make it feasible in different regions of the world, it would require digesters and reactors, which would make it costlier (Dhanya et al. 2020). The produced gas from the anaerobic degradation process contains different impurities. If the produced gas is used directly as a fuel, then it will corrode the metal parts of the engine and cause damage in it.

11.6.2 Conclusion

The motive of this chapter is to describe the methods involved in the production of methane as well as the advancement undergone in the technology to increases the purity of produced gas. The process of biomethane production is divided into two different groups: first, production of biogas using different substrates, like agricultural waste, sewage waste, and municipal waste as well as human waste, whereas secondly, purification of biogas in order to obtain biomethane, which can be used in transport sectors. The production of biogas occurs by anaerobic degradation of biomass with the help of different groups of microorganisms. As per the activities performed by the symbiotic microorganisms, they are classified into four different groups: (i) microbes responsible for hydrolysis, (ii) microbes responsible for fermentation and production of different organic acids, (iii) acidogenic microbes that degrade acids into hydrogen, acetate, and CO_2, and (iv) methanogenic microorganisms that generate biogas using CO_2, hydrogen, and acetic acids. These anaerobic biotechnologies of degrading biomass can be classified into three different groups as per the activities played by the microbes. These three different groups are as follows: hydrolysis (conversion of complex molecules to monomers), acidogenesis (conversion of hydrolyzed biomass to organic acids), and methanogenesis (conversion organic acids to methane gas). After production of biogas, it is required for purification of gases to remove the unwanted part from the biogas and concentrate the produced methane gas. Among these biological techniques, one is found to be more suitable for upgradation of biogas to obtain methane, but its maturity level is still low as compared to other mentioned technologies. These advanced techniques currently in use are water scrubbing, scrubbing by chemicals, pressure swing adsorption cryogenic separation, and membrane separation. Biomethane has lots of advantages, as it reduces the demand of fossil fuels. Methane gas is somewhat similar to different natural gases, such that it can easily replace the position of natural gases, oil, and coal to a large extent.

References

Adnan, A.I., Ong, M.Y., Nomanbhay, S. et al. (2019). Technologies for biogas upgrading to biomethane: a review. *Bioengineering* 6 (4): 92.

Agrafiotis, C., von Storch, H., Roeb, M., and Sattler, C. (2014). Solar thermal reforming of methane feedstocks for hydrogen and syngas production—a review. *Renew. Sust. Energ. Rev.* 29: 656–682.

Ahmad, F.B., Zhang, Z.W., Doherty, O.S., and O'Hara, I.M. (2019). The outlook of the production of advanced fuels and chemicals from integrated oil palm biomass biorefinery. *Renew. Sust. Energ. Rev.* 109: 386–411.

Alcaraz-Ibarra, S., Mier-Quiroga, M.A., Esparza-Soto, M. et al. (2020). Treatment of chocolate-processing industry wastewater in a low-temperature pilot-scale UASB: reactor performance and in-situ biogas use for bioenergy recovery. *Biomass Bioenergy.* 142: 105786.

Ali, S.S., Mustafa, A.M., Kornaros, M. et al. (2020). Construction of novel microbial consortia CS-5 and BC-4 valued for the degradation of catalpa sawdust and chlorophenols simultaneously with enhancing methane production. *Bioresour. Technol.* 301: 122720.

Angelidaki, I., Xie, L., Luo, G. et al. (2019). Biogas upgrading: current and emerging technologies. In *Biofuels: Alternative Feedstocks and Conversion Processes for the Production of Liquid and Gaseous Biofuels* (pp. 817–843). Academic Press.

Arutyunov, V.S., and Lisichkin, G.V. (2017). Energy resources of the 21st century: problems and forecasts. Can renewable energy sources replace fossil fuels? *Russ. Chem. Rev.* 86 (8): 777.

Atelge, M.R., Krisa, D., Kumar, G. et al. (2020). Biogas production from organic waste: recent progress and perspectives. *Waste. Biomass. Valori.* 11 (3): 1019–1040.

Awe, O.W., Zhao, Y., Nzihou, A. et al. (2017). A review of biogas utilisation, purification and upgrading technologies. *Waste. Biomass. Valori.* 8 (2): 267–283.

Banos, R., Manzano-Agugliaro, F., Montoya, F.G. et al. (2011). Optimization methods applied to renewable and sustainable energy: a review. *Renew. Sust. Energ. Rev.* 15 (4): 1753–1766.

Banu, J.R., Tamilarasan, K., Chang, S.W. et al. (2020). Surfactant assisted microwave disintegration of green marine macroalgae for enhanced anaerobic biodegradability and biomethane recovery. *Fuel.* 281: 118802.

Barbot, Y.N., Al-Ghaili, H., and Benz, R. (2016). A review on the valorization of macroalgal wastes for biomethane production. *Mar. Drugs.* 14 (6): 120.

Barcelos, M.C.S., Ramos, C.L., Kuddus, M. et al. (2020). Enzymatic potential for the valorization of agro-industrial by-products. *Biotechnol. Lett.*: 1–29.

Basu, S., Orr, S.A., and Aktas, Y.D. (2020). A geological perspective on climate change and building stone deterioration in London: implications for urban stone-built heritage research and management. *Atmos.* 11 (8): 788.

Bayrakdar, A., Sürmeli, R.Ö., and Çalli, B. (2017). Dry anaerobic digestion of chicken manure coupled with membrane separation of ammonia. *Bioresour. Technol.* 244: 816–823.

Bhagwat, G., Gray, K., Wilson, S.P. et al. (2020). Benchmarking bioplastics: A natural step towards a sustainable future. *J. Polym. Environ.*, 1–21.

Bharathiraja, B., Sudharsana, T., Jayamuthunagai, J. et al. (2018). Biogas production–a review on composition, fuel properties, feed stock and principles of anaerobic digestion. *Renew. Sust. Energ. Rev.* 90 (C): 570–582.

Bhatt, A.H., and Tao, L. (2020). Economic perspectives of biogas production via anaerobic digestion. *Bioengineering* 7 (3): 74.

Brémond, U., de Buyer, R., Steyer, J.-P. et al. (2018). Biological pretreatments of biomass for improving biogas production: an overview from lab scale to full-scale. *Renew. Sust. Energ. Rev.* 90: 583–604.

Canevesi, R.L.S., Andreassen, K.A., Silva, E.A. et al. (2019). Evaluation of simplified pressure swing adsorption cycles for bio-methane production. *Adsorption* 25 (4): 783–793.

Cano, P.I., Colon, J., Ramírez, M. et al. (2018). Life cycle assessment of different physical-chemical and biological technologies for biogas desulfurization in sewage treatment plants. *J. Clean. Prod.* 181: 663–674.

Chandra, R., Takeuchi, H., and Hasegawa, T. (2012). Methane production from lignocellulosic agricultural crop wastes: a review in context to second generation of biofuel production. *Renew. Sust. Energ. Rev.* 16 (3): 1462–1476.

Chojnacka, A., Szczęsny, P., Błaszczyk, M.K. et al. (2015). Noteworthy facts about a methane-producing microbial community processing acidic effluent from sugar beet molasses fermentation. *PLoS One* 10 (5): e0128008.

Choong, Y.Y., Chou, K.W., and Norli, I. (2018). Strategies for improving biogas production of palm oil mill effluent (POME) anaerobic digestion: a critical review. *Renew. Sust. Energ. Rev.* 82: 2993–3006.

Clarke, C.J., Wei-Chien, T., Levers, O. et al. (2018). Green and sustainable solvents in chemical processes. *Chem. Rev.* 118 (2): 747–800.

Collet, P., Flottes, E., Favre, A. et al. (2017). Techno-economic and life cycle assessment of methane production via biogas upgrading and power to gas technology. *Appl. Energy.* 192: 282–295.

Dhanya, B.S., Mishra, A., Chandel, A.K., and Verma, M.L. (2020). Development of sustainable approaches for converting the organic waste to bioenergy. *Sci. Total Environ.* 723: 138109.

Demirbas, A. (2008). Biofuels sources, biofuel policy, biofuel economy and global biofuel projections. *Energy Convers. Manag.* 49 (8): 2106–2116.

Dincer, I. (2000). Renewable energy and sustainable development: a crucial review. *Renew. Sustain. Energy Rev.* 4 (2): 157–175.

Devi, G.S., Vaishnavi, S., Srinath, S. et al. (2020). Energy recovery from biomass using gasification. In *Current Developments in Biotechnology and Bioengineering* (pp. 363–382). Elsevier.

Grau, B., Bernat, E., Antoni, R. et al. (2010). Small-scale production of straight vegetable oil from rapeseed and its use as biofuel in the Spanish territory. *Energy Policy* 38 (1), 189–196.

Farooq, M., Bell, A.H., Almustapha, M.N., and Andresen, J.M. (2017). Bio-methane from an-aerobic digestion using activated carbon adsorption. *Anaerobe.* 46: 33–40.

Florio, C., Fiorentino, G., Corcelli, F. et al. (2019). A life cycle assessment of biomethane production from waste feedstock through different upgrading technologies. *Energies.* 12 (4): 718.

Freitas, F.F., De Souza, S.S., Ferreira, L.R.A. et al. (2019). The Brazilian market of distributed biogas generation: overview, technological development and case study. *Renew. Sust. Energ. Rev.* 101: 146–157.

Hans, M. and Kumar, S. (2019). Biohythane production in two-stage anaerobic digestion system. *Int. J. Hydrog. Energy.* 44 (32): 17363–17380.

Herbes, C., Chouvellon, S., and Lacombe, J. (2018). Towards marketing biomethane in France—French consumers' perception of biomethane. *Ener. Sustain. Soci.* 8 (1): 1–17.

Hidalgo, D., Sanz-Bedate, S., Martín-Marroquín, J.M. et al. (2020). Selective separation of CH_4 and CO_2 using membrane contactors. *Renew. Energy* 150: 935–942.

Horschig, T., Adams, P.W.R., Röder, M. et al. (2016). Reasonable potential for GHG savings by anaerobic biomethane in Germany and UK derived from economic and ecological analyses. *Appl. Energy* 184: 840–852.

Höök, M., and Tang, X. (2013). Depletion of fossil fuels and anthropogenic climate change—a review. *Energy Policy* 52: 797–809.

Jiang, Y., Yang, X., Liang, P. et al. (2018). Microbial fuel cell sensors for water quality early warning systems: fundamentals, signal resolution, optimization and future challenges. *Renew. Sust. Energ. Rev.* 81: 292–305.

Jingura, R.M., and Kamusoko, R. (2017). Methods for determination of biomethane potential of feedstocks: a review. *Biofuel Res. J.* 4 (2): 573–586.

Jusoh, N., Yeong, Y.F., Lock, S.S.M. et al. (2020). Biomethane generation from biogas upgrading by means of thin-film composite membrane comprising Linde T and fluorinated polyimide: optimization of fabrication parameters. *RSC Adv.* 10 (6): 3493–3510.

Kailasa, S.K., Koduru, J.R., Vikrant, K. et al. (2020). Recent progress on solution and materials chemistry for the removal of hydrogen sulfide from various gas plants. *J. Mol. Liquids* 297: 111886.

Kamani, M.H., Eş, I., Lorenzo, J.M. et al. (2019). Advances in plant materials, food by-products, and algae conversion into biofuels: use of environmentally friendly technologies. *Green Chem.* 21 (12): 3213–3231.

Kamusoko, R., Jingura, R.M., Parawira, W., and Sanyika, W.T. (2019). Comparison of pretreatment methods that enhance biomethane production from crop residues-a systematic review. *Biofuel Res. J.* 6 (4): 1080–1089.

Kapoor, R., Ghosh, P., Kumar, M., and Vijay, V.K. (2019). Evaluation of biogas upgrading technologies and future perspectives: a review. *Environ. Sci. Pollut. Res.* 26 (12): 11631–11661.

Kim, S., Ko, D., Row, S.W., and Kim, J. (2016). Techno-economic evaluation of hybrid systems of pressure swing adsorption and membrane processes for coalbed methane separation. *Chem. Eng. Res. Des.* 115: 230–240.

Koupaie, E., Hosseini, S.D., Bazyar Lakeh, A.A. et al. (2019). Enzymatic pretreatment of lignocellulosic biomass for enhanced biomethane production-a review. *J. Environ. Manag.* 233: 774–784.

Kulkarni, M.B., and Ghanegaonkar, P.M. (2019). Methane enrichment of biogas produced from floral waste: a potential energy source for rural India. *Energy Sources Pt A Recov. Util. Environ. Effects* 41 (22): 2757–2768.

Kumar, R., Ghosh, A.K., and Pal, P. (2020). Synergy of biofuel production with waste remediation along with value-added co-products recovery through microalgae cultivation: a review of membrane-integrated green approach. *Sci. Total Environ.* 698: 134169.

Lee, S. and Kim., J.-K. (2020). Process-integrated design of a sub-ambient membrane process for CO2 removal from natural gas power plants. *Appl. Energy* 260: 114255.

Li, A., Chu, Y., Wang, X. et al. (2013). A pyrosequencing-based metagenomic study of methane-producing microbial community in solid-state biogas reactor. *Biotechnol. Biofuels* 6 (1): 3.

Li, C., Meckler, S.M., Smith, Z.P. et al. (2018). Engineered transport in microporous materials and membranes for clean energy technologies. *Adv. Mater.* 30 (8): 1704953.

Li, Y., Alaimo, C.P., Kim, M. et al. (2019). Composition and toxicity of biogas produced from different feedstocks in California. *Environ. Sci. Technol.* 53 (19): 11569–11579.

Li, Z., Wachemo, A.C., Yuan, H. et al. (2020). High levels of ammonia nitrogen for biological biogas upgrading. *Int. J. Hydrogen Energy* 45 (53): 28488–28498.

Liu, X., and Kokare, C. (2017). Microbial enzymes of use in industry. In *Biotechnology of Microbial Enzymes* (pp. 267–298). Academic Press.

Liu, X., Xu, Q., Wang, D. et al. (2019). Thermal-alkaline pretreatment of polyacrylamide flocculated waste activated sludge: process optimization and effects on anaerobic digestion and polyacrylamide degradation. *Bioresour. Technol.* 281: 158–167.

Logan, M., and Visvanathan, C. (2019). Management strategies for anaerobic digestate of organic fraction of municipal solid waste: current status and future prospects. *Waste Manage. Res.* 37 (1_suppl): 27–39.

Londhe, M.N., Sethy, S.K., Shettigar, S.K., and Ghosh, S.B. (2019). Metagenomics and physico-chemical analysis of the active mesophilic and mature phases of a fast composting system. *Curr. Sci.* 117 (11): 1842.

Luo, J., Zhang, Q., Wu, L. et al. (2018). Improving anaerobic fermentation of waste activated sludge using iron activated persulfate treatment. *Bioresour. Technol.* 268: 68–76.

Mani, S., Chowdhary, P., and Zainith, S. (2020). Microbes mediated approaches for environmental waste management. *Microorg. Sustain. Environ. Health*, 17.

Manyuchi, M., Ikhu-Omoregbe, D.I.O., and Oyekola, O. (2016). *Biogas Production from Saw Dust using Acti-zyme as Digestion Catalyst and its Upgrading to Bio methane using Chemical Absorption.*

Meetham, A.R., Bottom, D.W., and Cayton, S. (2016). *Atmospheric Pollution: Its History, Origins and Prevention*. Elsevier.

Mehrpooya, M., Rahbari, C., and Ali Moosavian, S.M. (2017). Introducing a hybrid multi-generation fuel cell system, hydrogen production and cryogenic CO2 capturing process. *Chem. Eng. Process.* 120: 134–147.

Miltner, M., Makaruk, A., and Harasek, M. (2017). Review on available biogas upgrading technologies and innovations towards advanced solutions. *J. Clean. Prod.* 161: 1329–1337.

Morgan, H.M. Jr., Xie, W., Liang, J. et al. (2018). A techno-economic evaluation of anaerobic biogas producing systems in developing countries. *Bioresou. Technol.* 250: 910–921.

Mostafazadeh, A.K., Drogui, P., Brar, S.K. et al. (2017). Microbial electrosynthesis of solvents and alcoholic biofuels from nutrient waste: a review. *J. Environ. Chem. Eng.* 5 (1): 940–954.

Naik, S.N., Goud, V.V., Rout, P.K., and Dalai, A.K. (2010). Production of first and second generation biofuels: a comprehensive review. *Renew. Sust. Energ. Rev.* 14 (2): 578–597.

Narwal, S.K., and Gupta, R. (2017). Biodegradation of xenobiotic compounds: an overview. *Handb. Res. Invent. Bioremed. Tech.* 186–212.

Nema, P., Nema, S., and Roy, P. (2012). An overview of global climate changing in current scenario and mitigation action. *Renew. Sust. Energ. Rev.* 16 (4): 2329–2336.

Nemestóthy, N., Bakonyi, P., Szentgyörgyi, E. et al. (2018). Evaluation of a membrane permeation system for biogas upgrading using model and real gaseous mixtures: the effect of operating conditions on separation behaviour, methane recovery and process stability. *J. Clean. Prod.* 185: 44–51.

Ning, Z., Ji, J., He, Y. et al. (2016). Effect of lipase hydrolysis on biomethane production from swine slaughterhouse waste in China. *Energy Fuel.* 30 (9): 7326–7330.

Ntaikou, I., Antonopoulou, G., and Lyberatos, G. (2010). Biohydrogen production from biomass and wastes via dark fermentation: a review. *Waste. Biomass. Valori.* 1 (1): 21–39.

Nwankwoala, H.N.L. (2015). Causes of climate and environmental changes: the need for environmental-friendly education policy in Nigeria. *J. Educ. Pract.* 6 (30): 224–234.

Olaniran, A.O., Balgobind, A., and Pillay, B. (2013). Bioavailability of heavy metals in soil: impact on microbial biodegradation of organic compounds and possible improvement strategies. *Int. J. Mol. Sci.* 14 (5): 10197–10228.

Pant, D., Singh, A., Van Bogaert, G. et al. (2012). Bioelectrochemical systems (BES) for sustainable energy production and product recovery from organic wastes and industrial wastewaters. *RSC Adv.* 2 (4): 1248–1263.

Park, S.-G., Rhee, C., Shin, S.G. et al. (2019). Methanogenesis stimulation and inhibition for the production of different target electrobiofuels in microbial electrolysis cells through an on-demand control strategy using the coenzyme M and 2-bromoethanesulfonate. *Environ. Int.* 131: 105006.

Peppers, J., Li, Y., Xue, J. et al. (2019). Performance analysis of membrane separation for upgrading biogas to biomethane at small scale production sites. *Biomass Bioenergy* 128: 105314.

Qian, Q., Asinger, P.A., Lee, M.J. et al. (2020). MOF-based membranes for gas separations. *Chem. Rev.* 120 (16): 8161–8266.

Quigg, A., Passow, U., Chin, W.-C. et al. (2016). The role of microbial exopolymers in determining the fate of oil and chemical dispersants in the ocean. *L&O Lett.* 1 (1): 3–26.

Rao, R., and Basak, N. (2020). Development of novel strategies for higher fermentative biohydrogen recovery along with novel metabolites from organic wastes: the present state of the art. *Biotechnol. Appl. Biochem.*

Rettenmaier, R., Schneider, M., Munk, B. et al. (2020). Importance of Defluviitalea raffinosedens for hydrolytic biomass degradation in co-culture with Hungateiclostridium thermocellum. *Microorgan.* 8 (6): 915.

Rotunno, P., Lanzini, A., and Leone, P. (2017). Energy and economic analysis of a water scrubbing based biogas upgrading process for biomethane injection into the gas grid or use as transportation fuel. *Renew. Energy.* 102: 417–432.

Ryue, J., Lin, L., Kakar, F.L. et al. (2020). A critical review of conventional and emerging methods for improving process stability in thermophilic anaerobic digestion. *Energy Sustain. Dev.* 54: 72–84.

Saadabadi, S.A., Thattai, A.T., Fan, L. et al. (2019). Solid oxide fuel cells fuelled with biogas: potential and constraints. *Renew. Energy* 134: 194–214.

Sahoo, K., and Mani, S. (2019). Economic and environmental impacts of an integrated-state anaerobic digestion system to produce compressed natural gas from organic wastes and energy crops. *Renew. Sustain. Energy Rev.* 115: 109354.

Salehabadi, A., Ahmad, M.I., Ismail, N. et al. (2020). *Energy, Society and the Environment: Solid-State Hydrogen Storage Materials*. Springer Nature.

Scapini, T., Camargo, A.F., Stefanski, F.S. et al. (2019). Enzyme-mediated enhanced biogas yield. In *Improving Biogas Production* (pp. 45–68). Springer, Cham.

Schiavon, M., Rada, E.C., Ragazzi, M. et al. (2017). Environmental advantages of treatment plants generating biomethane from food waste. *Energ. Sustain.* 224: 309–316.

Shah, A., Fayyaz, Q.M., Shah, M.M. et al. (2014). Microbial ecology of anaerobic digesters: the key players of anaerobiosis. *Sci. World J.*

Shelford, T., Gooch, C., Choudhury, A., and Lansing, S. (2019). *A Technical Reference Guide for Dairy-Derived Biogas Production, Treatment, and Utilization.*

Sigurnjak, I., Van Poucke, R., Vaneeckhaute, C. et al. (2020). Manure as a resource for energy and nutrients. *Bioref. Inorg. Recover. Mineral Nutr. Biomass Org. Waste* 65–82.

Sołowski, G., Shalaby, M.S., Abdallah, H. et al. (2018). Production of hydrogen from biomass and its separation using membrane technology. *Renew. Sustain. Energy Rev.* 82: 3152–3167.

Spitoni, M., Pierantozzi, M., Comodi, G. et al. (2019). Theoretical evaluation and optimization of a cryogenic technology for carbon dioxide separation and methane liquefaction from biogas. *J. Nat. Gas Sci. Eng.* 62: 132–143.

Villa, R., Rodriguez, L.O., Fenech, C., and Anika, O.C. (2020). Ensiling for anaerobic digestion: a review of key considerations to maximise methane yields. *Renew. Sustain. Energy Rev.* 134: 110401.

Vrbová, V., and Ciahotný, K. (2017). Upgrading biogas to biomethane using membrane separation. *Energy Fuels* 31 (9): 9393–9401.

Wainaina, S., Lukitawesa, M.K.A., and Taherzadeh, M.J. (2019). Bioengineering of anaerobic digestion for volatile fatty acids, hydrogen or methane production: a critical review. *Bioengin.* 10 (1): 437–458.

Wu, B., Lin, R., Kang, X. et al. (2020). Graphene addition to digestion of thin stillage can alleviate acidic shock and improve biomethane production. *ACS Sustain. Chem. Eng.* 8 (35): 13248–13260.

Yang, C., Miao, G., Pi, Y. et al. (2019). Abatement of various types of VOCs by adsorption/catalytic oxidation: a review. *Chem. Eng. J.* 370: 1128–1153.

Yousef, A.M., El-Maghlany, W.M., Eldrainy, Y.A., and Attia, A. (2018). New approach for biogas purification using cryogenic separation and distillation process for CO2 capture. *Energy.* 156: 328–351.

Yousef, A.M., El-Maghlany, W.M., Eldrainy, Y.A., and Attia, A. (2019). Upgrading biogas to biomethane and liquid CO2: a novel cryogenic process. *Fuel.* 251: 611–628.

Zhang, J., Loh, K.-C., Lee, J. et al. (2017a). Three-stage anaerobic co-digestion of food waste and horse manure. *Sci. Rep.* 7 (1): 1–10.

Zhang, X.D., ZhiGang, D., and Li, P.P. (2017b). Physical characteristics of high-rank coal reservoirs in different coal-body structures and the mechanism of coalbed methane production. *Sci. China Earth Sci.* 60 (2): 246–255.

Zheng, Z., Cai, Y., Zhang, Y. et al. (2021). The effects of C/N (10–25) on the relationship of substrates, metabolites, and microorganisms in "inhibited steady-state" of anaerobic digestion. *Water Res.* 188: 116466.

12

Biodiesel Production and Advancement from Diatom Algae

Abhishek Saxena and Archana Tiwari

Diatom Research Laboratory, Amity Institute of Biotechnology, Amity University, Noida, Uttar Pradesh, India

12.1 Introduction

Energy and power are the two major pillars of economic growth. The future of earth lies in the fastest route of development, but to keep the impetus high, a continuous supply of energy is the primary requisite. One needs fuel to stimulate the growth of the nation. In today's time, the consumption of fuel and environmental pollution is at its peak, and at this rate we will soon be under serious ecological complications (Arpia et al. 2021). Natural resources take thousands of years to develop, and once consumed, cannot be recovered in a quick time. Fossil fuels are depleting very quickly, considering their huge demand by the developing nations for faster economic growth. The main drawback of fossil fuels is their nonrenewability, which is the major cause of extreme climate change. No one could have predicted that crude oil price would be soaring, thus letting the world by surprise, because, despite the change in the environment, their availability is also uncertain in the future, and as a result, an energy crisis is inevitable (Popovich et al. 2019).

Earlier it was hypothesized that carbon-based fuel is inexhaustible, but this remains just a myth. Presently, fossil fuels, such as coal, petroleum, and natural gas, have been in constant use since the arrival of the nineteenth century, in order to fuel economic growth. However, excessive use of fossil fuels leads to climate deterioration through the emission of a large amount of carbon dioxide. The earth has experienced the warmest decade in 650 000 years according to a report by the National Aeronautics and Space Administration (NASA) and the National Oceanic and Atmospheric Administration (NOAA). Eventually, this has raised concerns over climate change, energy security, and exhaustion of hydrocarbon reserves. But energy efficiency is absolutely essential, and it is high time to switch to a renewable source of energy, thus pushing the boundaries of concern for energy substitutes (Qubeissi 2019). A biofuel is a renewable energy resource generated through contemporary processes from biomass rather than the natural geological process. Biomass can be converted into clean fuel via the physiochemical or fermentation process. The source of biofuels is crops, vegetable oils, animal fats, sugars, plant remains, and algae. They are the most suitable alternative to fossil fuels. This generation of biofuels create new economic and employment opportunities, expand green surroundings, and can be directly used in engines owing to their

Bioenergy Research: Evaluating Strategies for Commercialization and Sustainability, First Edition.
Edited by Neha Srivastava and Manish Srivastava.

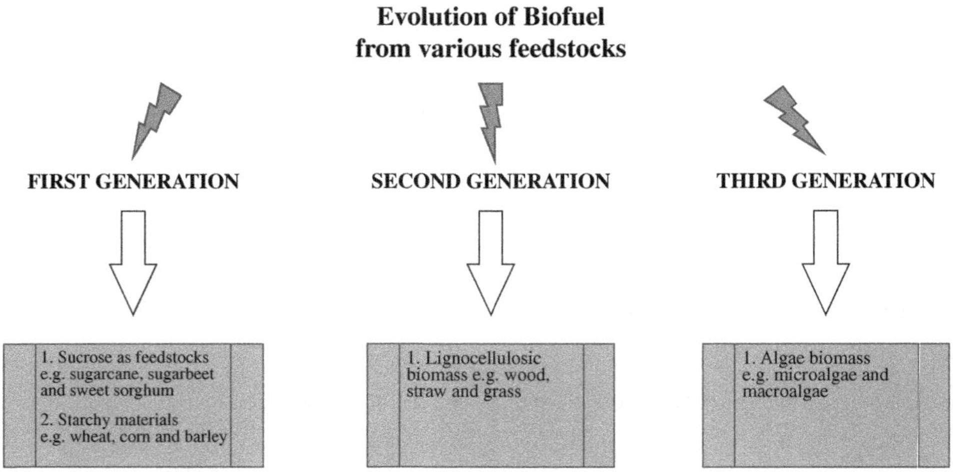

Figure 12.1 Schematic representation of various generation of biofuel. Source: Based on Chowdhury and Loganathan (2019).

ecofriendly actions. Assumptions like biofuel generation are harmful to the environment, for example, carrying palm oil in petrol/diesel-powered vehicles, and peat bog burning in order to prepare biomass can cause greenhouse emissions. These misleading postulations about ecological concerns, coupled with financial losses directed the European Parliament in 2018 to put an end to the import and use of palm oil by 2030 (Qubeissi 2019). Biofuel is divided into three categories: first-generation biofuel/conventional biofuel originated from food crops grown on arable land, second-generation biofuel generated from biomass produced from animal and plant source, and third-generation biofuel manufactured from algae and microbes as presented in Figure 12.1 (Chowdhury and Loganathan 2019). Michael Briggs, of the UNH Biofuel group, stated that fossil fuels can be replaced by using algae that have a more than 50% natural oil content. Briggs suggested growing algae at a larger scale to curb the energy crises. The oilgae, i.e., oil-rich algae, is extracted and further processed to produce a variety of biofuels. Algae culturing produces high yields of biofuel rather than crop-based biofuels since algae culturing requires neither farmland nor a large amount of water. Algae bioreactors are constantly being undertaken by various countries to scale up algae-based biofuels to commercial levels. Algae-based biofuels are highly advantageous because of their higher growth rate, wide range of aquatic habitats, short generation time, and high lipid production capacity make algae-based biofuels the most potential candidate for third-generation biofuel (Briggs 2004).

12.2 Diatom Algae as a Source of Lipids

Diatoms are the most significant group of unicellular, eukaryotic, and photosynthetic algae, and are decorated with a unique silicified cell wall, which is primarily engaged in growth and development. They occur universally both in freshwater and marine water sites, which belong to the stramenopile/heterokont phylum in class Bacillariophyceae. They evolve from higher plants, green microalgae, and red microalgae. They occur as secondary endosymbionts between red algae and a heterotrophic flagellate (Saxena et al. 2020). They fix up to 20% of the global carbon dioxide, which is partitioned into carbohydrate or lipids, particularly triacylglycerols (TAGs) making up to 25% dry biomass. The fatty acid profile is the most exclusive feature of diatoms containing fatty acid chain of various length, e.g., polyunsaturated fatty acids (PUFA), a feature deficient in chlorophytes or plants. The notable fatty acids, such as 14 : 0, 16 : 0, 16 : 1, and 20 : 5, originates from diatoms, whereas C 18 is found in traces. Medium-chain fatty acids produce less-viscous biodiesel (Bhatacharjya et al. 2020). Carotenoids, such as fucoxanthin and didinoxanthin, participate as natural food as well as feed in the pharmaceutical and cosmeceutical industry. A list of very long chain PUFAs, such as arachidonic acid (ARA) and eicosapentaenoic acid (EPA), serves as an indispensable source of fundamental omega-3 fatty acids, which is partitioned into TAGs and found in most diatoms for pharmaceutical and nutraceutical applications (Marella and Tiwari 2020).

Biofuels are biodegradable, renewable, harmless to the environment, and free from the contamination of sulfur and other pungent compounds. Biodiesel can deduct up to 90% of pollution from the environment compared to other fuels (d'Ippolito et al. 2015). Diatoms can produce a significant amount of energy-rich oils including a few strains having the potential to accumulate a great amount of oil in their total dry biomass. The diatom strain can be bioengineered for improving their specific traits and producing fine biochemicals as bioproducts and biofuels as a main component to compare them with petroleum economically. The maximum output from diatoms for different components can be achieved through a cascading approach. In a cascading approach, chemical and physical fractionation steps are held in a consecutive sequence. Diatoms components can be extracted through two-phase extraction, filtration, chromatography, and transesterification (Vinayak et al. 2014). Morphological identification of diatoms is based on their silica cell wall varied at the nanometer scale (Uthappa et al. 2018). Diatoms are freely accessible to water, carbon dioxide, and other nutrients, and thus allows the conversion of lipid up to 85% of their total weight thereby producing 30 times more oil per unit area than oilseed crops (Lebeau and Robert 2003). A liquid fuel energy equal to 1015 BTUs (~1018 kJ) could be produced from 200 kHa of open pond cultivation systems naturally (Sheehan et al. 2009).

Fundamentally, lipids, i.e., saturated and unsaturated fatty acids, were extracted to ascertain the exact amount present in diatoms (Hu et al. 2008). By manipulating nutrients and physicochemical factors, as well as cultivation methods, a significant variation in lipid production was found, and thus represents a solution to increase the lipid yield. Although such variation increases the growth period of diatoms by manipulating overall production per unit area. For instance, biofuels have provided better properties than traditionally used fuels in nitrogen starvation conditions (Milano et al. 2016). Under nitrogen starvation conditions, *Chaetoceros gracilis* can account for 70% of the total volume on a per weight basis (Wang and Seibert 2017). TAG and other lipid levels start increasing in microalgae with decreasing chlorophyll content (Shrestha and Hildebrand 2015). Many studies are utilizing both benthic and planktonic species of diatoms as a viable biofuel feedstock (Fitzer et al. 2019). The US Department of Energy (DOE) took a major step and launched an aquatic species program (ASP) during 1980–1996, in which approximately 3000 strains of microalgae were tested for their lipid productivity potential under the influence of different abiotic factors.

Out of all tested strains, 50 strains have provided outstanding performance in lipid productivity and survival capacity in harsh natural conditions. Sixty percent of those selected algal strains were diatoms (Guiry 2012). These findings have demonstrated the potentiality of diatoms to generate biofuel feedstocks. Diatoms are the ultimate cost-effective platform for generating feedstocks for biofuels generation. However, there are many challenges, such as isolation of potential strains of diatoms, to provide a proper source of nutrients and their consumption, diatoms gathering, development of coproduct, in order to refine and utilize residue biomass. Other important challenges are to engineer economically beneficial photobioreactors (PBRs) for commercial production of diatoms. The use of an integrated approach can exist with the help of engineers and biologists to develop such strains of diatoms that can grow perfectly in different cultivation systems, especially in economically feasible systems. A significant advancement in present technologies for the growth, production, and extraction of oil from algae is necessary to be made. There should be synchronized efforts to combine engineering advances with upgraded productive strains. Table 12.1 present various diatom species based on their lipid (%) dry cell weight (DCW) (Guiry 2012).

Table 12.1 Different diatom species and their relative lipid content.

S. No.	Diatom species	Lipid (%) DCW	References
1.	*Achnanthes delicatulum hauckiana*	29.8	Scholz and Liebezeit 2013
2.	*Achnanthes sp.*	27.7	Delgado et al. 2012; Zhao and Liang 2016
3.	*Aulacoseira ambigua*	19.7	Fields and Kociolek 2015
4.	*Bacillaria paradoxa*	33.61	Hausmann et al. 2016; Pena 2007
5.	*Cocconeis peltoides*	20.9	Scholz and Liebezeit 2013
6.	*Chaetoceros curvisetus,* *Chaetoceros muelleri,* *Chaetoceros calcitrans*	14.86, 33.6, 39.8	d'Ippolito et al. 2015; Rodolfi et al. 2009
7.	*Diatoma sp.*	15.76	Scholz and Liebezeit 2013; Taylor et al. 2007
8.	*Cocconeis sp.*	31.8	Chen et al. 2012
9.	*Cyclotella cryptica*	27.0	Sheehan et al. 1998
10.	*Melosira sp.*	14.75	Tan et al. 2017
11.	*Nitzschia dissipata var. media*	37.5	Tan et al. 2017; Sheehan et al. 1998
12.	*Phaeodactylum tricornutum*	10.7, 18.7	Branco-Vieira et al. 2017; Rodolfi et al. 2009
13.	*Synedra ulna*	7.58	Li et al. 2017
14.	*Tryblionella navicularis*	24.2	Scholz and Liebezeit 2013
15.	*Nitzschia punctata*	16.0	Saranya and Ramachandra 2020
16.	*Skeletonema costatum,* *Skeletonema sp.*	21.0, 31.8	Rodolfi et al. 2009
17.	*Thalassiosira pseudonana*	20.6	Rodolfi et al. 2009

Source: Based on Guiry (2012).

12.3 Biodiesel Production from Diatoms

Biodiesel is a renewable and efficient clean fuel that is produced from different bioresources (Tiwari and Marella 2018). Biodiesel has similar properties to traditional fossil fuels. Biodiesel, fatty acid methyl esters (FAME), are biodegradable and nontoxic, compared to traditional fuels. Biodiesel is generated by the conversion of the algae, plants, and animals oil and fat into (bio)diesel over a long period (Ali et al. 2017a, 2017b; Tiwari et al. 2019). The lipid content of diatoms has a great impact on biofuel costs (Joseph et al. 2016). The FAME profile establishes the suitability of the feedstock for biodiesel production ascertained by vital factors depicting the quality of biodiesel and its characteristics like cloud point, cetane number, plugging point, cold filter, iodine value, and kinematic viscosity (Davis et al. 2011). The fatty acid profile varies both qualitatively and quantitatively with the diatom's environmental conditions and physiological status. Consequently, it is imperative to isolate local species as a renewable source of feedstock, as they are more competitive, owing to their climatic, ecological, and geographical conditions (Rodolfi et al. 2017).

Diatoms can produce a significant amount of energy-rich oils, having the potential to accumulate a great amount of oil in their total dry biomass. The diatom strain can be bioengineered for improving their specific traits and for producing fine biochemicals as bioproducts and biofuels as a main component to compare them with petroleum economically. FAME is the derivative of oil during transesterification using chemical catalysts, such as acids, base, etc. Due to low cost, the methyl group is most preferred for transesterification reactions over ethanol. Figure 12.2 shows the flow chart of biodiesel production from diatoms. Diluted sulfuric acid (H_2SO_4) or hydrochloric acid (HCl) while sodium hydroxide (NaOH) or potassium hydroxide (KOH) are used as acid and base

Figure 12.2 Flow chart showing biodiesel production from diatoms.

catalyst in transesterification reactions, respectively (Moazeni et al. 2019; Saranya and Ramachandra 2020).

Synthesis of fatty acid in diatoms requires acetyl-CoA as the precursor that is similar to flowering plants. In eukaryotes, the acetyl-CoA is present in plastids, peroxisomes, and mitochondria (Leonardi et al. 2005). Plastidial acetyl-CoA is catalyzed by acetyl-CoA synthetase, whereas pyruvate decarboxylation is accelerated by plastidial pyruvate dehydrogenase complex (Lin and Oliver 2008). Mitochondrial acetyl-CoA is produced in the cytosol with citrate as the main substrate. The acetyl-CoA obtained from mitochondria/peroxisomes and fatty acid biosynthesized in plastid require movement toward these cell organelles. The acetyl-CoA is then transformed either to malate/pyruvate or directly require membrane-bound acetyl-CoA transporters. However, the four membrane plastids in diatoms made the matter more complicated. The outer membrane is directly linked to the endoplasmic reticulum (ER) membrane. In diatoms, acetyl-CoA changes into malonyl-CoA, a backbone for fatty acid synthesis. This step is carried out by ATP-dependent carboxylation of acetyl-CoA, which is catalyzed by acetyl-CoA carboxylase. This is the committed step for fatty acid biosynthesis in which acetyl-CoA carboxylase act as the initial enzyme (Saranya and Ramachandra 2020).

In *Cyclotella cryptic*, plastidial acetyl-CoA carboxylase is homomeric in nature (prokaryotic type). The upregulation of plastidial acetyl-CoA carboxylase in *C. cryptica* and *Navicula sapuvila* offer strong acetyl-CoA carboxylase activity, although lipid accumulation was not observed (Huerlimann and Heimann 2013). On the other hand, expression plastidial acetyl-CoA carboxylase from *P. tricornutum* in *E. coli* shows a dual increase in indeterminate lipid as well as a high level of monounsaturated fatty acids. In a recent study on *P. tricornutum*, the reduction of pyruvate dehydrogenase kinase (PDK)-encoding gene (PtPDK) resulted in a considerable rise of up to 82% neutral lipid without altering the fatty acid composition. This has confirmed that mitochondrial acetyl-CoA play an essential role in lipid metabolism in diatoms (Zulu et al. 2018). The very first proteomic investigation of oil-body–related proteins has been determined in *Fistulifera solaris,* thus presenting new insight into the accumulation of TAG in oil bodies (Nojima et al. 2013). Various species synthesizing LC-PUFA, such as *Chaetoceros* sp., *Cyclotella cryptica*, *Navicula saprophila*, *P. tricornutum*, *C. fusiformis*, *Fistulifera* sp., *T. pseudonana*, and *P. tricornutum*. *Chaetoceros gracilis* has been transformed by microparticle bombardment and electroporation technique, respectively (Sayanova et al. 2017). Naturally, diatoms produce novel bioactive compounds and fine chemicals, such as pigments, halogen-derived compounds, domoic acid and isomers, attractants and detergents, and polyamines. Among the diatom communities, *P. tricornutum* and *T. pseudonana* are the most exploited species, the genome of which is completely sequenced. Furthermore, many more species like *Amphora* sp., *Cyclotella cryptica*, *Fistulifera* sp., *Navicula* sp., and *Nitzschia* sp., have also been thoroughly studied to understand the lipid metabolism, but due to a lack of appropriate genetic tools, most of the species remain unsequenced. Both *T. pseudonana* and *P. tricornutum* can provide a great quantity of chief omega-3 fatty acids, wherein *P. tricornutum* build up 30% EPA and traces of docosahexaenoic acid (DHA), whereas *T. pseudonana* accumulates 26% EPA and a little high level of DHA than *P. tricornutum* out of the total lipid content, respectively.

Many plant species have been discovered as a highly suitable oil source for the production of biodiesel. However, they all need fertile land for cultivation, thereby directly competing with food crops for land. This is the reason why diatoms can provide a more useful tool than plants. Diatoms, on the other hand, can utilize nutrients from wastewater, and in return, remediate after purification this water, which can be used for various useful purposes, such as irrigation, industrial, and other domestic uses (Marella and Tiwari 2020). The residues of utilized biomass of diatoms can be further used for the formation of other bioproducts and fertilizers. Similarly, Khan et al. (2017) highlighted the same parameters of the substantial contribution of diatoms in biofuel and aquaculture purposes (Khan et al. 2017). The growth of diatoms in wastewater is significantly triggered by factors like silica, Fe, and trace metals. Nualgi, a nano-based micronutrient mixture can be used to optimize diatom growth. Culture grown in wastewater containing nualgi as the nutrient media produces a high biomass yield of 122.5 mg l^{--1} day^{-1} as well as lipid productivity of 37 mg l^{-1} day^{-1}. As the concentration of nutrient media varies, the fatty acid profile also changes accordingly. As a result, this makes diatoms an ideal candidate for CO_2 sequestration, biodiesel production, and wastewater phytoremediation (Marella et al. 2018).

12.4 Innovative Approaches toward Enhancement in Biodiesel Production and Challenges

It is believed that the main alternative to fossil fuels is biodiesel as it has received tremendous attention worldwide, which has become known to the public. Biodiesel is made from biologically renewable material that then transforms into fuel through a special process. Biodiesel is the monoalkyl esters of long-chain fatty acids, which are the derivative of transesterification of vegetable oil, animal fats, and alcohol through a catalyst or devoid of it. In 2005, total biodiesel production was estimated to be 3.8 billion liters, and approximately 85% was produced in the European Union (Ahmad et al. 2011). Biodiesel is a highly efficient renewable attractive energy resource because of its biodegradability, minimal toxicity, ecofriendly nature, and low combustion emission profile owing to its high oxygen content, and presence of a closed carbon cycle does not contribute to global warming. However, the challenge associated with biodiesel production is low yield, purity, side reactions, high energy input, high cost, etc. Use of biocatalyst for transesterification is the most recent technology as compared to acid or alkali catalysis. Lipolytic enzymes help in lipid mobilization, and turnover numbers, biosurfactants produced from bacteria, and fungi help in solubilizing lipids thus renders ecofriendly alternatives for third-generation algae-based fuels (Ahmad et al. 2011).

A large amount of biomass is required for the production of algae-based biofuel. But growing diatoms on large scale is quite an expensive affair as opposed to growing crops. Factors affecting diatoms growth are light, temperature, carbon dioxide, water, and inorganic salts. There are several strategies by which lipid accumulation can be directly enhanced in diatoms, such as improving light efficiency, carbon sequestration, and control of cell quiescence. Lipid accumulation can also be enhanced by giving stress to cells through nutrient depletion, variation in light intensity, temperature difference, salinity, and pH within the biological limits (Bartley et al. 2014; Chu et al. 2015; Suyono et al. 2015). Adopting strategies such as dual-stage cultivation and co-culture techniques assist in enhancing lipid accumulation (Singh et al. 2016). These factors require extra expenses and to minimize this, biodiesel production must depend on the free supply of sunlight despite periodic fluctuations. However, genetic manipulation of diatoms present specific control of the desired mechanism leading to enhancement in lipid accumulation under normal growth conditions (Sharma et al. 2018). Biodiesel production can be enhanced by genetic engineering and recombinant DNA technology, which has yielded a commanding approach toward optimization and maximizing oil production in a sustainable, technically feasible, and cost-effective manner. TAGs constitute 60–70% of the DCW in microalgae and each molecule of TAGs constitutes glycerol backbone with three fatty acids moieties. Depending upon the degree of unsaturation, fatty acids may be saturated fatty acids (SFA), monosaturated fatty acids (MUFA), or PUFA. Thus, the dominance of these fatty acids in microalgae cells determines the potential of TAG molecules for lipid accumulation (Sharma et al. 2018). Manipulation of enzymes and competitive parallel pathways can redirect the carbon and reductive equivalents flux toward lipid biosynthesis. Manipulation of single-gene encoding metabolic pathways is the most widely used technique but so far this strategy has generated only average success (Bajhaiya et al. 2017). Regulation of oil biosynthesis by transcription has controlled the activity of several components of the metabolic pathway (Courchesne et al. 2009).

Pathway databases, such as KEGG, dEMBF, and MetaCyc, and genome sequence databases, can be applied directly for genetic manipulation for high lipid biosynthesis (Sharma et al. 2018). So far, more than 40 microalgal species have been manipulated through nuclear transformation (Doron et al. 2016). Expression of target genes can be accomplished by tools like metabolic selection markers and "CHYSEL" to target plastid as well as the nuclear genome successfully (Rasala et al. 2014). With the advancement in microbial biotechnology, algonomics and integrated systems-biology modeling helps in understanding the linkage between genes, proteins, and metabolites (Benmoussa 2016). All these strategies improve the metabolite biosynthetic pathways, provide a clear view of the oil content, and pave the way through commercialization of diatoms-based biofuel (Benmoussa 2016).

Diatoms algae have the potential in revolutionizing biodiesel production without putting stress on agriculture and the forest ecosystem. The demand for diatoms-based biofuel is very promising in the future but a step toward feasible commercialization technology is distant from reality. Alleviating the production cost is a major challenge. Improvement of strain for enhanced oil production is the ultimate choice for industrial prospects. Scientific interference, such as genetic engineering and oil extraction procedure may cut down production cost by 15–20%, which will simplify the financial limitations (Chung et al. 2017). Successful implementation of these approaches determines the potential of biofuel production equivalence with fossil fuel. Therefore, it is necessary to put the best effort toward biocatalyst, synthetic biology, and genome control to develop novel strains according to the culture conditions. Integration of biofuel production with value-added products, for example, pharmaceuticals, antioxidants, and nutraceuticals, might generate a high return on investment (Jagadevan et al. 2018). Besides, biorefinery and phytoremediation are expected to utilize biomass for beneficial products (Rizwan et al. 2018). All these steps are intended for sustainable production aimed at the cost incurred during the culturing process. Culturing diatoms in open ponds is the most economic method for commercialization purposes (Kumar et al. 2018). Culturing genetically modified strains in an open pond is a

Table 12.2 Advantages and disadvantages of algal production methods.

S. No.	Algae production type	Advantages	Disadvantages	References
1.	Raceway pond	Cost-effective	Difficult to regulate, Contamination risk, difficult to scale up	Popovich et al. 2020
2.	Photobioreactor	Large surface to volume ratio	Difficult to scale up, Contamination risk, high capital cost, regular temperature monitoring, and periodic cleaning	Ozkan and Rorrer 2017
3.	Fermentation	Maintain high biomass concentration, easy to maintain under optimum condition	Compete for feedstock with other biofuel technology (sugarcane), difficult to scale up.	Choi et al. 2020

cost-effective strategy, although implications for health and the environment are a major concern: exposing transgenic strain can bring ecological calamity (Rastogi et al. 2018). For accurate monitoring and risk assessment analysis of transgenic species, biosafety regulations need to be designed by the concerned authority. The environment can also be made risk free by removing unwanted genes, which have no relevance for culture (Young and Purton 2016). Recently, it was reported that the culturing of microalgae in an open pond was not successful in outcompeting the native strains, whether transgenic or wild type. However, extensive studies must be carried out because the development of a cost-effective approach for diatoms-based fuel needs utmost certainty. The advantages and disadvantages of various strategies of algal production are given in Table 12.2 (Szyjka et al. 2017). A grand success in the demonstration of bio-jet fuel technology could open a new avenue in algae-based biofuel for commercial purposes (Chandra 2018).

12.5 Advancements in Diatoms-Based Biodiesel Production

The main objective of the producer is to put more emphasis on green products in the market in conjunction with recovery and recycling for economic benefits in a cost-effective manner. Biodiesel, i.e., FAME, are the most significant intention in the global markets for the generation of renewable fuel for the transportation of vehicles (Bajusová et al. 2019). Currently used feedstocks for biodiesel productions are edible as well as nonedible food sources, but the unit price is almost double than that of petroleum-based biofuels. This has favored the use of petroleum-based fuels in the market (Sati et al. 2019).

According to algaebase.org, researchers have identified more than 150 000 microalgal species and even now, new species are being identified across the world. By 2024, products derived from microalgae are expected to reach 1143 USD. The drive for algae-based biofuels in the market is showing a positive trend (Mehta et al. 2018). The biomass obtained from diatoms is a future alternative to petroleum-based feedstocks to ensure energy security, protect the environment from global warming, save foreign exchange, and promise socioeconomic benefits. Academia, industry, and several other organization are putting their best foot forward to come up with advanced technology for biofuel production. However, maintaining pure species, growth, and cultivation, biomass generation consistently still needs a significant scope for improvement. Sustainable diatoms-based fuel production throughout is quite challenging. The recent advancement in biodiesel production is nanocatalysts. Herein, the nanomaterial is used as a catalyst for homogeneous or heterogeneous catalysis. In the past, nanocatalyst have proved their mettle in purification of water, storage of energy, fuel cell, composite solid rocket propellants, production of biofuel, in medicine, wastewater treatment, photocatalytic activity, and many others (Akubude et al. 2019). The current research focuses on the use of nanoparticle in the form of nanocatalyst during the transesterification process. Calcium oxide nanoparticles (Cao-NP) have shown increased biodiesel production from 93% to 96% (Gupta and Agarwal 2016). Nanocalcium methoxide, i.e., Ca (OCH3)2 resulted in a FAME yield of more than 99%. The nanomagnetic solid base catalyst (CaO/Fe3O4) has successfully prepared biodiesel from date palm seed oil with a higher yield (Ali et al. 2017a,b). A high FAME yield of $97.7 \pm 2.14\%$ was achieved for the transesterification of canola oil using nano-γ-Al_2O_3 (Boz et al. 2009). Use of zinc oxide (ZnO) nanorods produced better results in biodiesel production from olive oil than conventional ZnO (Molina 2013). The nanocatalyst are highly efficient over other types of catalyst for biodiesel production, as presented in Table 12.3.

The unique approach of nanocatalysts is to extract oil from diatom algae without breaking their cell walls, which then converts into biodiesel without cell damage. Harvesting of fatty acids from diatom algae culture involves highly biocompatible mesoporous nanoparticles that can adsorb hydrophobic molecules. All this helps in the extraction of oil without actually killing the diatom algae. Mesoporous nanoparticles entrapped lipid molecules within the cell membrane in a highly efficient manner without disturbing the membrane structure and oil produced during lipid processing. Strontium and calcium oxides can be introduced into the porous structure of nanoparticles as the catalyst that facilitates transesterification of the captured lipid in vitro. Recently, nanoporous carbon and inorganic derivatives have been established as adsorbents for biofuel separation (Akubude et al. 2019). Thus, the use of nanocatalysts in the production of biodiesel from diatom algae has great potential in the commercialization of the biodiesel industry in the future.

12.6 Conclusion

Biomass produced by diatoms has numerous economic benefits. They are the most appropriate and sustainable oil feedstocks as the third-generation biofuels as they build up a great number of lipids composed of TAGs and fatty acid profile. Diatom oils have a huge demand in the global market for the sustainable production of biofuels. With the advancement of genetic engineering, strain improvement, cultivation strategies, and downstream process, the biodiesel obtained from diatoms may compete with the fossil fuels on a large scale. Nanocatalyzed transesterification is the upcoming promising strategy that deals with

Table 12.3 Different catalyst and their advantages and disadvantages.

S.No.	Type of catalyst	Advantage	Disadvantage	Examples	References
1.	Nanocatalyst	Cheaper, simpler and leaner process. Economical and recyclable. Low water consumption and environment contamination. Highly flexible. Large surface area, high yielding efficient process.	Toxic to human and the environment if exposed.	Iron, silver, gold, cobalt, zinc oxide, titanium dioxide, silicon dioxide, fullerenes, graphene	Akubude et al. 2019
2.	Ezymatic	Low energy input, no side reaction, pure products. Easy and efficient method require no downstream operation	Expensive and slow transesterification	Lipases	Akubude et al. 2019
3.	Ionic liquid	Noncombustible, thermally stable, negligible vapor pressure, recyclable, high catalytic activity, low operation cost and high yielding. High conversion ratio and favorable kinetic reactions.	Not so cost effective when compared to other catalysts. Complex process which can cause environment hazard.	Brønsted acidic ionic liquids Brønsted basic ionic liquids Lewis acidic ionic liquids	Akubude et al. 2019
4.	Homogeneous	Cost-effective at low temperature and pressure for brief period. Shows high reactivity and selectivity	Impurities may affect trans-esterification. High production and purification cost. Reusability and separation of catalyst	Sodium hydroxide (NaOH), potassium hydroxide (KOH), Carbonates, Methoxide, Sodium ethoxide, Sodium propoxide and Sodium butoxide	Akubude et al. 2019
5.	Heterogeneous	Simplify catalyst separation and reusability process.	Lower catalytic efficiency. Poor active sites and resistance toward mass transfer. Low yield and time consuming	Alkaline earth metal oxides, zeolites, KNO_3 loaded on Al2O3, BaO, SrO, CaO, and MgO	Akubude et al. 2019

bottleneck from conventional methods. Innovations can lead to shift toward diatoms-based biodiesel for future environment benefits.

Acknowledgments

We thank the Department of Biotechnology (DBT), New Delhi, India for providing financial assistance under project Grant No: BT/PR/15650/AAQ/3/815/2016 for conducting research.

References

Ahmad, A.L., Yasin, N.H.M., Derek, C.J.C., and Lim, J.K. (2011). Microalgae as a sustainable energy source for biodiesel production: a review. *Renew. Sust. Energ. Rev.* 15: 584–593.

Akubude, V.C., Nwaigwe, K.N., and Dintwa, E. (2019). Production of biodiesel from microalgae via nanocatalyzedtransesterification process: a review. *Mater. Sci. Energy Technol.* 2: 216–225.

Ali, M.A., Al-Hydary, I.A., and Al-Hattab, T.A. (2017a). Nano-magnetic catalyst CaO-Fe_3O_4 for biodiesel production from date palm seed oil. *Bull. Chem. React. Eng. Catal.* 12: 460–468.

Ali, S.F., Hassan, F.M., and Jabar, R.A. (2017b). Evaluation of water quality by trophic diatom index (TDI) in Tigris River within Wasit Province. *Indian J. Ecol.* 44 (71): 1–716.

Arpia, A.A., Chen, W.-H., Lam, S.S. et al. (2021). Sustainable biofuel and bioenergy production from biomass waste residues using microwave-assisted heating: a comprehensive review. *Chem. Eng. J.* 403: 126233.

Bajhaiya, A.K., Moreira, J.Z., and Pittman, J.K. (2017). Transcriptional engineering of microalgae: prospects for high-value chemicals. *Trends Biotechnol.* 35: 95–99. https://doi.org/10.1016/j.tibtech.2016.06.001.

Bajusová, Z., Ladvenicová, J., Bullová, T., and Gurčík, Ľ. (2019). Evaluation of economic impacts of biodiesel production in slovakia. *Visegrad J. Bioecon. Sustain. Dev.* 8 (2): 75–77. https://doi.org/10.2478/vjbsd-2019-0014.

Bartley, M.L., Boeing, W.J., Dungan, B.N. et al. (2014). pH effects on growth and lipid accumulation of the biofuel microalgae *Nannochloropsissalina* and invading organisms. *J. Appl. Phycol.* 26: 1431–1437. https://doi.org/10.1007/s10811-013-0177-2.

Benmoussa, M. (2016). Algomics for the development of a sustainable microalgae biorefinery. *Single Cell. Biol.* 5: 132. https://doi.org/10.4172/2168-9431.1000132.

Bhatacharjya, R., Marella, T.K., Tiwari, A.,Saxena, A., Singh, P.K., and Mishra, B. (2020). Bioprospecting of marine diatoms *Thalssiosira, Skeletonema*and *Chaetoceros*for lipids and other value-added products, Bioresour. Technol., 318:124073.

Boz, N., Degirmenbasi, N.M., and Kalyon, D. (2009). Conversion of biomass to fuel: Transesterification of vegetable oil to biodiesel using KF loaded nano-γ-Al_2O_3 as catalyst. *Appl Catal B* 89: 590–596.

Branco-Vieira, M., Martin, S.S., Agurto, C. et al. (2017). Potential of phaeodactylumtricornutum for biodiesel production under natural conditions in Chile. *Energies.* 11: 54. doi: 10.3390/en11010054.

Briggs, M. (2004). Widescale biodiesel production from algae, University of New Hampshire (US) (Biodiesel Group). www.resilience.org/stories/2004-10-03/widescale-biodiesel-production-algae (accessed August 31, 2018).

Chandra, J. (2018). SpiceJet Operates Country's First Biojet Fuel Flight. The Hindu. www.thehindu.com/business/Industry/spicejet-operates-indias-first-biojet-fuel-flight/article24790919.ece (accessed August 31, 2018).

Chen, L., Liu, T., Zhang, W. et al. (2012). Biodiesel production from algae oil high in free fatty acids by two-step catalytic conversion. *Bioresour. Technol.* 111: 208e214. https://doi.org/10.1016/J.BIORTECH.2012.02.033.

Choi, K.R., Sang, S.J., and Lee, Y. (2020). Metabolic engineering strategies toward production of biofuels. *Curr. Opin. Chem. Biol.* 59: 1–14.

Chowdhury, H. and Loganathan, B. (2019). Third-generation biofuels from microalgae: a review. *Curr. Opin. Green Sust. Chem.* 20: 39–44.

Chu, F.F., Shen, X.F., Lam, P.K.S., and Zeng, R.J. (2015). Polyphosphate during the regreening of *Chlorella vulgaris* under nitrogen deficiency. *Int. J. Mol. Sci.* 16: 23355–23368. https://doi.org/10.3390/ijms161023355.

Chung, Y.S., Lee, J.W., and Chung, C.H. (2017). Molecular challenges in microalgae towards cost-effective production of quality biodiesel. *Renew. Sust. Energ. Rev.* 74: 139–144. https://doi.org/10.1016/j.rser.2017.02.048.

Courchesne, N.M.D., Parisien, A., Wang, B., and Lan, C.Q. (2009). Enhancement of lipid production using biochemical, genetic and transcription factor engineering approaches. *J. Biotechnol.* 141: 31–41. https://doi.org/10.1016/j.jbiotec.2009.02.018.

Davis, R., Aden, A., and Pienkos, P.T. (2011). Techno-economic analysis of autotrophic microalgae for fuel production. *Appl. Energy.* 88: 3524–3531. https://doi.org/10.1016/j.apenergy.2011.04.018.

Delgado, C., Pardo, I., and García, L. (2012). Diatom communities as indicators of ecological status in Mediterranean temporary streams (Balearic Islands, Spain). *Ecological Indicators* 15 (1): 131–139.

d'Ippolito, G., Sardo, A., Paris, D. et al. (2015). Potential of lipid metabolism in marine diatoms for biofuel production. *Biotechnol. Biofuels* 8: 28. https://doi.org/10.1186/s13068-015-0212-4.

Doron, L., Segal, N., and Shapira, M. (2016). Transgene expression in microalgae—from tools to applications. *Front. Plant Sci.* 7: 505. https://doi.org/10.3389/fpls.2016.00505.

Fields, F.J. and Kociolek, J.P. (2015). An evolutionary perspective on selecting high-lipidcontent diatoms (Bacillariophyta). *J. Appl. Phycol.* 27: 2209–2220. https://doi.org/10.1007/s10811-014-0505-1.

Fitzer, S.C., Plancq, J., Floyd, C.J. et al. (2019). Increased pCO_2 changes the lipid production in important aquacultural feedstock algae *Isochrysisgalbana*, but not in *Tetraselmissue cicafitzer*. *Aqucul. Fish* 4: 12–148. https://doi.org/10.1016/j.aaf.2019.02.008.

Guiry, M.D. (2012). How many species of algae are there? *J. Phycol.* 48: 1057–1063. https://doi.org/10.1111/j.1529-8817.2012.01222.x.

Gupta, J. and Agarwal, M. (2016). Preparation and characterization of CaO nanoparticle for biodiesel production. *AIP Conf. Proceed.* 2016 (1724): 020066. https://doi.org/10.1063/1.4945186.

Hausmann, S., Charles, D.F., Gerritsen, J., and Belton, T.J. (2016). A diatom-based biological condition gradient (BCG) approach for assessing impairment and developing nutrient criteria for streams. *Sci. Total Environ.* 562: 914e927. https://doi.org/10.1016/j.scitotenv.2016.03.173.

Hu, Q., Sommerfeld, M., Jarvis, E. et al. (2008). Microalgaltriacylglycerols as feedstocks for biofuel production: perspectives and advances. *Plant J.* 54: 621–639. https://doi.org/10.1111/j.1365-313X.2008.03492.x.

Huerlimann, R. and Heimann, K. (2013). Comprehensive guide to acetyl-carboxylases in algae. *Crit. Rev. Biotechnol.* 33: 49–65.

Jagadevan, S., Banerjee, A., Banerjee, C. et al. (2018). Recent developments in synthetic biology and metabolic engineering in microalgae towards biofuel production. *Biotechnol. Biofuels* 11: 185. https://doi.org/10.1186/s13068-018-1181-1.

Joseph, M.M., Renjith, K.R., John, G. et al. (2016). Biodiesel prospective of five diatom strains using growth parameters and fatty acid profiles. *Biofuels* 8: 81–89. https://doi.org/10.1080/17597269.2016.1204585.

Khan, S., Siddique, R., Sajjad, W. et al. (2017). Biodiesel production from algae to overcome the energy crisis. *Hayati. J. Biosci.* 24: 163–167. https://doi.org/10.1016/j.hjb.2017.10.003.

Kumar, P., Kumar, D., Nehra, P., and Sharma, P.K. (2018). Green algae biomass cultivation, harvesting and genetic modifications for enhanced cellular lipids. In *Microbial Biotechnology*, Patra, J.K., Das, G., and Shin, H.S. (eds.), Singapore: Springer Nature, pp. 119–140.

Lebeau, T. and Robert, J.M. (2003). Diatom cultivation and biotechnologically relevant products. Part II: Current and putative products. *Appl. Microbiol. Biot.* 60: 62–632. https://doi.org/10.1007/s00253-002-1177-3.

Leonardi, R., Zhang, Y.-M., Rock, C.O., and Jackowski, S. (2005). Coenzyme a: Back in action. *Prog. Lipid Res.* 44: 125–153.

Li, X.L., Marella, T.K., Tao, L. et al. (2017). A novel growth method for diatom algae in aquaculture waste water for natural food development and nutrient removal. *Water Sci. Technol.* 75: 2777–2783. https://doi.org/10.2166/wst.2017.156.

Lin, M. and Oliver, D.J. (2008). The role of acetyl-coenzyme asynthetase in arabidopsis. *Plant Physiol.* 147: 1822–1829.

Marella, T.K. and Tiwari, A. (2020). Marine diatom *Thalassiosiraweissflogii* based biorefinary for co-production of eicosapentaenoic acid and fucoxanthin. *Bioresour. Technol.* 307: 123245. https://doi.org/10.1016/j.biortech.2020.123245.

Marella, T.K., Parine, N.R., and Tiwari, A. (2018). Potential of diatom consortium developed by nutrient enrichment for biodiesel production and simultaneous nutrient removal from waste water. *Saudi. J. Biol. Sci.* 25: 704–709. https://doi.org/10.1016/j.sjbs.2017.05.011.

Mehta, P., Singh, D., Saxena, R. et al. (2018). High-value coproducts from algae—an innovational way to deal with advance algal industry. In: *Waste to Wealth* (eds. R. Singhania, R. Agarwal, R. Kumar and R. Sukumaran), 343–363. Springer, Singapore https://doi.org/10.1007/978-981-10-7431-8_15.

Milano, J., Ong, H.C., Masjuki, H.H. et al. (2016). Microalgae biofuels as an alternative to fossil fuel for power generation. *Renew. Sust. Energ. Rev.* 58: 180–197. https://doi.org/10.1016/j.rser.2015.12.150.

Moazeni, F., Chen, Y.-C., and Zhang, G. (2019). Enzymatic transesterification for biodiesel production from used cooking oil, a review. *J. Clean. Prod.* 216: 117e128. https://doi.org/10.1016/J.JCLEPRO.2019.01.181.

Molina, C.M.M. (2013). Zno Nano-rods as Catalyts for Biodiesel Production from Olive Oil Ms.C. Thesis University of Louisville.

Nojima, D., Yoshino, T., Maeda, Y. et al. (2013). Proteomics analysis of oil bodyassociated proteins in the oleaginous diatom. *J. Proteome Res.* 12: 5293–5301. https://doi.org/10.1021/pr4004085.

Ozkan, A. and Rorrer, G.L. (2017). Effects of light intensity on the selectivity of lipid and chitin nanofiber production during photobioreactor cultivation of the marine diatom *Cyclotella* sp. *Algal Res.* 25: 216–227.

Pena, M.R. (2007). Cell growth and nutritive value of the tropical benthic diatom, amphora sp., at varying levels of nutrients and light intensity, and different culture locations, Springer. *Verlag, J. Appl. Phycol.*: 647e655. https://doi.org/10.1007/s10811-007-9189-0.

Popovich, C.A., Pistonesi, M., Hegel, P. et al. (2019). Unconventional alternative biofuels: quality assessment of biodiesel and its blends from marine diatom *Navicula cincta*. *Algal Res.* 39: 101438.

Popovich, C.A., Faraoni, M.B., Sequeira, A. et al. (2020). Potential of the marine diatom *Halamphoracoffeaeformis* to simultaneously produce omega-3 fatty acids, chrysolaminarin and fucoxanthin in a raceway pond. *Algal Res.* 51: 102030.

Qubeissi, M.A. (2019). Biofuels- challenges and opportunities. *Intech. Open*, www.intechopen.com/books/biofuels-challenges-and-opportunities/introductory-chapter-biofuels-challenges-and-opportunities. https://doi.org/10.5772/intechopen.84267.

Rasala, B.A., Chao, S.S., Pier, M. et al. (2014). Enhanced genetic tools for engineering multigene traits into green algae. *PLoS One* 9: e94028. https://doi.org/10.1371/journal.pone.0094028.

Rastogi, R.P., Pandey, A., Larroche, C., and Madamwar, D. (2018). Algal green energy – R & D and technological perspectives for biodiesel production. *Renew. Sust. Energ. Rev.* 82: 2946–2969. https://doi.org/10.1016/j.rser.2017.10.038.

Rizwan, M., Mujtaba, G., Memon, S.A. et al. (2018). Exploring the potential of microalgae for new biotechnology applications and beyond: a review. *Renew. Sust. Energ. Rev.* 92: 394–404. https://doi.org/10.1016/j.rser.2018.04.034.

Rodolfi, L., Zittelli, G.C., Bassi, N. et al. (2009). Microalgae for oil: strain selection, induction of lipid synthesis and outdoor mass cultivation in a low cost photobioreactor. *Biotechnol. Bioeng.* 102: 100–112.

Rodolfi, L., Biondi, N., Guccione, A. et al. (2017). Oil and eicosapentaenoic acid production by the diatom *Phaeodactylumtricornutum* cultivated outdoors in Green Wall panel (GWP®) reactors. *Biotechnol. Bioeng.* 114: 2204–2210. https://doi.org/10.1002/bit.26353.

Saranya, G. and Ramachandra, T.V. (2020). Novel biocatalyst for optimal biodiesel production from diatoms. *Renew. Energy* 153: 919e934.

Sati, H., Mitra, M., Mishra, S., and Baredar, P. (2019). Microalgal lipid extraction strategies forbiodiesel production: a review. *Algal Res.* 38: 101413.

Saxena, A., Prakash, K., Phogat, S. et al. (2020). Inductively coupled plasma nanosilica based growth method for enhanced biomass production in marine diatom algae. *Bioresour. Technol.* 314: 123747.

Sayanova, O., Mimouni, V., Ulmann, L. et al. (2017). Modulation of lipid biosynthesis by stress in diatoms. *Philos. Trans. R. Soc. B* 372: 20160407. https://doi.org/10.1098/rstb.2016.0407.

Scholz, B. and Liebezeit, G. (2013). Biochemical characterisation and fatty acid profiles of 25 benthic marine diatoms isolated from the Solth€orn tidal flat (southern North Sea). *J. Appl. Phycol.* 25: 453e465. https://doi.org/10.1007/s10811-012-9879-0.

Sharma, P.K., Saharia, M., Srivstava, R. et al. (2018). Tailoring microalgae for efficient biofuel production. *Front. Mar. Sci.* 5: 382. https://doi.org/10.3389/fmars.2018.00382.

Sheehan, J., Dunahay, T., Benemann, J., and Roessler, P. (1998). Look Back at the U.S. Department of Energy's Aquatic Species Program: Biodiesel from Algae, Close-Out Report, United States, p. 1998. doi: https://doi.org/10.2172/15003040

Sheehan, J., Dunahay, T., Benemann, J., and Roessler, P. (2009). *A Look Back at the U.S. Department of energy's Aquatic Species Program: Biodiesel from Algae*. National Renewable Energy Laboratory.

Shrestha, K.M. and Hildebrand, M. (2015). Characterization and manipulation of a DGAT2 from the diatom Thalassiosirapseudonana: improved TAG accumulation without detriment to growth, and implications for chloroplast TAG accumulation. *Algal Res.* 12: 239–248. https://doi.org/10.1016/j.algal.2015.09.004.

Singh, P., Kumari, S., Guldhe, A. et al. (2016). Trends and novel strategies for enhancing lipid accumulation and quality in microalgae. *Renew. Sust. Energ. Rev.* 55: 1–16. https://doi.org/10.1016/j.rser.2015.11.001.

Suyono, E.A., Haryadi, W., Zusron, M. et al. (2015). The effect of salinity on growth, dry weight and lipid content of the mixed microalgae culture isolated from Glagah as biodiesel substrate. *J. Life Sci.* 9: 229–233.

Szyjka, S.J., Mandal, S., Schoepp, N.G. et al. (2017). Evaluation of phenotype stability and ecological risk of a genetically engineered alga in open pond production. *Algal Res.* 24: 378–386. https://doi.org/10.1016/j.algal.2017.04.006.

Tan, X., Zhang, Q., Burford, M.A. et al. (2017). Benthic diatom based indices for water quality assessment in two subtropical streams. *Front. Microbiol.* 8: 601. www.frontiersin.org/article/10.3389/fmicb.2017.00601.

Taylor, J.C., Prygiel, J., Vosloo, A. et al. (2007). Can diatombased pollution indices be used for biomonitoring in South Africa? A case study of the Crocodile West and Marico water management area. *Hydrobiologia.* 592: 455e464. https://doi.org/10.1007/s10750-007-0788-1.

Tiwari, A. and Marella, T.K. (2018). Biofuels from microalgae. In: *Advances in Biofuels and Bioenergy*. Nageswara-Rao, M. and Jaya, R: Soneji, Intech Open https://doi.org/10.5772/intechopen.73012.

Tiwari, A., Marella, T.K., and Pandey, A. (2019). Algal Photobiohydrogen production. In: *Bioenergy and Biofuels* (ed. O. Konur), 313–330. CRC Press: Taylor and Francis.

Uthappa, U.T., Brahmkhatri, V., Sriram, G. et al. (2018). Nature engineered diatom biosilica as drug delivery systems. *J. Control. Release.* 281: 70–83. https://doi.org/10.1016/j.jconrel.2018.05.013.

Vinayak, V., Gordon, R., Gautam, S., and Rai, A. (2014). Discovery of a diatom that oozes oil. *Adv. Sci. Lett.* 20: 1256–1267. https://doi.org/10.1166/asl.2014.5591.

Wang, J.K. and Seibert, M. (2017). Prospects for commercial production of diatoms. *Biotechnol. Biofuels.* 10: 16. https://doi.org/10.1186/s13068-017-0699-y.

Young, R.E.B. and Purton, S. (2016). Codon reassignment to facilitate genetic engineering and bio-containment in the chloroplast of *Chlamydomonasreinhardtii*. *Plant Biotechnol. J.* 14: 1251–1260. https://doi.org/10.1111/pbi.12490.

Zhao, F. and Liang, J. (2016). Variations in the total lipid content and biological characteristics of diatom species for potential biodiesel production. *Fund. Renew. Energy Appl.* 6: 22e26. https://doi.org/10.4172/20904541.1000201.

Zulu, N.N., Zienkiewicz, K., Vollheyde, K., and Feussner, I. (2018). Current trends to comprehend lipid metabolism in diatoms. *Prog. Lipid Res.* 70: 1–16.

13

Biobutanol Production and Advancement

Enosh Phillips

Department of Biotechnology, St. Aloysius College (Autonomous), Jabalpur, M.P, India

13.1 Introduction

With the advent of industrial revolution in the eighteenth to nineteenth century, energy requirements increased everyday. Energy is essential for standard living. The present economy depends upon petroleum fuel (Lin et al. 2011). Petroleum is an efficient fuel for meeting our present-day energy needs. The concerns are rising worldwide with the depletion of fossil fuels, which includes petroleum. Moreover, combustion of petroleum has continuously affected the environment by raising the concentration of greenhouse gases, like CO, CO_2, and N_2O. Alternative sources of fuel for meeting energy needs are in demand. These alternative fuels are renewable in nature and have had a positive impact on the environment (Dominik et al. 2007; Kumar and Gayen 2011). Biofuels are the alternative fuels developed from biomass. They are defined as any gas or liquid fuel that can be used in the transport sector produced from biomass. Table 13.1 describes the variety of liquid fuel produced from biomass. Both the developing and developed nations encourages the production of biofuels, due to various reasons like environment, energy security, socioeconomic, and foreign exchange savings. Biofuels have oxygen levels up to 45%, whereas petroleum has none. Such properties make biofuels preferrable above petroleum (Demirbas 2008).

It is reported that 10% or more of energy demands are fulfilled by biofuels. Biofuels have been used from a long time in human history. In the nineteenth century, alcohol was reported to be used for energy purposes. In the 1860s, Nikolaus August Otto used ethanol in spark engine ignition. Heavy locomotives from Deutz Gas Engine Works ran on ethanol only in 1902. There are many such reports that suggest that the use of biomass-based fuels in the early nineteeth and twentieth century make up 90% of the present biofuel market, which is now governed by biodiesel and bioethanol, as they are complementary in their physical and chemical properties and utilized in ignition engines (Antoni et al. 2007).

13.2 Biobutanol

Although biodiesel and bioethanol dominate the biofuel market, they have several disadvantages. These are overcome by biobutanol. The use and development of biobutanol

Bioenergy Research: Evaluating Strategies for Commercialization and Sustainability, First Edition.
Edited by Neha Srivastava and Manish Srivastava.

Table 13.1 Here is the detail about the various types of liquid biofuel available and the source from which they are developed.

S. No.	Biofuel type	Biomass used	Reference
1	Bioethanol	Lignocellulosic biomass	Maurya et al. (2015)
2	Biodiesel	Algal biomass	Taher et al. (2014)
3	Biobutanol	Lignocellulosic biomass	Huzir et al. (2018a)

dates back to 1860 when Louis Pasteur reported its production from a mixed culture of *Clostridium butyricum* and *Clostridium acetobutylicum*. They produce biobutanol by the ABE (acetone-butanol-ethanol) fermentation process. Biodiesel and bioethanol are first-generation biofuels in which the former is produced from animal fats, vegetable oil, and restaurant greases, and later from sugary material by *Saccharomyce cerevisae*. Since they are first-generation biofuels, they challenge food crops. On the other hand, biobutanol offers many advantages listed here:

- Biobutanol can be used purely and in blended form with gasoline, whereas bioethanol can be blended up to 85% only.
- Biobutanol does not call for changes in combustion engines for combustion, whereas bioethanol need changes in combustion engines for better efficiency.
- It is safer than bioethanol and biodiesel due low vapor pressure.
- It is not hygroscopic, which gives surety of no contamination of water bodies over its spillage.
- It is noncorrosive.
- It is more efficient with increased mileage.
- It can be used as diesel also (Dürre 2007).

In Table 13.2, there is a brief view about the evolution of biobutanol production in the twentieth century.

Clostridium sp. has been vital in biobutanol production. It produces butanol via the ABE method. ABE has been in use for 100 years. Even Pasteur reported that butanol production from microorganisms was through ABE method. During World War II, the US reported 66% production of butanol was by fermentation processes. Other major countries like South Africa, Japan, Australia, and China also use fermentation for biobutanol production (Ranjan and Moholkar 2012). Looking at the US and Brazil incorporating bioethanol production in their economy, India also started production of biofuel and blended it with

Table 13.2 The table shows a short description how biofuels attained attention from initial days.

Year	Event	Reference
1861	Butanol production by microbial fermentation Louis Pasteur	Gabriel (1928)
1911	Fernbach patented culture involved in butanol production	Jones and Woods (1986)
1912	Chaim Weizmann isolated *Clostridium acetobutylicum*	Jones and Woods (1986)
1920	Industrial setup for butanol production	Zverlov et al. (2006)
1960	Plant in China reached annual production of 170 000 tons	Ni and Sun (2009)

gasoline. In India, plants for biofuel production have been established well before 1936. A report suggests that India produces 370 million tons of biomass from various sources, which can be used for biobutanol and other biofuel production. India requires a high amount of energy input with its ever-increasing population (Kumar and Gayen 2011).

13.3 ABE Process for Biobutanol Production

ABE is widely used at the industrial level for biobutanol production. There are challenges present in this method, but it provides the most efficient way of butanol production. In ABE, there is less by-product synthesis and more productivity (Patraşcu et al. 2017). ABE is basically an anaerobic metabolism carried out by microorganisms. The entire process can be divided into two phases: (i) acidogenesis and (ii) solventogenesis. During acidogenesis, microbe metabolizes available carbohydrate into acids. In the ABE process, conducted by Clostridia, the primary acids produced are butyric acid and acetic acid. The acids produced in acidogenesis are assimilated to produce acetone-butanol-ethanol. They are produced in the ratio 3 : 6 : 1 (acetone:butanol:ethanol). Many species of clostridia have been tested for efficient butanol production by the ABE method, which include the various *Clostridium* strains – *C. acetobutylicum, Clostridium beijerinckii, Clostridium saccharoperbutylacetonicu*, and *Clostridium saccharobutylicum* – and all have shown efficiency. The wild-type strains have a low yield, which can be enhanced by genetic engineering (Ellis et al. 2012; Li et al. 2019). From Figure 13.1, it is evident that glucose is converted to pyruvate, which is further metabolized to produce acetone-butanol-ethanol.

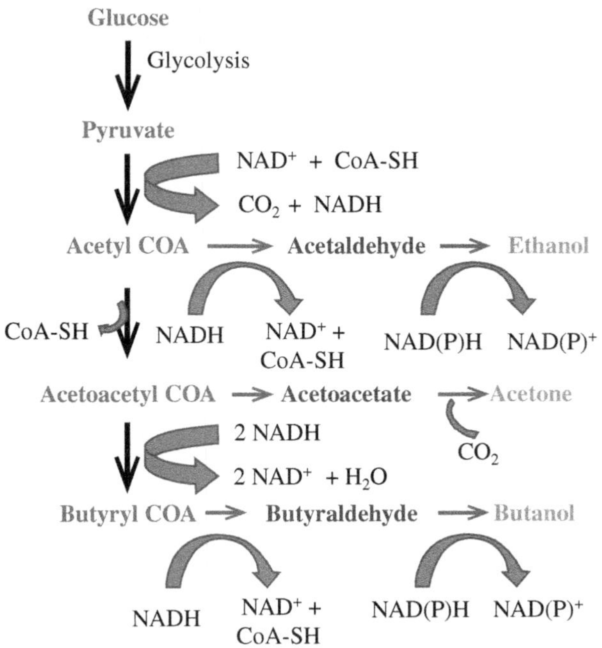

Figure 13.1 Schematic representation of the conversion of glucose into solvents. The ABE fermentation is adopted by bacteria under anaerobic conditions.

13.4 Biobutanol Production by ABE

Annually, biomass is produced in millions of tons by photosynthetic organisms, mainly by plants. Biomass may contain different amounts of polysaccharide useful for biobutanol production. Apart from cellulose and hemicellulose, lignin is also available in biomass. It has a network-like structure with random arrangements of phenlpropanoid units. It has other alcoholic structure units – coniferyl alcohol, sinapyl alcohol, and coumaryl alcohol. Monosaccharides are basically required for alcohol production. They can be produced by acid hydrolysis or by enzymatic treatment of cellulose and hemicellulose. However, due to the presence of lignin, yield is less. Yield is generally enhanced by pretreating the biomass. This step increases the cost of production. Both ethanol and butanol can be produced from biomass. Butanol is found to be more efficient than ethanol. It has higher efficiency and higher ignition temperature. It is also compatible with gasoline. In a general way, the biomass obtained is air- or heat-dried for up to 72 hours in the oven. This dried biomass is then mixed with concentrated sulfuric acid. Biomass is dried to avoid the presence of any water molecules that may alter product yield. Acid hydrolysis is the preferred method for the breakdown of polysaccharide, as it is more cost efficient and lowers production cost, necessary for encouraging the use of alternative fuel. After one hour of incubation in sulfuric acid, the solution is diluted with the help of boiling water (the temperature is kept around 105 °C). At this temperature, a second hydrolysis of biomass takes place. Hydrolysis is stopped and the solution is cooled with the help of ice bath. Different types of substrates are used as biomass for the production butanol as evident from Table 13.3. The hydrosylate produced neutralizes in the presence of calcium hydro-oxide.

Clostridium is used for the ABE method of butanol production. The fermentation medium contains yeast extract, KH2PO4, K2HPO4, p-aminobenzoic acid, biotin, MgSO4, MnSO4, FeSO4, and NaCl. Anaerobic condition is maintained for 24 hours at 37 °C (Han et al. 2013; Irimescu 2011; Tran et al. 2012). Pretreatment plays an important role in increasing the yield. Methods like physical, physiochemical, chemical, and biological methods

Table 13.3 Many substrates are tested for the production of biobutanol. Here is the list of few of the substrates used for the production of butanol.

S. No.	Substrate used	Amount of biobutanol produced	References
1	Sugar baggase	0.52 mol butanol/mol	Cheng et al. (2012)
2	Rice straw	$3.43\,g\,l^{-1}$	Gottumukkala et al. (2013)
3	Decanter cake	$3.42\,g\,l^{-1}$	Loyarkat et al. (2013)
4	Pea pod waste	$3.82\,g\,l^{-1}$	Nimbalkar et al. (2018)
5	Glycerol	$1.4\,g\,l^{-1}$	Yadav et al. (2014)
6	Waste acorn	$4\,g\,l^{-1}$	Heidari et al. (2016)
7	Felled oil palm trunk	$10\,g\,l^{-1}$	Komonkiat and Cheirsilp (2013)
8	Kraft black liquor	$7.3\,g\,l^{-1}$	Kudahettige-Nilsson et al. (2015)
9	Beech wood	$15\,g\,l^{-1}$ (Solvent concentration)	Tippkötter et al. (2014)
10	Sweet sorghum	$18\,g\,l^{-1}$ (ABE)	Mirfakhar et al. (2017)

are available. Methods involving efficient downstream processing are vital for lowering production cost. Many techniques are available, like adsorption, liquid–liquid extraction, gas stripping, perstraction, pervaporation, and reverse osmosis (Niemistö et al. 2014). Since the inception of the idea for the bioproduction of butanol as an alternative fuel, many developments have taken place to enhance production and make the process efficient.

13.5 Substrate Used in Biobutanol Production

Various substrates have been used for butanol production using fermentation by Clostridium sp. As early as the 1980s, sugarcane molasses is reported for butanol production (United States Patent (19) Levy g 54 Continuous Process for Producing, n.d.). Since then, other substrates are used for enhancing production. Agriculture waste and forest waste are seen as potential biomass for butanol production. In India, rice straw is one of the major agro-residues. In estimate, it is said that about 97 million metric tons of rice straw are produced annually. Rice straw contains 37% cellulose, 24% hemicellulose, and 14% lignin. Although rice straw is high in energy content, production efficiency decreases due to the presence of lignin. Pretreatment is therefore required for releasing simple sugars for fermentation. Rice straw can produce up to 13.5 g l^{-1} of butanol (Ranjan et al. 2013). Other substrates like wheat straw, corn stover, cassava waste, sugarcane baggase, and barley straw have been reported for biobutanol production. Fruits are also used for butanol production (Huzir et al. 2018a).

Novel substrates are always being searched out, just like in the case of municipal solid waste, which is helpful for reducing the cost of butanol production, as it is rich in pentose and hexose sugar. This is beneficial for increasing production, reducing pretreatment, and hence could bring the production costs down. With an increasing population, there will be an increase in the production of municipal production and thus, the problem of waste management will also increase. However, if municipal waste is to be utilized in butanol production, then it will also be an efficient way to manage the waste and lower the price of butanol. A work suggests that there could 50% decrease in the production price if municipal solid waste is used (Ashani et al. 2020).

Crops are a first-generation substrate for biofuel production and agriculture waste is the source for second-generation biofuel. However, both have certain constraints. First-generation sources compete with food crops and may pose a threat to an increasing population. On the other hand, second-generation biofuel sources have shown recalcitrant behavior. Looking at third-generation biofuels like microalgae, which are encouraged and overcome the shortcomings of the first and second generations has been shown to be the next step. As mentioned, microalgae is a third-generation biofuel source that is renewable and also has desirable characteristics, such as the microalgae are not recalcitrant and does not compete with food crops. At this point, only a few studies have been reported to involve algal biomass fermentation for biobutanol production. Most of these studies are confined to the laboratory stage. On comparing algae with other substrates, microalgae take less space (0.3–0.4 ha) than first- and second-generation substrates (1–2 ha). They also produce more energy as compared to the former generations. This establishes the significance of microalgae in biobutanol production (Yeong et al. 2018).

13.6 Advancement in Pretreatment Method

Lignocellulosic biomass has been vital in butanol production. The presence of lignin makes the fermentation process difficult, reducing the yield. In earlier methods, chemical reagents were used for treating lignocellulosic biomass and for enhancing the yield. Enzymes are encouraged for pretreatment (Wang et al. 2019). Commonly, acid and alkali treatment yields glucose from the biomass (Wang et al. 2016). Organosolv pretreatment, along with enzyme hydrolysis, has been reported to be beneficial. In this method, the biomass is mixed with ethanol-water waste and sulfuric acid. The mixture is then heated at high temperature for up to one hour. This treated biomass can be further treated with enzymes like cellulase and hemicellulose for generating more fermentable sugars. Such a combination is found to be beneficial, having a yield of about 150 g kg^{-1} (Farmanbordar et al. 2018). One of the major issues in pretreatment is the production of inhibitors, high energy requirement, and waste production, while other methods may require extreme conditions. Hydroxymethyl furfural, acetic acid, and furfural are the products of degradation and have shown to inhibit the enzyme activity during the fermentation process. A mild condition is always desirable. The following three aspects are desirable in pretreatment:

- Lignin removal
- Increase of surface area for better fermentation
- A decrease in cellulose crystallinity

DES, or deep eutectic solvents, are proposed to be efficient in the biomass pretreatment for biobutanol production. DES is a combination of two or three ionic compounds that are environment-friendly. DES increases the presence of cellulose in the medium to be further hydrolyzed by elimination lignin and hemicellulose. Procentese et al. (2017) reported the use of glycerol and choline chloride as DES for municipal solid waste (MSW) pretreatment. Agrofood residues, e.g., lettuce leaves are used. This treatment shows a positive response by increasing the fermentable elements. It shows a low energy input requirement (Procentese et al. 2017; Xu et al. 2016). Many feed stocks have been encouraging in biobutanol production like switch grass. Due to the presence of toxics in the hydrozylates after acid, pretreatment reduces the butanol production by *Clostridium beijerinckii*. Hydrothermolysis is also another way of treating biomass for enhancing butanol production. It provides the same amount of glucose and glycan when compared to acid and alkali hydrolysis. However, it has certain advantages over conventional methods, such as it does not require a catalyst and is noncorrosive to the reactor, bringing the production cost much lower (Liu et al. 2015).

13.7 Microbial Engineering for Production Enhancement

Clostridium is been widely used for biobutanol production. It is an anaerobic bacterium and is known for ABE fermentation. *C. acetobutylicum* and *C. beijerinckki* are well known. Other species, like *C. saccahrobutylicum, C. saccharoperbutylacetonicum, C. sporogenes,* and *C. pasteurinum,* have also been used for butanol production. During the fermentation carried out by *Clostridium,* acetic acid and butyric acid are produced as by-products. Over the sporulation the assimilation of acids solvent occurs. The biphasic nature of butanol production by clostridia is tightly regulated at the genetic level. The entire process of ABE can be divided into the following:

- Cold channel
- Hot channel

In cold channel, acetyl CoA is either reduced to ethanol or condensed to form butyryl CoA for butanol formation. On the other hand, hot channel condenses acetyl CoA to butyryl CoA, which is reduced to butanol. Thus, hot channel is a direct pathway, and for commercial scale production, many attempts are made to enhance butanol production (Gottumukkala et al. 2017).

Biosynthesis of butanol is shown to be confined to the clostridia species. With the development of genetic engineering, the metabolic activity of clostridia can be altered to obtain production benefits (Lütke-Eversloh and Bahl 2011). Separation of butanol from acetone is essential and is costly. Development of Clostridium species for high selectivity toward butanol is currently under study. *C. acetobutylicum ATCC 824* is a mutant strain that has lost a plasmid pSOL1, which led to the prevention of solvent production. RNA-based antisense technology has been used to downregulate the production of acetone in the wild-type strain. ctfB RNA antisense has been used for the purpose (Jin et al. 2009). Yu et al. in 2011 reported the overexpression of adhE2 gene (aldehyde/alcohol dehydrogenase) from *C. acetobutyricum ATCC* 25755. It is created by the engineering of *C. acetobutylicum ATCC 824*. The results were encouraging, as it produced a higher titer value of butanol (Yu et al. 2011). Butanol in itself at high concentration has inhibitory effects on microorganisms. A transcription regulator spo0A is suggested to control solvent production. Overexpression of it has shown that it increases the tolerance of microorganisms against butanol (Zheng et al. 2015). An anaerobic mesophile *C. cellulovorans* DSM 743B is been reported recently that is able to produce butyric acid as the major product from lignocellulose. Studies have suggested that *C. cellulovorans* can degrade the lignocellulosic material and metabolize it. About .42 $l\,g^{-1}$ of butanol is reported from this organism. However, there are certain restraints present, such as butanol concentration inhibition and process optimization, and still others have to be worked out. Recent work carried out on *C. cellulovorans* reports that Co-A–dependent ABE pathway genes from *C. acetobutyricum* are transferred. To overcome the butanol concentration, inhibition adaptive laboratory evolution approach is used. This approach is reported to enhance the production 138-fold (Wen et al. 2019). Such approaches imply that there may be the presence of better strains in nature yet to be identified. Availability of genetic tools can be utilized for the development of strains that could overcome the hindrances present in the wild-type strains.

13.8 Conclusion

Butanol is an effective biofuel when compared to bioethanol. Many studies have continuously stressed the fact that butanol can be blended with present-day fossil fuels and needs no upgrade in the combustion engine. Even switching from fossil fuel to a complete biofuel, like butanol, will need no change in the architecture of combustion engine. A huge amount of lignocellulosic waste is produced annually. Developing countries have focused on using the waste biomass for biobutanol production, which could help increase the living standards of their countries' populations. Becoming dependent on biobutanols, like biofuel, will not enhance these countries' socioeconomic statuses, but will conserve their environments. Biobutanol contains oxygen, which makes it ecofriendly. However, there are some hurdles, like biomass pretreatment, substrate utilization, process optimization, and metabolic engineering, in order to bring about financial competition against fossil fuel. Many projects are being conducted in this respect that are paving the way to a successful replacement of fossil-based fuels. *Clostridium acetobutylicum* has been vastly used in the ABE-modeled production of biobutanol. Many genetically altered strains of the *C. acetobutyricum* are available, and that has been successful in removing the obstructions available in the wild-type strains. Moreover, many other wild-type strains like *C. cellulovorans and E. coli* are also tested for this purpose and their genetically altered strains are developed for this purpose.

Acknowledgment

I would like to give my sincere thanks to the editors for giving support and time to writing this manuscript. Due to my medical condition, my work was delayed, but Dr. Neha Shrivastava encouraged me, for which I am heartly thankful to her. I thank my college, St. Aloysius (Autonomous), for providing all the support for the work.

References

Antoni, D., Zverlov, V.V., and Schwarz, W.H. (2007). Biofuels from microbes. *Appl. Microbiol. Biotechnol.* 77 (1): 23–35. https://doi.org/10.1007/s00253-007-1163-x.

Ashani, P.N., Shafiei, M., and Karimi, K. (2020). Biobutanol production from municipal solid waste: technical and economic analysis. *Bioresour. Technol.* 308: 123267. https://doi.org/10.1016/j.biortech.2020.123267.

Cheng, C.L., Che, P.Y., Chen, B.Y. et al. (2012). Biobutanol production from agricultural waste by an acclimated mixed bacterial microflora. *Appl. Energy* 100: 3–9. https://doi.org/10.1016/j.apenergy.2012.05.042.

Demirbas, A. (2008). Biofuels sources, biofuel policy, biofuel economy and global biofuel projections. *Energy Convers. Manage.* 49 (8): 2106–2116. https://doi.org/10.1016/j.enconman.2008.02.020.

Dominik, A., Zverlov, V.V, and Schwarz, W.H. (2007). *Biofuels from microbes*. doi: https://doi.org/10.1007/s00253-007-1163-x

Dürre, P. (2007). Biobutanol: an attractive biofuel. *Biotechnol. J.* 2 (12): 1525–1534. https://doi.org/10.1002/biot.200700168.

Ellis, J.T., Hengge, N.N., Sims, R.C., and Miller, C.D. (2012). Acetone, butanol, and ethanol production from wastewater algae. *Bioresour. Technol.* 111: 491–495. https://doi.org/10.1016/j.biortech.2012.02.002.

Farmanbordar, S., Amiri, H., and Karimi, K. (2018). Simultaneous organosolv pretreatment and detoxification of municipal solid waste for efficient biobutanol production. *Bioresour. Technol.* 270: 236–244. https://doi.org/10.1016/j.biortech.2018.09.017.

Gabriel, C.L. (1928). Butanol fermentation process. *Ind. Eng. Chem.* 20 (10): 1063–1067. https://doi.org/10.1021/ie50226a020.

Gottumukkala, L.D., Parameswaran, B., Valappil, S.K. et al. (2013). Biobutanol production from rice straw by a non acetone producing *Clostridium sporogenes* BE01. *Bioresour. Technol.* 145: 182–187. doi: 10.1016/j.biortech.2013.01.046.

Gottumukkala, L.D., Haigh, K., and Görgens, J. (2017). Trends and advances in conversion of lignocellulosic biomass to biobutanol: microbes, bioprocesses and industrial viability. *Renewable Sustainable Energy Rev.* 76: 963–973. https://doi.org/10.1016/j.rser.2017.03.030.

Han, S.H., Cho, D.H., Kim, Y.H., and Shin, S.J. (2013). Biobutanol production from 2-year-old willow biomass by acid hydrolysis and acetone-butanol-ethanol fermentation. *Energy* 61: 13–17. https://doi.org/10.1016/j.energy.2013.04.069.

Heidari, F., Asadollahi, M.A., Jeihanipour, A. et al. (2016). Biobutanol production using unhydrolyzed waste acorn as a novel substrate. *RSC Adv.* 6 (11): 9254–9260. https://doi.org/10.1039/c5ra23941a.

Huzir, N.M., Aziz, M.M.A., Ismail, S.B. et al. (2018a). Agro-industrial waste to biobutanol production: eco-friendly biofuels for next generation. *Renewable Sustainable Energy Rev.* 94: 476–485. https://doi.org/10.1016/j.rser.2018.06.036.

Irimescu, A. (2011). Fuel conversion efficiency of a port injection engine fueled with gasoline-isobutanol blends. *Energy* 36 (5): 3030–3035. https://doi.org/10.1016/j.energy.2011.02.047.

Jin, Y.L., Jang, Y.S., Lee, J. et al. (2009). Metabolic engineering of *Clostridium acetobutylicum* M5 for highly selective butanol production. *Biotechnol. J.* 4 (10): 1432–1440. https://doi.org/10.1002/biot.200900142.

Jones, D.T. and Woods, D.R. (1986). Acetone-butanol fermentation revisited. *Microbiol. Rev.* 50 (4): 484–524. American Society for Microbiology (ASM). doi: https://doi.org/10.1128/mmbr.50.4.484-524.1986.

Komonkiat, I. and Cheirsilp, B. (2013). Felled oil palm trunk as a renewable source for biobutanol production by *Clostridium* spp. *Bioresour. Technol.* 146: 200–207. https://doi.org/10.1016/j.biortech.2013.07.067.

Kudahettige-Nilsson, R.L., Helmerius, J., Nilsson, R.T. et al. (2015). Biobutanol production by *Clostridium acetobutylicum* using xylose recovered from birch Kraft black liquor. *Bioresour. Technol.* 176: 71–79. https://doi.org/10.1016/j.biortech.2014.11.012.

Kumar, M. and Gayen, K. (2011). Developments in biobutanol production: new insights. *Appl. Energy.* 88 (6): 1999–2012. https://doi.org/10.1016/j.apenergy.2010.12.055.

Li, Y., Tang, W., Chen, Y. et al. (2019). Potential of acetone-butanol-ethanol (ABE) as a biofuel. *Fuel* 242: 673–686. https://doi.org/10.1016/j.fuel.2019.01.063.

Lin, L., Cunshan, Z., Vittayapadung, S. et al. (2011). Opportunities and challenges for biodiesel fuel. *Appl. Energy.* 88 (4): 1020–1031. https://doi.org/10.1016/j.apenergy.2010.09.029.

Liu, K., Atiyeh, H.K., Pardo-Planas, O. et al. (2015). Butanol production from hydrothermolysis-pretreated switchgrass: quantification of inhibitors and detoxification of hydrolyzate. *Bioresour. Technol.* 189: 292–301. https://doi.org/10.1016/j.biortech.2015.04.018.

Loyarkat, S., Cheirsilp, B., and Umsakul, K. (2013). Decanter cake waste as a renewable substrate for biobutanol production by *Clostridium beijerinckii*. *Process Biochem.* 48 (12): 1933–1941. https://doi.org/10.1016/j.procbio.2013.09.013.

Lütke-Eversloh, T. and Bahl, H. (2011). Metabolic engineering of *Clostridium acetobutylicum*: recent advances to improve butanol production. *Curr. Opin. Biotechnol.* 22 (5): 634–647. https://doi.org/10.1016/j.copbio.2011.01.011.

Maurya, D.P., Singla, A., and Negi, S. (2015). An overview of key pretreatment processes for biological conversion of lignocellulosic biomass to bioethanol. *3 Biotech.* 5 (5): 597–609. https://doi.org/10.1007/s13205-015-0279-4.

Mirfakhar, M., Asadollahi, M.A., Amiri, H., and Karimi, K. (2017). Phenolic compounds removal from sweet sorghum grain for efficient biobutanol production without nutrient supplementation. *Ind. Crops Prod.* 108: 225–231. https://doi.org/10.1016/j.indcrop.2017.06.028.

Ni, Y. and Sun, Z. (2009). Recent progress on industrial fermentative production of acetone-butanol-ethanol by *Clostridium acetobutylicum* in China. *Appl. Microbiol. Biotechnol.* 83 (3): 415–423. https://doi.org/10.1007/s00253-009-2003-y.

Niemistö, J., Saavalainen, P., Isomäki, R. et al. (2014). Biobutanol production from biomass. In: *Biofuel Technologies: Recent Developments*, 443–470. Berlin Heidelberg: Springer-Verlag https://doi.org/10.1007/978-3-642-34519-7_17.

Nimbalkar, P.R., Khedkar, M.A., Chavan, P.V., and Bankar, S.B. (2018). Biobutanol production using pea pod waste as substrate: impact of drying on saccharification and fermentation. *Renewable Energy* 117: 520–529. https://doi.org/10.1016/j.renene.2017.10.079.

Patraşcu, I., Bîldea, C.S., and Kiss, A.A. (2017). Eco-efficient butanol separation in the ABE fermentation process. *Sep. Purif. Technol.* 177: 49–61. https://doi.org/10.1016/j.seppur.2016.12.008.

Procentese, A., Raganati, F., Olivieri, G. et al. (2017). Low-energy biomass pretreatment with deep eutectic solvents for bio-butanol production. *Bioresour. Technol.* 243: 464–473. https://doi.org/10.1016/j.biortech.2017.06.143.

Ranjan, A. and Moholkar, V.S. (2012). Biobutanol: science, engineering, and economics. *Int. J. Energy Res.* 36 (3): 277–323. https://doi.org/10.1002/er.1948.

Ranjan, A., Khanna, S., and Moholkar, V.S. (2013). Feasibility of rice straw as alternate substrate for biobutanol production. *Appl. Energy* 103: 32–38. https://doi.org/10.1016/j.apenergy.2012.10.035.

Taher, H., Al-Zuhair, S., Al-Marzouqi, A.H. et al. (2014). Effective extraction of microalgae lipids from wet biomass for biodiesel production. *Biomass Bioenergy* 66: 159–167. https://doi.org/10.1016/j.biombioe.2014.02.034.

Tippkötter, N., Duwe, A.M., Wiesen, S. et al. (2014). Enzymatic hydrolysis of beech wood lignocellulose at high solid contents and its utilization as substrate for the production of biobutanol and dicarboxylic acids. *Bioresour. Technol.* 167: 447–455. https://doi.org/10.1016/j.biortech.2014.06.052.

Tran, L.S., Sirjean, B., Glaude, P.A. et al. (2012). Progress in detailed kinetic modeling of the combustion of oxygenated components of biofuels. *Energy* 43 (1): 4–18. https://doi.org/10.1016/j.energy.2011.11.013.

United States Patent (19) Levy g 54 Continuous Process For Producing. (n.d.).

Wang, Y., Guo, W., Cheng, C.L. et al. (2016). Enhancing bio-butanol production from biomass of *Chlorella vulgaris* JSC-6 with sequential alkali pretreatment and acid hydrolysis. *Bioresour. Technol.* 200: 557–564. https://doi.org/10.1016/j.biortech.2015.10.056.

Wang, P., Chen, Y.M., Wang, Y. et al. (2019). Towards comprehensive lignocellulosic biomass utilization for bioenergy production: efficient biobutanol production from acetic acid pretreated switchgrass with *Clostridium saccharoperbutylacetonicum* N1-4. *Appl. Energy* 236: 551–559. https://doi.org/10.1016/j.apenergy.2018.12.011.

Wen, Z., Ledesma-Amaro, R., Lin, J. et al. (2019). Improved n-butanol production from clostridium cellulovorans by integrated metabolic and evolutionary engineering. *Appl. Environ. Microbiol.* 85 (7) https://doi.org/10.1128/AEM.02560-18.

Xu, G.C., Ding, J.C., Han, R.Z. et al. (2016). Enhancing cellulose accessibility of corn stover by deep eutectic solvent pretreatment for butanol fermentation. *Bioresour. Technol.* 203: 364–369. https://doi.org/10.1016/j.biortech.2015.11.002.

Yadav, S., Rawat, G., Tripathi, P., and Saxena, R.K. (2014). A novel approach for biobutanol production by *Clostridium acetobutylicum* using glycerol: a low cost substrate. *Renewable Energy* 71: 37–42. https://doi.org/10.1016/j.renene.2014.05.004.

Yeong, T.K., Jiao, K., Zeng, X. et al. (2018). Microalgae for biobutanol production – technology evaluation and value proposition. *Algal Res.* 31: 367–376. https://doi.org/10.1016/j.algal.2018.02.029.

Yu, M., Zhang, Y., Tang, I.C., and Yang, S.T. (2011). Metabolic engineering of *Clostridium tyrobutyricum* for n-butanol production. *Metab. Eng.* 13 (4): 373–382. https://doi.org/10.1016/j.ymben.2011.04.002.

Zheng, J., Tashiro, Y., Wang, Q., and Sonomoto, K. (2015). Recent advances to improve fermentative butanol production: genetic engineering and fermentation technology. *J. Biosci. Bioeng.* 119 (1): 1–9. doi: https://doi.org/10.1016/j.jbiosc.2014.05.023.

Zverlov, V.V., Berezina, O., Velikodvorskaya, G.A., and Schwarz, W.H. (2006). Bacterial acetone and butanol production by industrial fermentation in the Soviet Union: use of hydrolyzed agricultural waste for biorefinery. *Appl. Microbiol. Biotechnol.* 71 (5): 587–597. https://doi.org/10.1007/s00253-006-0445-z.

Index

Note: Page numbers followed by *f* and *t* represent figures and tables respectively.

Bioenergy Research: Evaluating Strategies for Commercialization and Sustainability, First Edition.
Edited by Neha Srivastava and Manish Srivastava.

C